Control of Respiration

Edited by D.J. Pallot

CROOM HELM
London & Canberra

©1983 David J. Pallot
Croom Helm Ltd, Provident House, Burrell Row,
Beckenham, Kent BR3 1AT

British Library Cataloguing in Publication Data

Control of respiration.
    Respiration—Regulation
    I. Pallot, David J.
    612'.2   Q123

    ISBN 0-7099-2015-6

Typeset by Elephant Productions, London
Printed and bound in Great Britain

CONTENTS

CONTRIBUTORS

H. Acker: Max-Planck-Institut für Systemphysiologie, Rheinlandamm 201, D-4600 Dortmund 1, Federal Republic of Germany

P.A. Barnard, PhD: Institute for Environmental Medicine and University of Pennsylvania School of Medicine, Philadelphia PA 19104, USA

Peter J. Fleming, MB, ChB, MRCP, FRCP(C): Consultant in Neonatal Paediatrics, Bristol Maternity Hospital, Southwell Street, Bristol 2, England

Sukhamay Lahiri, DPhil: Professor of Physiology, University of Pennsylvania School of Medicine, Philadelphia PA 19104, USA

Hans H. Loeschcke: Institut fur Physiologie, Ruhr – Universitat Bochum, Lehrstuhl 1, 4630 Bochum – Queremburg, Postfach 2148, Federal Republic of Germany

D.S. McQueen, BPharm, PhD: Reader in Pharmacology, University of Edinburgh, 1 George Square, Edinburgh EH8 9JZ, Scotland

A.S. Paintal: ICMR Centre for Respiratory Physiology and Department of Physiology, Vallabhai Patel Chest Institute, University of Delhi, Delhi – 110007, India

David J. Pallot, BSc, PhD: Senior Lecturer in Anatomy, Medical Sciences Building, University of Leicester, University Road, Leicester LE1 7RH, England

J.M. Patrick, MA, DPhil, BM, BCh: Senior Lecturer in Physiology, University of Nottingham Medical School, Queens Medical Centre, Clifton Boulevard, Nottingham NG7 2UH, England

José Ponte, MD, PhD, FFARCS: Senior Lecturer in Anaesthetics, King's College Hospital Medical School, Denmark Hill, London SE5 9RS, England

Andrzej Trzebski: Department of Physiology, Institute of Physiological Sciences, Medical Academy, Warsaw 00-927, Poland

Christopher B. Wolff, MB, ChB, MRCP: Senior Lecturer in Physiology, Guy's Hospital Medical School, London SE1 9RT, England

R. Zhang, MD: Institute of Biophysics, Chinese Academy of Sciences, Beijing, People's Republic of China

PREFACE

The scientific literature has expanded dramatically in recent years, making entry into the structure of any given area extremely difficult; concurrent with this explosion more people are required to become acquainted with information outside their main line of expertise. For this reason there is a need for review articles which give an overall review of circumscribed areas. This volume reviews the subject of respiratory control mechanisms; the authors of each chapter are active research workers engaged in the area covered by their chapter.

The first four chapters are concerned with the basic physiological mechanisms which sense changes in the respiratory system, in the standard physiology textbook parlance chemical and neural sensory receptors. The peripheral arterial chemoreceptors sense changes in arterial oxygen tension, carbon dioxide and pH. The first chapter describes the basic responses in the organ produced by changes in blood chemistry. Later chapters discuss changes in activity produced by exercise, chronic hypoxia and the possible role of the chemoreceptors in initiation of respiration in the new-born. In Chapter 1, a section considers the action of drugs on the peripheral chemoreceptors, and finally there is a discussion of the possible mechanisms whereby the organs sense changes in blood chemistry. This pattern is followed in subsequent chapters wherever possible; first a discussion of the basic physiological properties, followed by any clinical application and discussion of the mechanism whereby the receptor might operate.

The remaining chapters are of a more applied nature. In Chapter 5 Acker discusses a most important field, often neglected in textbooks, that of the mechanisms controlling the distribution of oxygen to the tissues at the level of the microcirculation. Included in this chapter is a discussion of oxygen and blood supply to tumours, a topic of considerable clinical importance. In Chapter 7 there is a discussion of the periodic nature of O_2, CO_2 and pH changes during the respiratory cycle, and of the use such information is put to in controlling respiration and the changes in respiratory oscillations produced by some disease states.

Other chapters are concerned more directly with the clinical situation and adaptations of the respiratory system. Chapter 6 on respiratory control in man discusses the method and problems of collecting data in human subjects, the respiratory control system in disease and drugs affecting the respiratory control system. Chapter 8 discusses the onset of respiration in the new-born whilst the final chapter is concerned with adaptation to high altitude.

The volume, in setting out reviews of these areas, will be of use to those requiring either an up-to-date view of respiratory control mechanisms or to people requiring an entrée to a massive literature. Clinicians, in particular those with an interest in chest medicine and anaesthesia, will be able to update their knowledge, while advanced students of biology can obtain an introduction to respiratory mechanisms.

1 PERIPHERAL ARTERIAL CHEMORECEPTORS

D.S. McQueen and D.J. Pallot

This chapter describes the influence of the peripheral arterial chemo-receptors on respiration and also examines the transduction mechanism of these receptors. It is our intention to concentrate mainly on the recent literature; readers requiring information about historical aspects, or more detailed evidence than it is possible to give in this chapter, are referred to reviews on chemoreceptors (e.g. Schmidt & Comroe, 1940; Heymans & Neil, 1958; Dejours, 1962; Anichkov & Belen'kii, 1963; Torrance, 1968; Biscoe, 1971; Eyzaguirre & Fidone, 1980) and the proceedings of recent international meetings on arterial chemoreceptors (Purves, 1975; Paintal, 1976; Acker *et al.*, 1977; Belmonte *et al.*, 1981).

The peripheral arterial chemoreceptors are sensory receptors which monitor the oxygen and carbon dioxide tensions and pH of arterial blood, and form part of the integrated system that controls respiration. A fall in the partial pressure of oxygen in arterial blood (PaO_2), a rise in $PaCO_2$ or a fall in pH, increases the activity of these receptors, increases discharge frequency in their sensory (afferent) nerves, which terminate centrally mainly in the nucleus tractus solitarius, and ultimately may stimulate respiration. Two main groups of arterial chemoreceptors exist in mammals, namely the aortic and carotid bodies located, respectively, near the arch of the aorta and in the bifurcation of the common carotid arteries. The carotid body is very vascular and is supplied with arterial blood by one or more arteries originating from the bifurcation or the external carotid artery; venous blood drains into the internal jugular vein. The aortic bodies are supplied with blood from small branches of the aortic arch. In addition, pulmonary chemoreceptors have been described (Coleridge, Coleridge & Howe, 1967) and tissue similar in structure to that of chemoreceptors is found along the course of the abdominal aorta (abdominal para-ganglia, Abbot & Howe, 1972). This review will only be concerned with chemoreceptors of the aortic and carotid bodies because it is these which have been most extensively studied and they appear to be of greatest importance.

Each carotid and aortic body receives nerve fibres from two sources. The glossopharyngeal (cranial IX) nerve supplies sensory fibres to the carotid body (glomus caroticum) via a branch known variously as the carotid nerve, the intercarotid nerve, the carotid sinus nerve, the sinus nerve or Hering's nerve, and the cell bodies of these sensory fibres are located in the sensory (petrosal) ganglion of the IXth nerve. In addition, the carotid body is also innervated by sympathetic fibres (ganglioglomerular nerves) from the superior cervical ganglion (see Figure 1.1). Some sympathetic fibres also course down the sinus nerve. The aortic bodies are innervated by the (aortic) vasodepressor nerve (a branch of the vagus) which has its cell bodies in the nodose ganglion; they also receive a sympathetic nerve supply from the stellate ganglion.

Figure 1.1: A Diagrammatic Representation of the Innervation of the Cat Carotid Body. Note the dual innervation from the sinus nerve and ganglioglomerular nerve. The aortic bodies also receive a dual innervation.

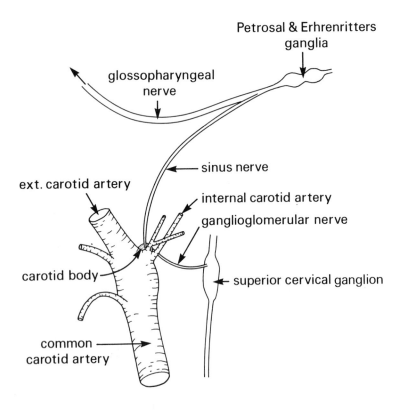

Histologically the carotid and aortic bodies consist of groups of cells collected around numerous large capillaries. A connective tissue stroma is found between groups of cells and contains numerous nerve fibres; on closer examination each cell group is seen to contain two cell types which have been given a variety of names (see Biscoe, 1971). We shall apply the term Type I to the most common type of cell, and Type II for the less frequently encountered cell which is often located at the periphery of a cell group. The general histological features of the carotid body are illustrated in Figure 1.2; and we shall make a more detailed examination of the important structural features of peripheral chemoreceptors in a later section.

The embryological origin of the carotid body has only recently been established. Using a heterospecific grafting technique (Le Dovarin, Le Lievre & Fontaine, referred to by Verna, 1979) which enabled donor cells to be identified in the recipient, Pearse and his colleagues (see Verna, 1979) established that in birds Type I cells were definitely derived from neural crest tissue, and Type II cells may also originate from the same source. A number of studies have described the development of the carotid body (see Verna, 1979); all of these demonstrate that the carotid body is structurally mature well before birth, a somewhat surprising finding in view of the physiological immaturity of the organ prior to the onset of respiration (see Chapter 8).

Two particularly interesting features of the carotid chemoreceptors emerged from studies performed on cats. They have a very high oxygen consumption of $\simeq$ 9ml $(100g^{-1})min^{-1}$, indicative of intense metabolic activity, together with a very substantial blood flow of 2 litres $(100g^{-1})$ min^{-1}. The difference between arterial and venous blood oxygen content is, however, fairly low (0.2 - 0.5 ml/100ml) and the venous blood is quite red; this could result from low oxygen extraction from the blood, or from direct passage of blood from arteries to veins via extensive anastamoses in the carotid body.

Study of Peripheral Arterial Chemoreceptor Activity

Various techniques are employed for studying chemoreceptor activity, including the classical approach of using reflexly induced respiratory changes as an indicator of receptor activity (see Heymans & Neil, 1958). Electrophysiological techniques are now commonly used because they avoid some of the complications associated with reflex studies, such as the influence of the anaesthetic on the reflex pathway, or

Figure 1.2: The Histology of the Carotid Body. Note how the cells are set in a connective tissue stroma (ct) amidst numerous blood vessels (c). Some bundles of nerve fibres are also seen (small arrow head). Even at this magnification two cell types are visible on the basis of nuclear morphology (T1 and large arrow heads), and position in the cell groups.

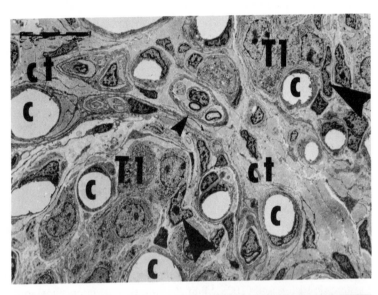

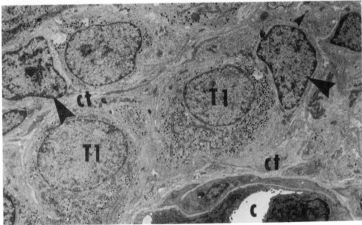

secondary changes such as those initiated by alterations in blood gas tensions or by the lung inflation reflex. Such techniques are generally regarded as providing more information about activity at the receptor; what they actually provide is information from a small sample of chemosensory fibres which is taken to represent the response of the whole carotid body complex.

Recordings of single units (single active fibres) obtained using extracellular recording electrodes show that action potentials associated with chemoreceptor fibres occur irregularly. This distinguishes them from the regular pressure-related discharge associated with baroreceptor fibres (Figure 1.3). The interval between successive chemoreceptor potentials is in fact random, following a Poisson distribution (see Biscoe, 1971). Furthermore, unlike the baroreceptors which cease firing when the arterial blood pressure falls below a certain level, chemoreceptor afferents do not exhibit a firing threshold.

Figure 1.3: Electrical Recording of Sensory Activity in a Filament Dissected from the Peripheral End of a Carotid Sinus Nerve which had been Cut Centrally (A). Note the irregular discharge of the large biphasic chemoreceptor unit and the regular discharge of the smaller monophasic baroreceptor unit which is synchronous with the pulse seen in the blood pressure record (B) (McQueen, unpublished record). In (C) the lower part of the chemoreceptor action potential was used to trigger 50 successive sweeps of a storage oscilloscope during intense chemoreceptor stimulation evoked by sodium cyanide (5 μg intra-carotid); it can be seen from the trace on the right-hand side of the figure that no chemoreceptor action potential occurred within 7 ms of a triggering potential. This observation, together with the shape and amplitude of the action potential, enables it to be classified as a single unit.

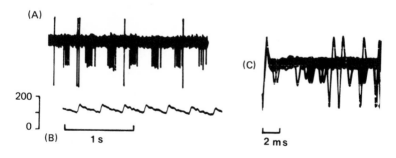

The effects on chemoreceptor discharge and on respiration of breathing gas mixtures containing different concentrations of CO_2 and O_2 have been studied by various workers using both respiratory and electrophysiological techniques (e.g. Lahiri & Delaney, 1975).

Function

As already mentioned, the arterial chemoreceptors respond to changes in arterial blood gas tensions and pH and reflexly influence respiration. The chemoreceptors are responsible for about 15 per cent of the respiratory drive during eupnoeic (normal) respiration, most of this contribution appears to come from the carotid chemoreceptors in man. The carotid body is the only mechanism in the body for detecting hypoxia and initiating rapid reflex respiratory adjustment. The central chemoreceptors are not stimulated by hypoxia; rather breathing is depressed by the direct effect of hypoxia on the CNS. This effect is in contrast to that of CO_2-H^+ in the physiological range which can act centrally to increase respiration, as well as reflexly via the arterial chemoreceptors. Very high levels of CO_2-H^+ can cause a central depression of ventilation. In this chapter we are concerned exclusively with the influence of arterial chemoreceptors on respiration, but it should be appreciated that these receptors reflexly influence other body functions, such as control of the cardiovascular system (Heymans & Neil, 1958; Anichkov & Belen'kii, 1963).

Peripheral chemoreceptor discharge increases when either PaO_2 falls or the $PaCO_2$ rises. The two stimuli interact in the peripheral chemoreceptors in such a way that hypoxia enhances the discharge associated with hypercapnia, and vice versa. The interaction is additive or more than additive (multiplicative) at lower discharge frequencies. Clearly as the response to either stimulus alone approaches maximum, the potentiation produced by the two stimuli given together is less marked. There is not much change in ventilation when PaO_2 falls during normocapnia in steady-state conditions, until a PaO_2 of about 8.7 kPa ($1 kPa \simeq 7.5$ mmHg) is reached when breathing increases. Peripheral chemoreceptor activity, however, increases when slight falls in oxygen tension occur even at levels of PaO_2 well in excess of the respiratory 'threshold'; chemoreceptor activity is not linearly related to PaO_2, the curve of discharge against PaO_2 steepens as the PaO_2 falls. The explanation for the lack of correlation between chemoreceptor discharge and ventilation is that the peripheral chemoreceptor drive is opposed by direct depression of the respiratory centre by hypoxia; additionally the fall in $PaCO_2$ that occurs in response to the hyperventilation decreases the drive from the central chemoreceptors. When a critical PaO_2 is reached, these factors are unable to negate the increased input from the peripheral chemoreceptors, and respiration increases.

There is an almost linear relationship between carotid chemoreceptor discharge and $PaCO_2$ over the physiological range, the slope of the curve steepening as PaO_2 is lowered. A 'fan' of lines relates respiration to $PaCO_2$ at different oxygen tensions in man (see Cunningham, 1974). Below a certain $PaCO_2$, known as the 'CO_2 threshold', ventilation appears to be independent of $PaCO_2$ (even though peripheral chemoreceptor activity is not) and the overall CO_2-ventilation curve is often described as being 'dog-legged' in shape, something which is most apparent under hypoxic conditions. Experiments in anaesthetised animals show that ventilation is not totally independent of $PaCO_2$ in the range 2.7 – 5.3 kPa, and apnoea can occur even in severe hypoxia when the $PaCO_2$ is very low. However, anaesthesia interferes with the responses, as is shown by the fact that anaesthesia in man raises the CO_2 threshold and the 'dog-leg' is not seen. Respiratory thresholds vary from individual to individual, and even within individuals; it is thus very difficult to obtain meaningful data from conscious animals and man when breathing is subject to minimal drive because ventilation becomes very irregular (see Cunningham, 1974). Hence it is difficult to be sure what the CO_2-respiration dog-leg signifies, but it may result from CO_2-H^+ acting in the medulla to modify afferent input from the peripheral chemoreceptors.

Experiments in man using the single breath O_2 test (see Dejours, 1962) enable assessment of peripheral chemoreceptor respiratory drive to be made since the single breath will influence the arterial chemoreceptors before the CNS, and changes in respiration will not be complicated by the alterations in blood gas tensions and pH which accompany alterations in ventilation under steady-state conditions. It is found that a single breath of 100 per cent O_2 causes a transient fall of about 15 per cent in respiration. This response has a latency of 1–3 breaths due mainly to the circulation time from alveoli to arterial chemoreceptors, and is not obtained when the peripheral arterial chemoreceptors are denervated. If oxygen breathing continues, respiration returns to normal because although removal of peripheral chemoreceptor drive initially reduces ventilation, CO_2 retention occurs and the rise in $PaCO_2$ stimulates respiration, so that within one minute breathing is back to normal. This can be seen in Figure 1.5, where respiration during breathing 100 per cent O_2 in steady state is not much different from that during air-breathing, although carotid chemoreceptor discharge continues to be reduced by the high oxygen tension.

Figure 1.4: Discharge of a Single Chemoreceptor Unit Showing the Effects of Intracarotid Injections of Different Substances at the Marker. Also shown is the maximal discharge recorded 90 s after starting to ventilate the animal with 100 per cent N_2 instead of air. (McQueen, unpublished cat data.)

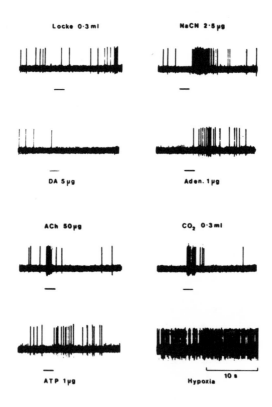

Peripheral chemoreceptors can respond to CO_2 or H^+, but the response to H^+ is not as fast or intense as that to a transient CO_2 stimulus producing the same pH change. This is probably because molecular CO_2 crosses biological membranes faster than H^+. Carbonic anhydrase inhibitors such as acetazolamide slow the hydration of CO_2

and reduce the speed and intensity of the chemoreceptor response to CO_2, which suggests that at least part of this response involves conversion of CO_2 into H^+/HCO_3^-. Part of the effect has been attributed to molecular CO_2 because the chemoreceptor response to CO_2 under steady conditions is greater than that to the equivalent change in H^+. The precise stimulus (CO_2 and/or H^+-HCO_3^-) at the peripheral chemoreceptor remains to be determined, and it will be necessary to ensure that drugs like acetazolamide act specifically as inhibitors of carbonic anhydrase in the chemoreceptors, and do not have other actions which affect the response of these receptors to CO_2.

There has been some controversy concerning the ability of CO_2-H^+ to stimulate the aortic chemoreceptors, and Hanson, Rao & Torrance (1979) who investigated the CO_2 sensitivity of the cat aortic chemoreceptors concluded that although some of the aortic chemoreceptors do respond to CO_2, the responses are very variable in the steady state. This is in contrast to the marked effect CO_2 has on the carotid chemoreceptors, and the authors consider that the difference may be due to a greater adaptation of the aortic receptors to CO_2. The fact that single carotid chemoreceptor units can respond to both CO_2 and hypoxia, as well as other stimuli, is illustrated in Figure 1.4.

It is generally agreed that activation of the carotid chemoreceptors increases the tidal volume and, to a lesser extent, the frequency of respiration. Aortic chemoreceptor stimulation has less effect on tidal volume, but does increase respiratory frequency. The overall increase in respiratory minute volume in response to the same stimulus is about seven times greater for the carotid chemoreceptors than it is for the aortic chemoreceptors in anaesthetised dogs (Daly & Ungar, 1966). However, the overall contribution to respiration made by the aortic and carotid bodies varies from species to species and depends on the techniques used. Humans with denervated carotid bodies but intact aortic bodies do not show an increase in tidal volume when exposed to hypoxia whereas those with intact carotid bodies do (Wasserman & Whipp, in Paintal, 1976). Hypoxia depresses the CNS and might mask the input from aortic chemoreceptors, but people who had undergone bilateral carotid body resection and were not hypoxic showed no significant increase in ventilation following i.v. injection of doxapram, a drug which can stimulate the peripheral chemoreceptors, whereas normal individuals responded to the same dose with a significant increase in ventilation (Honda *et al.*, 1979). Smith & Mills (1980) however have demonstrated that cats which have undergone bilateral section of the sinus nerve develop a renewed sensitivity to hypoxia after 2–3 months, presumably due to stimulation of the aortic bodies.

Chemoreceptor stimulation also indirectly affects respiration by reflexly affecting bronchial airflow. Nadel & Widdicombe (1962) concluded that carotid chemoreceptor stimulation in dogs reflexly constricts the upper and lower airways, thus leading to changes in blood gas tensions. Reflex changes in cardiac output and catecholamine secretion from the adrenal gland may also indirectly affect blood gas tensions. There is evidence that sudden application of stimuli (i.e. abrupt changes in blood gas tensions) evoke respiratory responses only if they reach the peripheral chemoreceptors at the appropriate point of the respiratory cycle. Thus, chemoreceptor stimulation during inspiration leads to an augmented breath and an increase in respiration, whereas the same stimulus arriving during expiration may have no effect on respiration (Black & Torrance, 1967). This type of stimulus is not very physiological, but oscillations of PO_2, PCO_2 and pH related to respiratory periodicity do occur in arterial blood (see Chapter 7).

Cat carotid chemoreceptors can follow oscillations up to a frequency of about 20 Hz for PaO_2 and 70 Hz for $PaCO_2$, the faster the frequency, the smaller the oscillation in chemoreceptor discharge. The receptors are, therefore, capable of signalling breath to breath changes in blood gas tensions and pH. The phase relationship between chemoreceptor afferent discharge and activity of the inspiratory neurones may well be important in adjusting respiration during speaking, eating, swallowing, yawning, laughing, crying and, perhaps, exercise. Temporal differences between aortic and carotid chemoreceptor input to the CNS could theoretically be used to compute blood/gas flow to the brain, but recent work by Lahiri *et al.* (1980a) revealed that the latency to onset of response following changes in the gas mixture breathed, or intravenous injection of chemoreceptor stimulants such as cyanide, is greater for the aortic than it is for the carotid chemoreceptors in cats. Whether the CNS makes use of the temporal difference between carotid and aortic chemoreceptor input, or input from myelinated and unmyelinated fibres, remains to be established. It has also been found that rapid adaptation of chemoreceptor discharge occurs when CO_2 is used as a stimulus, but not when hypoxia is used (Black, McCloskey & Torrance, 1971) and the response to CO_2 is generally more rapid in onset than that to hypoxia. The rapid or dynamic nature of the response to peripheral chemoreceptor stimulation is a particularly marked feature of these receptors and is illustrated in Figure 1.4.

Hypoxia can be induced in tissues receiving a normal blood flow by lowering the oxygen tension of the blood (i.e. hypoxic hypoxia affecting the amount of oxygen dissolved in the plasma) or, by causing histotoxic hypoxia, as does cyanide, by inhibiting cytochrome oxidase. Examples of both these methods have already been given. An additional method of producing hypoxia is to reduce the amount of oxyhaemoglobin in the blood, leading to anaemic hypoxia, and this can be achieved by lowering the number of red blood cells or by using carbon monoxide to form carboxyhaemoglobin, thereby reducing the oxygen content of the blood without affecting the oxygen tension. One of the reasons why carbon monoxide is so dangerous is that no hyperventilation occurs during poisoning with this gas. There has been considerable controversy over the question of whether or not carbon monoxide stimulates arterial chemoreceptors, some workers finding carbon monoxide to be without effect on carotid chemoreceptors, others that it stimulates the aortic chemoreceptors and/or the carotid chemo-receptors. Recent work by Lahiri *et al.* (1981) appears to clarify matters by showing that in cats in which carotid and aortic chemo-receptor discharge are recorded simultaneously, carbon monoxide inhalation during normoxia always stimulated aortic chemoreceptors before it affected the carotid receptors, and that the steady-state response of aortic chemoreceptors to carboxyhaemoglobin was greater than that of most carotid chemoreceptors; only about 10 per cent of the carotid chemoreceptor fibres tested showed any increase in discharge frequency in response to moderate increases in carboxyhaemoglobin. The authors hypothesise that the aortic bodies have a much lower perfusion relative to their oxygen utilisation compared to the carotid bodies, and as a consequence are able to monitor oxygen delivery and initiate circulatory reflexes for oxygen homeostasis; whilst carotid chemoreceptors monitor oxygen tension and initiate strong reflex effects on respiration.

Thus, the carotid body chemoreceptors seem to be satisfied by the oxygen dissolved in the plasma, and carbon monoxide poisoning is unlikely to cause any reflex respiratory changes via the peripheral chemoreceptors until the levels of carboxyhaemoglobin are quite high. Results obtained from experiments in which the haematocrit was reduced are consistent with the carbon monoxide results, a decrease in the number of red blood cells during normoxia stimulated the aortic chemoreceptors, but much larger reductions in haematocrit seldom affected the carotid chemoreceptors provided the PaO_2 was above 8.0 kPa.

Blood Flow

The carotid body blood flow rate is very high, as already mentioned. Purves (1970) performed experiments in anaesthetised cats from which he concluded that carotid blood flow of the intact innervated carotid body is linearly related to mean arterial pressure over the range 13.3 - 22.7 kPa, and hypoxia or hypercapnia can cause a small increase in carotid body blood flow. Others have presented evidence that autoregulation of blood flow occurs in the carotid bodies of some species. In many vascular beds trauma, or extremes of perfusion pressure, abolishes autoregulation, and great care has to be taken not to damage what can be a fairly delicate mechanism. Changes in arterial blood pressure within the physiological range do not cause any significant alteration in carotid chemoreceptor discharge, but do affect the aortic chemoreceptors; a fall in pressure increases discharge. It is interesting to note that when a cat carotid body is removed from the animal and superfused *in vitro*, chemoreceptor discharge recorded from the carotid sinus nerve increases rapidly in response to a reduction in flow of the superfusing fluid. Oxygen usage is also reduced. The flow-sensitivity of the *in vitro* carotid body thus makes it more akin to the *in vivo* aortic body.

The capillaries of the carotid body are fenestrated, and numerous A-V shunts exist. There has been much discussion and experimental work concerning the distribution of blood within the carotid and aortic bodies and the suggestion has been made that 'plasma skimming' may occur. This arises from non-Newtonian fluid flow in narrow tubes, and the suggestion is that as the speed of the current is greatest and the pressure lowest at the centre of the blood vessel, red cells will migrate to the central axis and travel faster than the plasma at the side walls. A small branch from a larger artery would receive blood with a lower proportion of red cells, i.e. the plasma has been skimmed. Such a mechanism would explain the observations by Acker & Lübbers (in Acker *et al.*, 1977) that the decrease in tissue PO_2 is the same under zero flow conditions whether the organ is perfused by blood or saline equilibrated with air. These authors propose that during progressively increasing hypoxia an increasing number of red cells enter the carotid body. The idea is attractive for it enables the tissue PO_2 to be maintained. Unfortunately, however, preliminary quantitative histological studies on the area occupied by red cells within the carotid body vasculature at different levels of PaO_2 in rabbits do not support this concept (Pallot & Verna, unpublished observations).

Experiments with oxygen-sensitive microelectrodes (e.g. Acker & Lübbers in Acker *et al.*, 1977) have established that the PO_2 in carotid body tissue is about 5.3 kPa, much lower than the PaO_2. However, differing results have been obtained concerning the oxygen gradient within the carotid body, and further refinement of the technique will be needed before one can obtain reliable information concerning the oxygen tension in the vicinity of the nerve endings and in the cells.

Influence of Exercise on Chemoreceptor Discharge

There has been considerable debate concerning the question of whether or not arterial chemoreceptors contribute to the hyperpnoea of exercise (see Cunningham (1974) for a detailed discussion of breathing in exercise). There is some evidence which favours peripheral chemo-receptor involvement. Thus, administration of oxygen reduces the hyperpnoea of exercise, the respiratory response to exercise in ordinary individuals is potentiated by hypoxia and it is absent in people who have undergone bilateral carotid resection, as is their late, but not early, respiratory response to exercise (Wasserman & Whipp, in Paintal 1976). If arterial chemoreceptors are involved in exercise hyperpnoea, what is it that activates the receptors? PaO_2, $PaCO_2$, pH and body temperature change only slightly in the early stages, and although the possibility exists that the sensitivity of the peripheral chemoreceptors to these variables may alter during hypoxia, other influences need to be considered. Sympathetic activity increases during exercise and increased sympathetic drive to the peripheral chemoreceptors can increase ventilation. Some results in man accord with this idea, but Eisele, Ritchie & Severinghaus (1967) concluded that the steady-state ventilatory response to moderate exercise in man is not influenced by sympathetic innervation of the chemoreceptors since blocking the stellate ganglia had no marked effect on the hyperpnoea of exercise. Experiments in anaesthetised cats showed that moving the hind limbs caused an immediate increase in carotid chemoreceptor discharge which was abolished by removing the sympathetic innervation of the carotid body or cutting the femoral and sciatic nerves (see Biscoe, 1971). Could it be that in exercise increased activity in muscle afferents reflexly increases chemoreceptor discharge? Not according to Davies & Lahiri (1973), who also performed experiments on cats (anaesthetised or decerebrated), since they were unable to show any change in carotid chemoreceptor activity during exercise, although hyperpnoea occurred.

The present position concerning the involvement of peripheral chemoreceptors in exercise hyperpnoea is somewhat confusing, and further studies appear to be necessary in order to establish their role. It could be that muscle afferent activity acts centrally to increase the influence on respiration of a given chemoreceptor input. It is also possible that during exercise, changes in the oscillations of blood gas tensions, cardiac output, circulating catecholamines or other hormones, and carotid sinus nerve efferent activity, might influence arterial chemoreceptor activity by acting either separately or interactively (see also Chapter 7).

Chronic Resection of the Carotid Bodies

Many patients in Japan some 25 years ago underwent bilateral carotid body resection for asthma. This procedure only temporarily relieved their symptoms (see Honda *et al.*, 1979). Treated patients have a normal $PaCO_2$ and pH, but a slightly reduced PaO_2, do not show any reflex increase in respiration while breathing hypoxic gas mixtures in the steady-state; in single breath studies there is evidence of some chemosensitivity. It appears that although aortic chemoreceptors can prevent the central respiratory depression of hypoxia, they do not elicit hyperventilation and so, according to Wasserman & Whipp (in Paintal, 1976), are relatively insensitive respiratory control organs in man. Using few-breath oxygen tests in conscious dogs, Ungar & Bouverot (1980) have demonstrated that even moderate hypoxia can cause a marked depression of respiration. Therefore, it may be that aortic chemoreceptors do make an important contribution to respiration, but this is not apparent during hypoxia in the absence of carotid chemoreceptors because of the substantial depressant effect hypoxia exerts on central respiratory neurones.

Animal experiments show that different species vary in their ability to recover peripheral chemosensitivity following bilateral carotid body resection. Slight sensitivity returns in ponies, but in cats after about 260 days peripheral chemoreflex activity apparently returns to near normal. No chemoreceptor activity was recorded from the regenerated carotid sinus nerve, and bilateral vagotomy abolished the reflex respiratory response to physiological and pharmacological stimuli, implying that the aortic receptors were being stimulated (Smith & Mills, 1980). The suggestion is that, about 30–40 days after bilateral carotid body removal, aortic chemoreceptors increase their influence on respiration, possibly as a consequence of changes in gain of the central component of the reflex pathway.

High Altitude

This topic will be dealt with in detail in Chapter 9 and we shall confine ourselves to noting here that during adaptation to high altitude, respiration is stimulated by hypoxia acting via the peripheral chemoreceptors; hypocapnia and alkalosis limit the hyperventilation. However, acclimitisation can still occur after chronic denervation of the aortic and carotid chemoreceptors which suggests that although the peripheral chemoreceptors do contribute to the process they are not absolutely essential. It is interesting that carotid bodies taken from high altitude-dwelling man and animals are greatly enlarged. This hypertrophy is also observed in chronically hypoxic individuals at sea level and can be mimicked in the laboratory. Thus, Dhillon, Barer & Walsh (in Pallot, 1983) have shown a sixfold increase in carotid body volume after exposure of rats to 10 per cent O_2 for three weeks. Barer, Chiocchio & Pallot (unpublished) found a massive increase in catecholamine concentration in rat carotid bodies after similar treatment.

Influence of Peripheral Chemoreceptors on Respiration in Neonates

The role of arterial chemoreceptors in initiating respiration, and the influence of sympathetic nerves on peripheral chemoreceptors in the new-born is considered in Chapter 8. We should like to concentrate here on the question of whether failure of peripheral chemoreceptors contributes to the sudden infant death syndrome (SIDS) or 'cot death', in which apparently healthy infants die when aged between about 1 and 10 months (average 2-4 months in the study of Naeye *et al.*, 1976), while asleep. Histological studies of carotid bodies taken from infants who had died of SIDS show abnormalities when compared with carotid bodies taken from age-matched infants dying from other causes, including congenital heart disease (Naeye *et al.*, 1976; Cole *et al.*, 1979).

In general the carotid bodies in SIDS showed a reduction in the size and number of cells and also a decrease in the dense cytoplasmic granules in the cells, something which was not seen in the carotid bodies of infants who had been chronically hypoxic thoughout life and eventually died of congenital heart disease. This suggests that degranulation in SIDS is not secondary to hypoxia.

A correlation has been suggested between prolonged apnoea during sleep, particularly REM sleep, and SIDS; infants considered at risk for SIDS (Shannon, Kelly & O'Connell, 1977) showed hypoventilation during sleep and a decreased ventilatory response to breathing CO_2. While it is obvious that peripheral chemoreceptor involvement in SIDS is far from established, and other causes have been suggested, there seems sufficient evidence to warrant serious study of the possibility. It could be that in this disorder arterial chemoreceptors do not function normally because of some defect in the receptor mechanism, or alternatively the CNS fails to respond to chemoreceptor input during hypoxia; perhaps excessive efferent activity in the sinus nerve during hypoxia is a contributing factor. It will be necessary to determine how sleep influences the arterial chemoreceptors and respiratory control. There may be a case for using the single breath oxygen test to examine whether peripheral chemoreceptor function is normal in young infants, and this should be done while the child is asleep. Such a test may help to identify those infants who may be at risk, and could give information concerning the role of peripheral chemoreceptors in SIDS which is otherwise going to be difficult to obtain from animal studies.

Peripheral Arterial Chemoreceptors and Respiratory Disorders

Although the arterial chemoreceptors may not be essential for ordinary life in healthy individuals, the situation is very different in unhealthy people. For example, patients with chronic obstructive lung disease (emphysema, chronic bronchitis) are usually hypoxic, hypercapnic and acidotic; such patients have chronically enlarged carotid bodies (see Heath, Smith and Jago in Pallot, 1983).

The consequence of chronic CO_2 retention is that the central chemoreceptors become less responsive to CO_2, respiratory depression occurs, and the balance of respiratory drive shifts from the central to the peripheral chemoreceptors; hypoxaemia depresses the CNS, but the peripheral chemoreceptors are stimulated. A similar situation can arise when the CNS has been depressed by drugs such as barbiturates or opiates.

Injudicious administration of oxygen to raise the PaO_2 in patients such as those described above can, by removing peripheral chemoreceptor drive, stop ventilation, thereby causing a further rise in $PaCO_2$

and a fall in PaO_2. However, there is a definite need to increase the PaO_2 in patients with respiratory depression, particularly when hypoxaemic episodes associated with periods of hypoventilation occur during REM sleep (compare SIDS) in 'blue and bloated' patients who are already hypoxic (Douglas *et al.*, 1979), and this requires controlled oxygen therapy, with or without the assistance of mechanical ventilation or respiratory stimulant drugs.

Drugs which Stimulate the Peripheral Chemoreceptors

Drugs are chemicals and so it is not surprising that a large number of them can affect the arterial chemo (chemical)-receptors. We shall confine ourselves to considering here only those drugs which are fairly specific chemoreceptor stimulants and which may have a clinical use as respiratory stimulants. In passing, however, it is worth noting that one widely used non-specific chemoreceptor stimulant, nicotine, evokes reflex changes in B.P., heart rate, and respiration (see Ginzel, 1975). It remains to be established whether repeatedly stimulating the arterial chemoreceptors by smoking cigarettes causes changes in the receptors, which in turn leads to pathophysiological alterations in the respiratory and cariovascular systems.

Suberyldicholine is a nicotinic receptor agonist which has been advocated for use as a respiratory stimulant (Anichkov & Belen'kii, 1963) as it has advantages over centrally acting drugs. Doxapram appears to act as a peripheral chemoreceptor stimulant (Mitchell & Herbert, 1975) and has been used to test peripheral chemoreceptor functioning in man. Some drugs may appear to be peripheral chemoreceptor stimulants, but in fact act centrally to potentiate input from the chemoreceptors. It is necessary to be cautious when interpreting results from experiments in which respiration is recorded and drug effects studied. Abolition of the respiratory-stimulating property of a drug by peripheral chemoreceptor denervation is not, on its own, sufficient evidence that the drug stimulates the chemoreceptors and more direct evidence, e.g. electrophysiological, is needed. The drug might, for example, have been acting centrally, and changes in the pattern of respiration following denervation may mask this action, or it could have been indirectly affecting ventilation via the baroreceptors since peripheral chemoreceptor denervation generally involves denervating the baroreceptors. Also, evidence from animals, anaesthetised or conscious, is not necessarily directly applicable to man.

A new chemoreceptor stimulant drug, which may be useful in replacing hypoxic drive from the peripheral arterial chemoreceptors during oxygen breathing, is almitrine. Experiments in cats (Figure 1.5) show that intravenous infusion of almitrine increases carotid chemoreceptor discharge and respiration, even when the animal breathes 100 per cent O_2; it does not seem to be a nicotinic agonist since its action was not affected by hexamethonium or mecamylamine. Almitrine may prove to be clinically useful for replacing hypoxic drive during breathing oxygen-rich mixtures, and it will be interesting to determine its mechanism of action on the peripheral chemoreceptors.

Mechanism of Chemoreception

It seems fair to state at the outset that there is no generally accepted explanation of the mechanisms involved in chemoreception, and there is still considerable debate concerning which of the elements present in peripheral chemoreceptor tissues (cells, nerve fibres, blood vessels, connective tissue) is the primary receptor. Many hypotheses have been advanced to explain how the physiological stimulus to the peripheral arterial chemoreceptors (hypoxia and/or hypercapnia) is transduced into the neural signal carried by sensory fibres to the CNS. The topic is a complex one, partly because of technical difficulties inherent in studying small structures such as the carotid and aortic bodies, and partly because a lot of conflicting evidence has been amassed from experiments, often performed on different species, using a variety of experimental techniques. We propose to examine the different components in the chemoreceptors and then will review some of the main hypotheses, finally finishing this chapter by considering some of the more recent evidence which may suggest new interpretations of existing information, or lines for future research.

Cells of the Arterial Chemoreceptors

Specific chemoreceptor tissue comprises about 50 per cent of the carotid body in cat and rat. The tissues involved are the Type I and Type II cells, blood vessels, nerve fibres and nerve terminals. The Type I cell (see Figure 1.6) is present in the greatest number and is about 10μm in diameter. It has a large dense nucleus, vesicles containing electron-dense granules, an endoplasmic reticulum, Golgi apparatus and microtubules, i.e. all the features associated with secretory cells. The cells are arranged in groups, with a single nerve fibre

Figure 1.5: The Effects of i.v. Infusion of 50 μg (kg^{-1}) min^{-1} Almitrine. Panels show, from above downwards: respiratory half-minute volume 60s after switching to breathing the gas indicated (each breath being represented by a step in the pneumotachograph record); femoral B.P.: counter output in counts s^{-1} for a single carotid chemoreceptor unit. The neurograms to the right of the figure were recorded during oxygen breathing. A is the control, B is 2 and 4 min after starting the infusion, C is 6 and 8 min and D 10 and 12 min after starting the infusion.

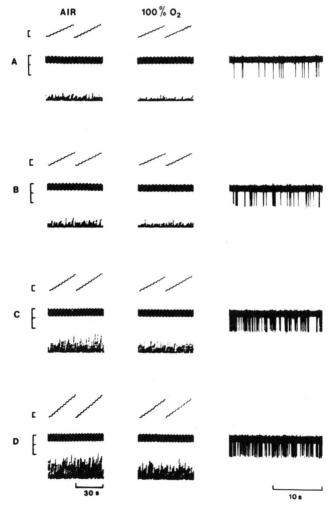

Source: McQueen, Price and Ungar, unpublished data.

Figure 1.6: The Ultrastructure of the Carotid Body. T1 — Type I cell; Type II cell; ➤ nerve fibres; ● endoplasmic reticulum; ✱ mitochondria; ★ electron-dense cored vesicles; n.e. nerve endings. Fig 1.6a shows two adjacent Type I cells. Note the mitochondria and electron-dense cored vesicles. Most of the free surface of the Type I cells is covered by attenuated processes of Type II cells (arrows); in some areas however (small arrow heads) the Type I cells communicate directly with the extracellular space. Fig 1.6b and 1.6c illustrate nerve endings. Note how in 1.6b the ending contains mainly mitochondria, whilst those in 1.6c contain many clear cored vesicles. In all three figures (and in Fig 1.2) note the peripheral position of the Type II cells, and the numerous unmyelinated nerve fibres.

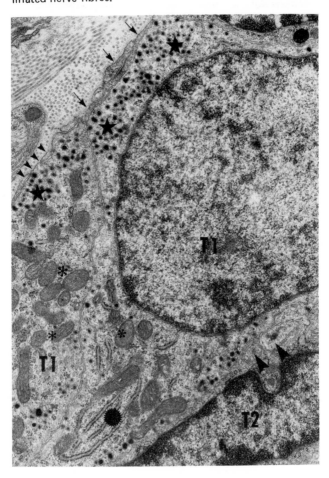

Figure 1.6b and c (*contd*)

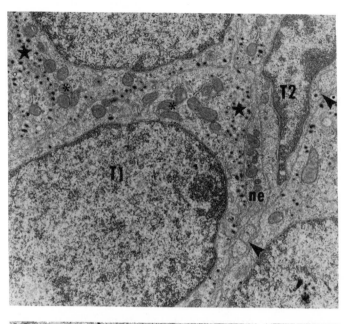

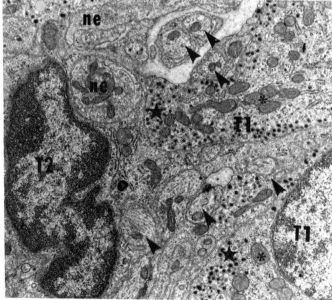

sending branches to between 10 and 20 Type I cells; a fibre can inner-vate more than one cell group. The nerve terminals are intimately associated with the cells, there being a synapse or gap of about 30 μm between nerve terminal and Type I cell membrane. Cutting the inner-vation to cat and rat carotid bodies does not affect the appearance of Type I cells, although the sensory fibres degenerate. In the rabbit, however, section of the sinus nerve causes regression of the carotid body (Kienecke *et al.*, 1981; Tan, Pallot & Purves, 1981 both in Bel-monte *et al.*, 1981). Type I cells are also intimately associated with the blood supply, being exposed to the pericapillary space where the cells are only covered by a basement membrane. The Type II cells which partially surround 4–5 Type I cells have not been studied as extensively as Type I cells. They appear to contain carbonic anhydrase and possibly other enzymes. Unlike the Type I cells, Type II cells do not appear to be innervated, but they do encircle numerous small unmyelinated nerve fibres. Pallot (1976 in Paintal, 1976) has demonstrated that some of these unmyelinated fibres in the cat and mouse carotid bodies terminate without forming Type I cell junctions. The major ultrastructural features of the carotid body are illustrated in Figure 1.6.

Various substances have been identified in carotid body cells and the list includes dopamine, noradrenaline, adrenaline, 5-hydroxytrypta-mine (5-HT, serotonin) and probably acetylcholine in amounts that vary from species to species and even within a species. The fact that these substances are neurotransmitters in other parts of the nervous system has led to their being considered as putative transmitters in arterial chemoreceptors. However, for a long time the carotid body was regarded as a gland, particularly since it is morphologically similar to the chromaffin cell of the adrenal medulla, and there have been sugges-tions that it may have an endocrine or paraneurone (Fujita & Kobayashi, 1979) function. Pearse (1969) classified the Type I cell in his APUD (amine precursor uptake and decarboxylation) cell series and predicted that they secrete a polypeptide which he named glomin. Recent work using immunocytochemical techniques have established that VIP (vasoactive intestinal polypeptide), substance P, methionine and leucine enkephalin are present in the carotid body and it would not be surpris-ing if more polypeptides are found there. Substance P like material has been identified in cells and nerve fibres of the cat carotid body (Cuello & McQueen, 1980). Whether more than one of the putative neuro-transmitters coexist in the cells, and whether the different peptides are stored and released with particular substances requires more detailed investigation as does the identity of the nerves with which the peptides are associated.

The existence of so many substances in the carotid body, together with evidence based on vesicle size and density measurements, has led to the suggestion that Type I cell sub-types exist. This subject is reviewed by Verna (1979) who casts doubt on the hypothesis that cells may be subdivided on the basis of vesicle size.

Nerve Supply to the Arterial Chemoreceptors

The nerve supply to the arterial chemoreceptors has been most extensively studied in carotid body, but there is no evidence to suggest that innervation of the aortic bodies is appreciably different. There is some evidence that in rats a small proportion of nerve endings associated with Type I cells are sympathetic in origin, (McDonald & Mitchell, 1975) but the overwhelming majority of sympathetic nerve fibres from the superior cervical ganglion innervate the carotid body vasculature. Most of the Type I cell nerve endings are derived from nerve fibres in the carotid sinus nerve, and the classical opinion was that they are the terminations of sensory axons (De Castro, 1928). The evidence for this rested on degeneration studies as illustrated diagrammatically in Figure 1.7.

Figure 1.7: Diagram to Explain the Degeneration Studies. 1 and 2 represent extra- and intra-cranial sections respectively (T1 — Type I cell; n.e. nerve ending; CNS brain stem).

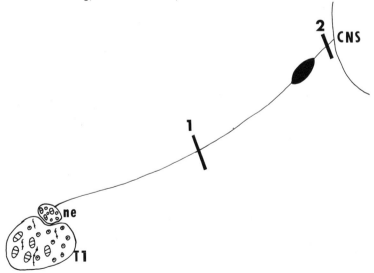

If, after section of the carotid sinus nerve at 1, the nerve endings associated with Type I cells degenerate, then the parent axons run in that nerve. Section of the nerve at 2, central to its sensory ganglion, should differentiate between afferent and efferent nerve endings since the peripheral process of an axon degenerates when severed from its cell body. According to De Castro (1928) section of the carotid sinus nerve at 1 results in degeneration of Type I cell nerve endings whereas section at 2 does not affect the endings.

Physiological studies by Biscoe's group in the late 1960s (reviewed by Biscoe, 1971) have demonstrated efferent or centrifugal activity in the carotid sinus nerve which can be increased by hypoxia, hypercapnia and i.v. adrenaline. In the intact preparation, activation of this efferent pathway decreases afferent activity, and section of the carotid sinus nerve increases chemosensory discharge (see Biscoe, 1971; O'Regan, 1977).

The presence of such a centrifugal pathway in the sinus nerve, and the effects of section of the sinus nerve on chemoreceptor activity, led Biscoe, Lall & Sampson (1970) to repeat the degeneration experiments referred to above (Figure 1.7). They found that three months after section of the IXth nerve between the ganglia and brainstem there was a 60–70 per cent reduction in the number of Type I cell nerve endings; furthermore normal chemoreceptor afferent activity could still be recorded from the sinus nerve. The authors suggested that many of the Type I cell nerve endings represented the terminations of the efferent pathway referred to above. These experiments have been the object of considerable criticism, and subsequent attempts to repeat them have failed (see Verna, 1979). In spite of this it is very difficult to explain the degeneration unless one is to assume that one or other (petrosal or Ehrenritters) ganglion was damaged by Biscoe *et al.* (1970).

This suggestion, which seemed unlikely in view of the normal afferent discharge, may with hindsight prove to be correct. In this respect it is interesting that Willshaw & McAllen (1981 in Belmonte *et al.*, 1981) have provided evidence that the cell bodies of the sinus efferents are located outside the brainstem, in some as yet unidentified site.

Each nerve terminal is characterised by an accumulation of mitochondria and/or clear vesicles (see Figure 1.6). In addition, some endings contain dense-cored vesicles, neurofilaments, neurotubules and glycogen particles. It is presumed that the clear and dense-cored vesicles contain unidentified transmitter substance(s).

The evidence from degeneration studies which purports to establish the nature (sensory or motor) of nerve fibres opposed to the Type I cell has been conflicting, and part of the difficulty seems to arise from the lack of quantitative data. Morphological studies are unable to establish the function of synapses, but the appearance of the nerve-cell junction sometimes is in accord with a motor (efferent) synapse, with vesicles in the nerve terminal apposed to the cell, sometimes a sensory synapse (vesicles in cell), and occasionally a mixture of the two, referred to as reciprocal synapses (McDonald & Mitchell, 1975; Blakeman & Pallot, 1983 in Pallot, 1983). Experiments performed on mutant mice (Pallot, 1978) have shown quantitatively that in these animals, which have a motor disorder, few endings are found on Type I cells in the carotid body as compared with normal mice, although the mutants have normal chemoreceptor reflexes and this implies that a high proportion of endings are motor, supporting the concept of an efferent pathway. Against this is more recent evidence with radioactive amino acids such as proline which, after injection into the petrosal (sensory) ganglion where the cell bodies are situated, is transported to the carotid body. It is claimed that a substantial proportion of nerve endings on Type I cells in cats are labelled, implying that these are sensory nerve terminals. Such a technique however does not lend itself to quantitative studies.

There is evidence from horseradish peroxidase studies which suggests that only very few efferent fibres course from the CNS in the carotid sinus nerve. Again, part of the problem in interpreting the results arises from difficulties with the methodology and the quantification of results. It will be necessary to establish where the cell bodies of the efferent fibres are located.

Pallot and his colleagues (Morgan, Pallot & Willshaw, 1981; Pallot, Morgan & Willshaw in Belmonte *et al.*, 1981) have recently provided evidence for the existence of an efferent innervation of Type I cells in the cat. They ventilated animals with either 10 per cent or 100 per cent O_2 with one sinus nerve intact and the other cut and then examined the vesicle content of the nerve endings. In all cases after ventilation with 100 per cent O_2 there were more vesicles in the endings from the intact nerve carotid body than the contralateral (sectioned nerve) organ. After 10 per cent O_2 there were fewer vesicles in the nerve endings of the intact nerve carotid body than in the contralateral one. As the stimulus to the carotid bodies themselves is similar, regardless of the state of the nerve, this result indicates that the stimulus acted central to the point of section, and is hence a direct efferent effect on these nerve endings. Parallel purely structural studies (Pallot & Blakeman, 1983 in Pallot, 1983) suggest that on the basis of nerve ending polarity some 50 per cent of endings might be efferents, assuming that the methods for assessing synaptic polarity in the CNS are applicable to the periphery (see Verna, 1979).

The balance of evidence then, seems to suggest that both afferent and efferent fibres terminate on Type I cells. In view of the serial reconstruction studies (Biscoe & Pallot, 1972; Nishi & Stensaas, 1974) which suggest that most cells receive innervation from a single axon, it may be that afferent and efferent endings are found on different Type I cells. Efferent fibres may also supply blood vessels since it has been shown that sinus nerve stimulation affects carotid body blood flow, an effect which can be prevented by atropine. The question arises as to how physiologically important the efferent pathway to the peripheral chemoreceptors is, as there is certainly no doubt that the sensory pathway can continue operating in the absence of all efferent nerve input to the chemoreceptors. The argument about afferent/ efferent fibres in the carotid sinus nerve has led to the compromise suggestion (McDonald, 1980) that carotid sinus nerves can carry impulses in both directions, being antidromically activated by potential changes in the central terminals under certain conditions, but such a mechanism cannot explain the observation referred to above. It is also possible that some of the inhibition of chemosensory discharge seen following stimulation of the sinus nerve results from substances being released from sensory endings by antidromic nerve impulses (e.g. polypeptides).

There are myelinated and unmyelinated fibres in the carotid and aortic nerves and conduction velocity studies show about 66 per cent of carotid chemoreceptor fibres conduct at 5-50 m/s (A fibres — presumed to be myelinated), with the remainder conducting at 0.5-2 m/s (C fibres — unmyelinated). Fidone & Sato (1969) found that the fibres differed in their sensitivity to various stimuli, with A fibres being the most sensitive to acetylcholine (ACh). In contrast, the C fibres of the aortic chemoreceptors have been reported to be much more sensitive to ACh than the A fibres in cats. There may be differences in the two chemo-receptor organs in the responsiveness of their myelinated and unmye-linated fibres to ACh, but Paintal (1971) has criticised the technique used by Fidone & Sato for estimating the conduction velocity of fibres in the very short length of carotid sinus nerve available for experimenta-tion, arguing that their results may be misleading.

Sympathetic Nerve Supply to Arterial Chemoreceptors

Arterial chemoreceptors receive a sympathetic nerve supply, as pre-viously mentioned, and there is general agreement that sympathetic nerve stimulation can increase carotid chemoreceptor discharge and respiration although this is a somewhat variable effect. Blood flow through the carotid body is reduced by the same procedure and oxygen consumption falls. Cutting the ganglioglomerular (sympathetic) nerves leads to an increase in blood flow through the carotid body. The fact that sympathetic nerves have no effect on chemosensory discharge of the *in vitro* carotid body preparation supports the notion that the sympathetic nerves regulate blood flow in the carotid body.

Apart from effects on chemoreceptor discharge which are secondary to changes in blood flow, there may also be some direct influence on Type I cells. There is morphological evidence that Type I cells in rats receive a pre-ganglionic sympathetic innervation. Sympathectomy has no effect on the electron-dense granules in the Type I cells or on the catecholamine content of cat carotid body (Zapata *et al.*, 1969; Mir, Al Neamy, Pallot & Nahorski, 1983 in Pallot, 1983) but in the rat carotid body sympathectomy causes a 90 per cent reduction in noradrenaline levels (Mir *et al.*, 1983, in Pallot, 1983).

Blood Vessels

There has already been some discussion concerning blood flow through the carotid body, and it will be necessary to perform further studies *in vivo* in order to compare flow in the aortic and carotid bodies, and to determine how flow is regulated. Meanwhile, it is important to realise that many physiological and pharmacological stimuli are likely to influence blood flow in these organs. Performing experiments *in vitro* may avoid these difficulties, but does introduce other complications, such as the reduced O_2 usage and the inability to sustain chemoreceptor discharge in response to a prolonged stimulus. It has been suggested that responses alter if the carotid body is perfused by saline rather than blood, and this may be due to the greater oxygen-carrying capacity of blood compared with saline, or to the presence of some essential factor in the blood.

Mechanisms of Chemoreception

Various hypotheses have been advanced to explain the mechanism involved in chemoreception, and these will be considered in relation to recent evidence. Biscoe (1971) proposed that free nerve endings in the pericellular space are the chemosensors and the endings on Type I cells are efferent, being involved in modulating sensory activity. The suggestion is that these fine nerve endings have a high metabolic rate, which is needed to maintain their polarisation, and a large surface to volume ratio. They respond rapidly to hypoxia, a fall in PaO_2 leading to depolarisation of the fibres as a consequence of reduced ion pump (e.g. sodium pump) activity. Sympathetic nerve stimulation reduces oxygen usage by the carotid body, but increases chemoreceptor discharge, and this could be interpreted as meaning that the oxygen supply is locally rate limiting to metabolism, since sympathetically-induced vasoconstriction would reduce flow, leading to chemoreceptor stimulation and reduced oxygen consumption. However, the oxygen consumption of the individual elements of the carotid body is not known, and changes in the oxygen usage of the whole organ may not reflect consumption of the nerve terminals. Further work is needed to characterise these 'free endings'.

Evidence showing that many of the fibres terminating on Type I cells are sensory, has already been referred to, and is obviously not compatible with the free ending hypothesis. Freezing the carotid body causes a permanent loss of Type I and Type II cells, and although the carotid sinus nerve regenerates and the vasculature appears normal, no respiratory reflexes are elicited from the carotid body and nor can chemoreceptor activity be recorded from the carotid sinus nerve. Electrical stimulation of the regenerated nerve does evoke reflex respiratory changes, so demonstrating that the afferent pathway is normal (Verna, Roumy & Leitner, 1981, in Belmonte *et al.*, 1981). The difficulty with such experiments lies in assessing the damage to the nerve endings. Crushing a carotid sinus nerve causes the sensory fibres to degenerate and chemosensory activity is lost. The fibres regenerate, but no chemosensory activity is recorded until the fibres have grown fairly close to the Type I cells (Zapata, Stensaas & Eyzaguirre, 1976), a situation which is similar to that seen in the skin where, following a crush of the sensory nerve innervating the Merkel cells in cats, the regenerating sensory endings show unspecialised sensitivity while the axon is growing, and typical response characteristics are not restored until the Merkel cell–nerve complex is reformed.

It appears from this evidence that chemoreception can only occur when the regenerating fibres come close to the Type I cell – they do not have to be as closely apposed as normal for function to be restored (Zapata *et al.*, 1976) – and this can be interpreted as meaning either that the cells release a chemical (trophic factor) which activates the sensory ending, or the cell causes the ending to become chemosensitive, or vice versa. Kienecker, Knoche & Binmann (1981, in Belmonte *et al.*,

1981) have reported that in rabbits chemosensitivity appears long before the Type I cells are innervated, and it has been claimed that carotid sinus nerve fibres allowed to regenerate into the adventitia of the external carotid artery are chemosensitive (Bingmann *et al.* in Belmonte *et al.*, 1981; Tan, Pallot & Purves, 1981 in Belmonte *et al.*, 1981), but slight chemosensitivity may be a non-specific property of neuromas (see Smith & Mills, 1981 in Belmonte *et al.*, 1981). From this it will be appreciated that the evidence is equivocal and different species have been used by different workers in the studies. Tan *et al.* (1981) seem to provide the best evidence for specific chemosensitivity in neuromas as they demonstrate a recording from a neuroma which is indistinguishable from that of a normal chemoreceptor afferent. Even in this work there is the possibility of reinnervation of miniglomera by sprouts from the regenerating sinus nerve. A useful experiment would be to study the neuroma *in vitro*, when subsequent histology could exclude this possibility.

A further hypothesis based on the sensory endings being the chemosensors is that of Mitchell & McDonald (in Purves, 1975) which proposes that sensory fibres apposed to Type I cells are directly affected by hypoxia or hypercapnia and afferent activity leads to release of a transmitter from the sensory nerve terminal, and the transmitter activates the Type I cell (which is considered to be a dopaminergic interneurone) and dopamine then inhibits the sensory terminal. Morphological evidence for such reciprocal synapses has already been discussed and, again, much hinges on the species and whether or not the nerve endings are chemosensitive. However, the fact that cutting the carotid sinus nerve does not change the levels of known putative transmitters in the carotid body, something which might be expected to occur if sensory fibres contain such a substance argues against this idea.

The hypothesis that physiological stimuli cause the Type I cell to release a chemical (transmitter) which activates the sensory nerve fibres (i.e. the Type I cell is the chemosensor) has been the subject of a great many studies. The problem lies in determining whether chemical transmission occurs and, if it does, establishing the identity of the hypothetical transmitter(s). The oldest candidate, one might even call it classical, is acetylcholine (ACh), and the evidence for and against its role in chemoreception has been reviewed recently (Eyzaguirre & Fidone, 1980; Eyzaguirre, in Belmonte *et al.*, 1981). Briefly, ACh is released in response to physiological stimuli and there appears to be a choline uptake mechanism in Type I cells. The random nature of spontaneous chemosensory discharge could, by analogy with miniature end-plate potentials of the neuromuscular junction, be due to release of quanta of ACh. However, the properties of fine nerve terminals in the chemoreceptors (cf. Biscoe, 1971) could equally account for the randomness of the normal discharge.

Figure 1.8: Response of Carotid Chemoreceptors (three units counted) to Intra-carotid Injection of 100% CO_2-equilibrated Locke Solution, NaCN and ACh Before and After Administering the Nicotinic Blocking Drug, Pentolinium. The increase in discharge above baseline values ($\triangle \epsilon x$) for each of the responses is submaximal and the baseline discharge in ct/s is given at the foot of each column. After pentolinium the dose of ACh had to be increased by a factor of 10 in order to approximate the pre-pentolinium response, whereas responses to CO_2 and NaCN were not appreciably affected by the nicotinic antagonist.

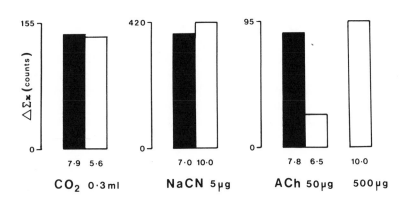

RESPONSES EVOKED BEFORE ■ AND AFTER □ PENTOLINIUM (10 mg I.C)

Source: McQueen, unpublished record from 3.5 kg cat.

ACh levels in the cat carotid body are not affected by cutting either the carotid sinus or the sympathetic nerves, implying that ACh is not stored in nerves. ACh excites all chemoreceptor fibres (e.g. see Figures 1.4, 1.8, Fidone & Sato, 1969), although Paintal (1971) considers it does not affect aortic myelinated chemoreceptor afferents, whereas physiological stimuli do. The major argument against ACh being a chemosensory transmitter in the chemoreceptors has been failure of antagonists to cause any significant reduction in responses to physiological stimuli, although they virtually abolish the excitatory response evoked by exogenous ACh, and whereas anticholinesterases, such as physostigmine, potentiate exogenous ACh, they have little or no effect on more physiological stimulants (Figure 1.8, and McQueen, 1977, 1980).

Although the pharmacological evidence does not favour a chemo-excitatory role for ACh, it can be argued that, if the concentration of ACh released locally within the carotid body was high compared with that which reaches the site following exogenous administration, or if exogenous ACh acts at an 'extra-synaptic' site, concentrations of drugs used to modify the action of ACh released by physiological stimuli at the intrinsic site may not be adequate. This is difficult to refute entirely, but the fact that very high doses of drugs which penetrate tissues fairly readily, and which are allowed to act for prolonged periods, affect responses to exogenous ACh without having much effect on responses to physiological stimuli, taken together with the observation that large-molecular-weight substances (e.g. horseradish peroxidase) can readily penetrate the chemoreceptor tissues, makes the argument appear improbable.

Results from autoradiographic experiments with labelled α-bungaro-toxin (Dinger *et al.*, 1981) show binding sites located on Type I, and possible Type II cells, with no evidence of binding to nerves. The authors equate the binding site for α-bungarotoxin with the acetylcholine nicotinic receptor site, but does bungarotoxin in fact label that nicotinic receptor which when activated causes an increase in chemosensory discharge? The reason for caution is that α-bungarotoxin in the doses used by McQueen (1977) was not very effective at antagonising the excitatory action of ACh on the cat carotid body. It would be interesting to see whether labelled hexamethonium and mecamylamine, established antagonists of the ACh response, bind to the same sites as α-bungarotoxin. If it emerges that there are really no nicotinic receptors on chemosensory nerve endings, or only on those from unmyelinated fibres, it will be necessary to revise considerably the 'cholinergic' hypothesis that ACh released from cells by physiological stimuli depolarises the sensory fibres. The presence of ACh receptors on Type I or Type II cells may mean that ACh released from one cell can influence another cell in the group. In this way, ACh might release one or more chemoexcitatory substances via actions on nicotinic and possibly muscarinic receptors on cells, and these actions would be susceptible to pharmacological antagonists. Physiological stimuli may be acting through several mechanisms to cause chemoreception, such that loss (or potentiation) of the cholinergic component does not greatly affect the overall response.

ACh may have more than one action on the chemoreceptor complex for there is some evidence that it has an inhibitory role, and indeed chemoinhibition is the predominant effect of ACh on the rabbit carotid body (see Docherty and McQueen, 1979). Some of the inhibitory effects of efferent nerve stimulation of the chemosensory discharge can be prevented by atropine, and this suggests that ACh may be involved in this pathway. Cholinergic receptors may also be involved in control of blood flow through the carotid body.

Intracellular electrophysiological studies have shown that Type I cells are variably affected by physiological and chemical stimuli and fairly insensitive to changes in ionic composition of their environment, although they do respond to changes in temperature and osmotic pressure. Hypoxia, ACh and NaCN have no consistent effect on the membrane potential, sometimes hyperpolarising, sometimes depolarising the cells, yet they invariably increase sensory discharge (see Eyzaguirre & Fidone, 1980). One might suspect that these small cells are rather badly damaged by the electrode because the average resting membrane potential of -20mV seems low and as techniques improve, values nearer to -60mV are being obtained. The variability of cells in their responses to various stimuli may be a consequence of the technical difficulties associated with recording from small cells, or from the fact that they are *in vitro*, but it may also mean that individual cells in a glomerulus differ in their sensitivity to stimuli, perhaps according to the type of receptor(s) they carry. If graded (receptor) potentials are a requirement for the primary receptor element, then the Type I cell does not appear to fit the bill. However, the extent to which the cell membrane potential and resistance are important for the primary transducer element remains to be established. It is also noteworthy that the cells of the arterial chemoreceptors in cats and rats do not seem to be affected by chronic denervation of the sensory nerve, appearing morphologically similar to 'normal' cells and showing similar changes in membrane potential in response to stimulants, although voltage noise is reduced. This contrasts with gustatory chemoreceptors of the tongue, and touch receptors in the skin, where the taste bud cells and Merkel cells disappear following sensory denervation, although in the latter case this is species dependent. In this respect it is interesting that the rabbit carotid body does appear to involute following section of the sinus nerve (Pallot, unpublished; Kienecker, personal communication).

The other putative neurotransmitters in the carotid body (noradrenaline, adrenaline, dopamine and 5-HT) have all been investigated but as yet their role, if any, in chemoreception remains, like that of ACh, to be established. We shall briefly consider evidence relating to dopamine, which is present in the carotid body and released during physiological stimulation of the chemoreceptors. Although there may be species differences, in general low doses of dopamine inhibit spontaneous chemoreceptor activity, whereas higher doses, or lower doses after administration of drugs which block the inhibitory effect, cause chemoexcitation (see Docherty & McQueen, 1978). Dopamine receptors in the carotid body may, therefore, be of two types, inhibitory (DAi) and excitatory (DAe). Respiration in man is reduced by dopamine (Bainbridge & Heistad, 1980). It was suggested by Osborne & Butler (1975) that in the carotid body dopamine is continually released from Type I cells during normoxia and suppresses the tendency of sensory endings to depolarise spontaneously. In hypoxia, dopamine release is attenuated and the chemosensory activity increases. But dopamine release is found to increase during hypoxia, and there is also pharmacological evidence which is not in accord with the hypothesis (see Docherty & McQueen, 1978).

The fact that administration of exogenous dopamine causes a predominately inhibitory effect on carotid chemosensors has led to various proposals concerning the involvement of dopamine in negative feedback and the consensus is that it has a 'modulatory' role, modifying ongoing discharge without being directly responsible for that discharge. In contrast to the inhibitory action, it was suggested over 20 years ago that catecholamines might stimulate sensory nerves in the carotid body (see Biscoe, 1971), and Docherty (1980) recently presented pharmacological evidence which is compatible with endogenous dopamine having an excitatory influence on cat carotid chemoreceptors. Recently, interest in the possible importance of noradrenaline to chemoreceptors has revived and the debate continues as to whether more noradrenaline than dopamine is present in the carotid body; it seems to be species-dependent. One complication of studying the actions of catecholamines *in vivo* is the well-known vascular effects of these substances, and this has led to experiments being performed *in vitro*, where vascular effects are eliminated. However, the extent to which results from the *in vitro* blood-free preparations are directly comparable to those obtained *in vivo* is a moot point as has already been mentioned.

Various other hypotheses exist concerning how the peripheral chemoreceptors work. These include the following: intracellular bicarbonate ions regulate sensory discharge (see Torrance in Belmonte *et al.*, 1981); Type II cells are the chemosensors, with changes in a low affinity cytochrome enzyme leading to the establishment of a K^+ gradient which can depolarise nerve terminals situated near the Type II cells (Mills & Jöbsis, 1970); Type II cells are chemosensors, which respond to changes in oxygen tension, by mechanically affecting the sensory nerve ending and causing a depolarisation (see Paintal, 1977); changes in cyclic nucleotides, particularly ATP, within chemoreceptor tissue are involved in regulation of chemoreceptor discharge (Anichkov & Belen'kii, 1963) — ATP and adenosine do affect chemosensory activity (see Figure 1.4). These various hypotheses are supported, to a greater or lesser degree by experimental evidence, but to date no single hypothesis has emerged as being generally acceptable, and the involvement of chemical transmission in chemoreception remains to be established.

The variety of hypotheses may mean that different mechanisms within the receptor complex are capable of increasing chemosensory discharge, with the final stage of all the processes being the same. Thus, hypoxia and hypercapnia might act via different cells and/or mechanisms, and there is some evidence to suggest this may be the case, but the final pathway which leads to increased chemosensory discharge could be common, perhaps involving adenylate cyclase or Ca^{2+}. Further studies using ion-sensitive electrodes, tissue slices, cultured cells, and microanalytical techniques should enable a better understanding of the biophysical and biochemical processes to emerge in the near future. It is often assumed that aortic and carotid chemoreceptors are quantitatively similar, but we have already seen some evidence which implies that this may not be the case. Results obtained with one set of receptors, or in one species, should be applied with great caution to the other set, or to other species, lest the basic mechanisms of chemoreception differ.

We have not so far considered the role polypeptides might play in the peripheral chemoreceptors. They do influence carotid chemoreceptor discharge when injected (see McQueen in Belmonte *et al.*, 1981), but because the role of polypeptides in the body has yet to be established, one hesitates to suggest their function in the peripheral chemoreceptors. They may be neurotransmitters or neuromodulators, but could equally well have a neuroendocrine, trophic or other function on cells and/or nerve endings. It will probably emerge that some of the peptides are stored in chemoreceptor cells together with the 'classical' putative transmitters, and may be released along with these substances. Drugs which can specifically affect polypeptides (e.g. antagonists, enzyme inhibitors) are needed in order to facilitate study of the physiological role of polypeptides. In the present absence of such drugs, some information may be obtained using specific antibodies to interfere with the peptides. The presence of these various substances in the chemoreceptor tissue may mean that the chemoreceptors are not homogeneous, but rather made up of cell clusters in which the branches of a single nerve fibre are influenced by cells with different characteristics, perhaps releasing different substances in response to physiological and pharmacological stimuli, these substances having excitatory or inhibitory actions, either directly or indirectly mediated (by influencing sensitivity to physiological stimuli), on chemosensory activity. The substances released may, besides influencing the sensory nerves, also affect cells and blood vessels in the vicinity. The integrated input to the CNS from such a glomerulus would be the algebraic summation of the various components and might be subject to efferent feedback.

The role of Type I and Type II cells in chemoreception is problematic — whether the cells release one or more chemical transmitters to affect the sensory nerve endings, or whether cellular activity influences the sensitivity of the nerve endings some other way, remains to be established, as indeed does the more general question concerning which of the chemoreceptor elements is the chemosensor. An alternative proposal is that some of the cells are not directly involved in chemoreception, but have a separate function, perhaps endocrine. In this context it should be remembered that the carotid body was, for a long time, considered to be a gland, and it could be that it has a dual function — sensory organ and gland.

Although two types of chemosensory fibre exist, namely myelinated and unmyelinated, it is not known whether they subserve different functions and act by different mechanisms. Preliminary studies in anaesthetised rats have demonstrated a significant reduction in basal respiration and a loss of chemoreceptor responsiveness in anaesthetised animals which were treated with capsaicin neonatally (Figure 1.9). This drug causes a permanent loss of the majority of unmyelinated sensory afferent nerve fibres and the results observed could be the consequence of a destruction of the unmyelinated chemoreceptor afferents. The reduced respiratory responsiveness may be due to a reduced peripheral chemoreceptor input to the CNS, or some effect in the CNS consequent to destruction of the primary afferent terminals. It has been suggested that substance P is a neurotransmitter of chemoreceptor afferents in the nucleus tractus solitarius of the rat (Gillis *et al.*, 1980), and capsaicin does cause a destruction of SP-containing fibres. However, it is also able to affect unmyelinated sensory fibres containing other polypeptides, so the results cannot be interpreted exclusively in terms of a loss of SP-containing fibres. The study of animals treated neonatally with capsaicin may allow an assessment of the relative contribution made to respiratory reflexes by myelinated and unmyelinated chemoreceptor afferents, and may also be useful for investigating the physiological role of polypeptides in the peripheral chemoreceptors.

Figure 1.9: Averaged Increase in Respiratory Minute Volume in Pento-barbitone Anaesthetised Rats During the 20s Period Following i.v. Injection of NaCN ($\log_{10}$ scale) in Vehicle-treated (–, n=3) and Capsaicin-treated (---, n=3) Animals. Lines were fitted to the data by eye. The rats were seven months old when used for the study and had been injected with either capsaicin (50 mg kg^{-1} s.c.) or drug vehicle 50:50 polyethylene glycol/saline) two days after birth. The body weights of the two groups were not significantly different, but capsaicin-treated animals had a significantly lower ($P < 0.01$) respiratory minute volume than the controls (93±8.1 compared with 154±6.7 ml min^{-1}) and their respiratory frequency was also reduced (51±1.2 against 59±2.1 breaths min^{-1}).

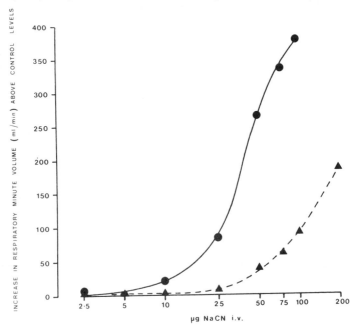

Source: McQueen and Cervero, unpublished observations.

In conclusion, the influence of the peripheral arterial chemoreceptors on respiration has been investigated and is fairly well understood, whereas the identity of the chemosensor(s) and the mechanism(s) of transduction within the peripheral chemoreceptors remain to be established. The pathophysiology of the aortic and carotid bodies should be studied since respiratory and other disorders might follow from changes in peripheral chemoreceptor function, perhaps as a result of 'sclerosis of the organs' (Gomez, 1908) or as a consequence of repeated stimulation.

References

Abbot, C.P. & Howe, A. (1972) 'Ultrastructure of Aortic Body Tissue in the Cat', *Acta anatomica, 81*, 609-19

Acker, H., Fidone, S., Pallot, D., Eyzaguirre, C., Lubbers, D.W. & Torrance, R.W. (eds) (1977) *Chemoreception in the Carotid Body*, Springer-Verlag, Heidelberg

Anichkov, S.V. & Belen'kii, M.L. (1963) *Pharmacology of the Carotid Body Chemoreceptors*, Pergamon Press, Oxford

Bainbridge, D.W. & Heistad, D.D. (1980) 'Effect of Haloperidol on Ventilatory Response to Dopamine in Man', *J. Pharmacol. Exp. Ther., 213*, 13-17

Belmonte, C., Pallot, D.J., Acker, H. & Fidone, S. (eds) (1981) *Arterial Chemoreceptors*, Leicester University Press, Leicester

Biscoe, T.J. (1971) 'Carotid Body: Structure and Function', *Physiol. Rev., 51*, 437-95

Biscoe, T.J., Lall, A. & Sampson, S.R. (1970) 'Electronmicroscopic and Electrophysiological Studies in the Carotid Body Following Intracranial Section of the IXth Nerve', *J. Physiol., 208*, 132-52

Biscoe, T.J. & Pallot, D.J. (1972) 'Serial Reconstruction with the Electronmicroscope of Carotid Body Tissue: The Type I Cell Nerve Supply', *Experientia, 28*, 222-3

Black, A.M. & Torrance, R.W. (1967) 'Chemoreceptor Effects in the Respiratory Cycle', *J. Physiol., 189*, 59-61P

Black, A.M.S., McCloskey, D.I. & Torrance, R.W. (1971) 'The Response of Carotid Body Chemoreceptors in the Cat to Sudden Changes of Hypercapnic and Hypoxic Stimuli', *Respiration Physiol., 13*, 36-49

Cole, S., Lindenberg, L.B., Galioto, F.M. (Jr.), Howe, P.A., de Graaf, A.C., Davis, J.M., Lubka, R. & Gross, E.M. (1979) 'Ultrastructural Abnormalities of the Carotid Body in Sudden Infant Death Syndrome', *Pediatrics, 63*, 13-17

Coleridge, H., Coleridge, J.G.C. & Howe, A. (1967) 'A Search for Pulmonary Arterial Chemoreceptors in the Cat with a Comparison of the Blood Supply of the Aortic Bodies in the Newborn and Adult Animal', *J. Physiol. Lond., 191*, 353-74

Cuello, A.C. & McQueen, D.S. (1980) 'Substance P: a Carotid Body Peptide', *Neurosci. Lett., 17*, 215-9

Cunningham, D.J.C. (1974) 'Integrative Aspects of the Regulation of Breathing: a Personal View' in J.G. Widdicombe (ed.), *Respiratory Physiology*, MTP International Review of Science. Physiology, Series 1, vol. 1, Butterworth, London, pp. 303-69

Daly, M. de B. & Ungar, A. (1966) 'Comparison of the Reflex Responses Elicited by Stimulation of the Separately Perfused Carotid and Aortic Body Chemoreceptors in the Dog', *J. Physiol., 182*, 379-403

Davies, R.O. & Lahiri, S. (1973) 'Absence of Carotid Chemoreceptor in the Cat', *Respiratory Physiology, 18*, 92-100

De Castro, F. (1928) 'Sur la structure et l'innervation du sinus carotidien de l'homme et des mammiferes. Nouveaux faits sur l'innervation et la fonction du glomus caroticum', *Tras. Lab. Invest. Biol. Univ. Madrid, 25*, 331-84

Dejours, P. (1962) 'Chemoreflexes in Breathing', *Physiol. Rev., 42*, 335-58

Dinger, B., Gonzalez, C., Yoshizaki, K. & Fidone, S. (1981) 'Alpha-bungarotoxin Binding in Cat Carotid Body', *Brain Res., 205*, 187-93

Docherty, R.J. (1980) 'A Quantitative Pharmacological Study of Some Putative Neurotransmitters in the Carotid Body of the Cat and the Rabbit', *Ph.D. Thesis, University of Edinburgh*

Docherty, R.J. & McQueen, D.S. (1978) 'Inhibitory Action of Dopamine on Cat Carotid Chemoreceptors', *J. Physiol., 279*, 425-36

Docherty, R.J. & McQueen, D.S. (1979) 'The Effects of Acetylcholine and Dopamine on Carotid Chemosensory Activity in the Rabbit', *J. Physiol., 288*, 411-23

Douglas, N.J., Calverley, P.M.A., Leggett, R.J.E., Brash, H.M., Flenley, D.C. & Brezinova, V. (1979) 'Transient Hypoxaemia during Sleep in Chronic Bronchitis and Emphysema', *Lancet, i*, 1-4

Eisele, J.H., Ritchie, B.C. & Severinghaus, J.W. (1967) 'Effect of Stellate Ganglion Block on the Hyperpnoea of Exercise', *J. Appl. Physiol., 22*, 966-9

Eyzaguirre, C. & Fidone, S.J. (1980) 'Transduction Mechanisms in Carotid Body: Glomus Cells, Putative Neurotransmitters and Nerve Endings', *Am. J. Physiol., 239*, C 135-52

Fidone, S.J. & Sato, A. (1969) 'A Study of Chemoreceptor and Baroreceptor A- and C-fibres in the Cat Carotid Nerve', *J. Physiol., 205*, 527-48

Fujita, T. & Kobayashi, S. (1979) 'Current Views on the Paraneurone Concept', *Trends in Neuroscience, 2*, 27-30

Gillis, R.A., Helke, C.J., Hamilton, B.L., Norman, W.P. & Jacobowitz, D.M. (1980) 'Evidence that Substance P is a Neurotransmitter of Baro- and Chemoreceptor Afferents in Nucleus Tractus Solitarius', *Brain Res., 181*, 476-81

Ginzel, K.H. (1975) 'The Importance of Sensory Nerve Endings as Sites of Drug Action', *Naunyn-Schmiedeberg's Arch. Pharmacol., 288*, 29-56

Gomez, L.P. (1908) 'The Anatomy and Pathology of the Carotid Gland', *Am. J. Med. Sci., 136*, 98-111

Hanson, M.A., Rao, P.S. & Torrance, R.W. (1979) 'Carbon Dioxide Sensitivity of Aortic Chemoreceptors in the Cat', *Respiration Physiol., 36*, 301-10

Heymans, C. & Neil, E. (1958) *Reflexogenic Areas of the Cardiovascular System*, Churchill, London

Honda, Y., Watanabe, S., Hashizume, I., Satomura, Y., Hata, N., Sakakibara, Y. & Severinghaus, J.W. (1979) 'Hypoxic Chemosensitivity in Asthmatic Patients Two Decades after Carotid Body Resection', *J. Appl. Physiol., 46*, 632-8

Lahiri, S. & Delaney, R.G. (1975) 'Relationship between Carotid Chemoreceptor Activity and Ventilation in the Cat', *Respiration Physiol., 24*, 267-86

Lahiri, S., Mulligan, E., Nishino, T., Mokashi, A. & Davies, R.O. (1981) 'Relative responses of aortic body and carotid body chemoreceptors to carboxyhemoglobinemia', *J. Appl. Physiol., 50*, 580-6

Lahiri, S., Nishino, T., Mokashi, A. & Mulligan, E. (1980b) 'Relative Responses of Aortic and Carotid Bodies to Hypotension', *J. Appl. Physiol., 48*, 781-8

Lahiri, S., Nishino, T., Mulligan, E. & Mokashi, A. (1980a) 'Relative Latency of Responses of Chemoreceptor Afferents from Aortic and Carotid Bodies', *J. Appl. Physiol., 48*, 362-9

McDonald, D.M. (1980) 'Regulation of Chemoreceptor Sensitivity in the Carotid

Body: the Role of Presynaptic Sensory Nerves', *Fed. Proc.*, *39*, 2627-35

McDonald, D.M. & Mitchell, R.W. (1975) 'The Innervation of Ganglion Cells, Glomus Cells and Blood Vessels in the Rat Carotid Body', *J. Neurocytol.*, *4*, 177-230

McQueen, D.S. (1977) 'A Quantitative Study of the Effects of Cholinergic Drugs on Carotid Chemoreceptors in the Cat', *J. Physiol.*, *273*, 515-32

McQueen, D.S. (1980) 'Effects of dihydro-βerythroidine on the cat carotid chemoreceptors', *Q.J. Exp. Physiol.*, *65*, 229-37

Mills, E. & Jobsis, F.F. (1970) 'Simultaneous Measurement of Cytochrome-a_3 Reduction and Chemoreceptor Afferent Activity in the Carotid Body', *Nature*, *225*, 1147-9

Mitchell, R.A. & Herbert, D.A. (1975) 'Potencies of Doxapram and Hypoxia in Stimulating Carotid Body Chemoreceptors and Ventilation in Anesthetized Cats', *Anesthesiology*, *42*, 559-66

Morgan, S.E., Pallot, D.J. & Willshaw, P. (1981) 'The Effect of Ventilation with Different Concentrations of Oxygen upon the Synaptic Vesicle Density in Nerve Endings of the Cat Carotid Body', *Neuroscience*, *6*, 1461-66

Nadel, J.A. & Widdicombe, J.G. (1962) 'Effect of Changes in Blood Gas Tensions and Carotid Sinus Pressure on Tracheal Volume and Total Lung Resistance to Airflow', *J. Physiol.*, *163*, 13-33

Naeye, R.L., Fisher, R., Ryser, M. & Whalen, P. (1976) 'Carotid Body in the Sudden Infant Death Syndrome', *Science*, *191*, 567-8

Nishi, K. & Stensaas, L.J. (1974) 'The Ultrastructure & Source of Nerve Endings in the Carotid Body', *Cell Tiss. Res.*, *154*, 303-19

O'Regan, R.G. (1977) 'Control of Carotid Body Chemoreceptors by Autonomic Nerves', *Irish J. Med. Sci.*, *146*, 199-205

Osborne, M.P. & Butler, P.J. (1975) 'New Theory for Receptor Mechanism of Carotid Body Chemoreceptors', *Nature*, *254*, 701-3

Paintal, A.S. (1971) 'Action of Drugs on Sensory Nerve Endings', *Ann. Rev. Pharmacol.*, *11*, 231-40

Paintal, A.S. (ed.) (1976) *Morphology and Mechanisms of Chemoreceptors*, Navchetan Press Ltd., New Delhi

Paintal, A.S. (1977) 'Effects of Drugs on Chemoreceptors Pulmonary and Cardio-vascular Receptors', *Pharmac. Ther. Bull.*, *3*, 41-63

Pallot, D.J. (1978) 'The Innervation of Type-I Cells in a Mutant Mouse' in R.E. Coupland & W.G. Frossmann (eds), *Peripheral Neuroendocrine Interaction*, Springer-Verlag, Heidelberg, pp. 112-7

Pallot, D.J. (ed.) (1983) *The Peripheral Arterial Chemoreceptors*, Croom Helm, London

Pearse, A.G.E. (1969) 'The Cytochemistry and Ultra-structure of Polypeptide Producing Cells of the APUD Series and the Embryologic, Physiologic and Pathologic Implications of the Concept', *J. Histochem. Cytochem.*, *17*, 303-13

Purves, M.J. (1970) 'Effects of Hypoxia, Hypercapnia on Carotid Body Blood Flow', *J. Physiol.*, *215*, 33-47

Purves, M.J. (ed.) (1975) *The Peripheral Arterial Chemoreceptors*, Cambridge University Press, Cambridge

Schmidt, C.F. & Comroe, J.H. (Jr.) (1940) 'Functions of the Carotid and Aortic Bodies', *Physiol. Rev.*, *20*, 115-51

Shannon, D.C., Kelly, D.H. & O'Connell, K. (1977) 'Abnormal Regulation of Ventilation in Infants at Risk for Sudden-Infant-Death Syndrome', *N. Engl. J. Med.*, *297*, 747-50

Smith, P.G. & Mills, E. (1980) 'Restoration of Reflex Ventilatory Response to Hypoxia after Removal of Carotid Bodies in the Cat', *Neuroscience*, *5*, 573-80

Torrance, R.W. (ed.) (1968) *Arterial Chemoreceptors*, Blackwell, Oxford

Ungar, A. & Bouverot, P. (1980) 'The Ventilatory Responses of Conscious Dogs to Iscocapnic Oxygen Tests. A Method for Exploring the Central Component of Respiratory Drive and its Dependence on O_2 and CO_2', *Resp. Physiol.*, *39*, 183-97

Verna, A. (1979) 'Ultrastructure of the Carotid Body in Mammals', *Int. Rev. Cytol.*, *60*, 271-330

Verna, A., Roumy, M. & Leitner, L-M. (1981) 'Role of the Carotid Body Cells; Long-term Consquences of their Cryodestruction', *Neurosci. Lett.*, *16*, 281-5

Zapata, P., Hess, A., Bliss, E.L. & Eyzaguirre, C. (1969) 'Chemical, EM and Physiological Observations on the Role of Catecholamines in the Carotid Body', *Brain Res.*, *14*, 473-96

Zapata, P., Stensaas, L.J. & Eyzaguirre, C. (1976) 'Axon Regeneration Following a Lesion of the Carotid Nerve: Electrophysiological and Ultrastructural Observations', *Brain Res.*, *113*, 235-53

2 CENTRAL CHEMORECEPTORS

Hans H. Loeschcke

Central Chemosensitivity

The aim of this chapter is to analyse the stimulus driving ventilation and the adaptation of this drive to metabolic needs. The main questions are what is the adequate stimulus, where is the receptor, what is the mechanism of stimulation, what are the conditions surrounding the receptor and how does the feedback operate?

After a period when CO_2 and O_2 (or lack of O_2) were considered as the chemical stimuli, Winterstein (1911) proposed a unifying theory suggesting that PO_2 and PCO_2 would act on one receptor by one mechanism and that the common denominator was the hydrogen ion which was either dissociated from H_2CO_3 or from the acids formed in oxygen deficiency.

In his latest formulation of the reaction theory Winterstein (1956) accepted that there are two kinds of chemoreceptors which have to be considered separately, the peripheral and central chemoreceptors. The part of Winterstein's theory concerning oxygen and CO_2 in the peripheral chemoreceptors is still controversial; though the sites of the oxygen receptor have been identified in the carotid and aortic glomera.

For the central chemoreceptor, which is not sensitive to oxygen (except what might be called modulations by PO_2), the question is restricted to the alternative of the hydrogen ion or of molecular CO_2 as the adequate stimulus. How difficult this is to answer becomes clear when the dramatic effect of inhaled CO_2 on ventilation is compared to the effect of acid injection which is relatively small. This discrepancy suggested a specificity of CO_2 (Nielsen, 1936). Much effort was necessary to show that this view was incorrect.

The quantitative investigation of the blood components and their action on ventilation led to the general conclusion that all three possible stimuli PCO_2, PO_2 and pH would to some extent contribute to the ventilatory drive. This so-called multiple factor theory (Gray, 1950) necessitated that the single partial drives and their interactions be investigated. The following discussion will come out with the conclusion that H^+ is the sole stimulus of the central chemoreceptors.

41

The Reaction Theory of Central Chemosensitivity

In the next step information was gathered closer to the receptor of the central chemosensitivity to which the further discussion will now be restricted before generalisations again may be possible. This step became feasible as soon as techniques were developed to perfuse the subarachnoid spaces and the effects of CO_2 or of H^+ on ventilation were observed (Leusen, 1954a, b). Loeschcke, Koepchen & Gertz (1958) used such a perfusion experiment to differentiate between H^+ and PCO_2 by perfusion either with solutions with constant PCO_2 and varied pH (Figure 2.1) or with constant pH and varied PCO_2. Since PCO_2 at constant pH had no effect or even a depressing effect, while pH changes at constant PCO_2 drove ventilation, this excluded CO_2 as a stimulus in an experiment like this. Of course in the Henderson-Hasselbalch equation there are three parameters pH, PCO_2 and HCO_3^-, constancy of one factor implies, necessarily, variation of the other two, and this must still be considered a shortcoming of the argument that H^+ is *the* stimulus.

These observations were however confirmed and extended by Mitchell *et al.* (1963a, b) in the cat, Fencl, Miller & Pappenheimer (1966), Pappenheimer (1967) in the goat and dog and Hori, Roth & Yamamoto (1970) in the rat. The conclusions stayed the same. Pappenheimer (1967) as well as Loeschcke (1969) postulated that if there is a unique receptor to H^+ there must be a way to explain the experimental observations assuming the same chemical signal in respiratory as in metabolic acidosis in such a way that the reactions of the two types of acidosis can be quantitatively explained. Of course the contribution of the peripheral chemoreceptors must also be considered. This contribution in the steady state, however, is small.

Not all investigators were able to confirm the above experiments. Cragg, Patterson & Purves (1977) did not see increases of ventilation when applying acid pH on the ventral side of the medulla. It must, however, be stated that the experiment is not easy. The operation has to be done in such a way that after opening the dura and arachnoid there is not the slightest bleeding during the entire experiment. If there is bleeding there will be either a gross, or sometimes invisible, deposition of fibrin which can diminish or abolish the response. Also the anaesthesis has to be light; barbiturates especially depress or abolish central chemosensitivity. This usually is not recognised at first glance because the respiratory drive from peripheral chemoreceptors is still active; the phenomenon is well documented however (Benzinger,

Figure 2.1: Ventilation during Perfusion of the Fourth Ventricle and Subarachnoid Space with Different pH. PCO_2 was kept constant. Anaesthetised cat.

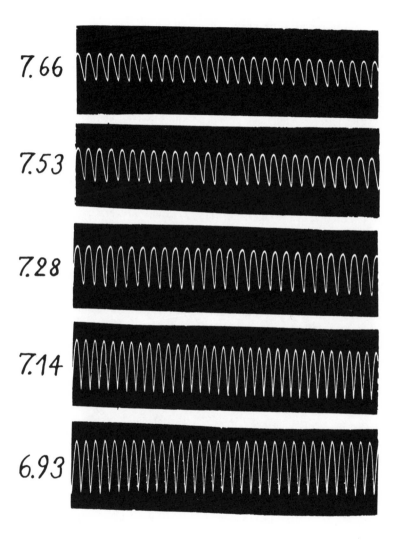

Source: Loeschcke, Koepchen, Gertz, 1958; courtesy of Springer-Verlag.

Opitz & Schoedel, 1938; Aström, 1952) that sensitivity to inhaled CO_2 is very easily anaesthetised. Of course in addition the perfusing fluid temperature, osmolality and calcium concentration must be maintained carefully.

In the experiments cited above pH was measured in the outflow of the perfusion or superfusion fluids and while Pappenheimer's group used ventricular perfusions our group adopted the technique of superfusion of the ventral medullary surface. Recently pH has been measured with microelectrodes in the tissue of the medulla oblongata (Cragg *et al.*, 1977) and with macroelectrodes on the surface of the medulla (Ahmad, Berndt & Loeschcke, 1976; Shams, 1981).

The latter technique is based on the observation that there is free access to the glass electrode from the intercellular compartment. This is demonstrated by the free entry of horseradish peroxidase (Figure 2.2) from the subarachnoid into the intercellular spaces (Dermietzel, 1976). The former technique (Cragg *et al.*, 1977) has the advantage that the electrode is in the tissue and the second one (Ahmad *et al.*, 1976) that it is non-invasive. Cragg *et al.* (1977) could answer the old question (Lambertsen *et al.*, 1961) whether or not the time course of the pH in the medullary tissue is compatible with the time course of ventilation if CO_2 is inhaled. It was answered positively indicating that indeed the electrode measures a pH which is representative of the pH stimulus applied to the medullary surface.

With the older technique several problems were tackled. Ahmad & Loeschcke (1983a) obtained the same result as Cragg *et al.* (1977) and established the similarity of the time courses of ventilation and change of extracellular pH (pH$_e$) by demonstrating that in an approximately stepwise PCO_2 change the on- and off-transients of pH$_e$ coincided with little or no hysteresis. The tidal volume (V_T), especially after denervation of the peripheral chemoreceptors, followed clearly the pH as measured on the surface (Ahmad & Loeschcke, 1981b). There was, for example, no conformity of pH$_e$ and V_T if pH was measured on the surface of the cortex (Figure 2.3).

The slope of the response of V_T to changes of pH$_e$ if CO_2 was inhaled (under oxygen) was determined by Ahmad & Loeschcke (1983a) in anaesthetised cats with intact sino-aortic nerves as -246 ml min^{-1} pH^{-1}. In decerebrated cats with intact chemoreceptors it was -633 ml min^{-1} pH^{-1} and in decerebrated chemodenervated cats -650 ml min^{-1} (Figure 2.4).

Shams (1981) has studied the effects of intravenously injected HCl and H_2SO_4. Whilst the latter acid is not commonly used for inducing

Figure 2.2: (a) Distribution in the Extracellular Spaces of the Marker
Horseradish Peroxidase in an Electron Microscopic Section of the
Ventral Medullary Surface (VMS). The marker was applied to the
subarachnoid space of an anaesthetised cat. Recesses clad with basal
membrane are filled with marker. Also the marker has entered the
extracellular spaces. (b) The marker accumulates around a blood
capillary in the perivascular space (PVS).

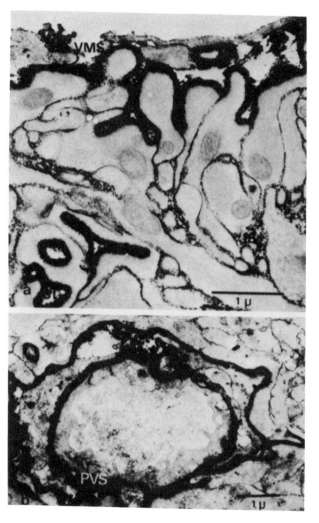

Source: Dermietzel (1976); courtesy of Thieme Verlag.

metabolic acidosis it is better tolerated than HCl producing less depression of the arterial blood pressure. The acid-base effects are the same.

Figure 2.3: Transients of V_T Plotted against Extracellular pH on the Cortex (left) and on the Medullary Surface (right) in Anaesthetised Cats. Data obtained during approximately rectangular increase and subsequent decrease of endtidal PCO_2. Points plotted are 12s apart. On and off transients fall approximately on the same line for the medulla oblongata, while for the cortex there is marked hysteresis.

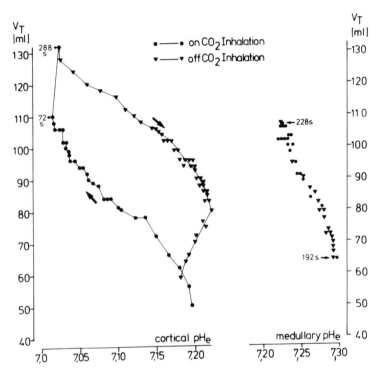

Source: Ahmad & Loeschcke (1983c); courtesy of Springer-Verlag.

Shams' results came partly as a surprise. As long as the changes of extracellular pH (pH_e) after acid injection were only slight there was a steep response of ventilation or of tidal volume and this response coincided with the response to pH if CO_2 was inhaled. So far this is a full confirmation, and the best documentation, of the identity of the metabolic and respiratory response of the receptor. The high gain

Figure 2.4: V_T Plotted against pH, Averages of four (5) Experiments in two Decerebrated Cats (bars = standard error). Sinus nerves intact (right), sinus nerves cut (left). On transients after approximately rectangular variation of endtidal PCO_2. These curves represent the pH (PCO_2) response of tidal volume. The parallel shift after cutting the sinus nerve is interpreted as a loss of a basic (not CO_2-dependent) drive.

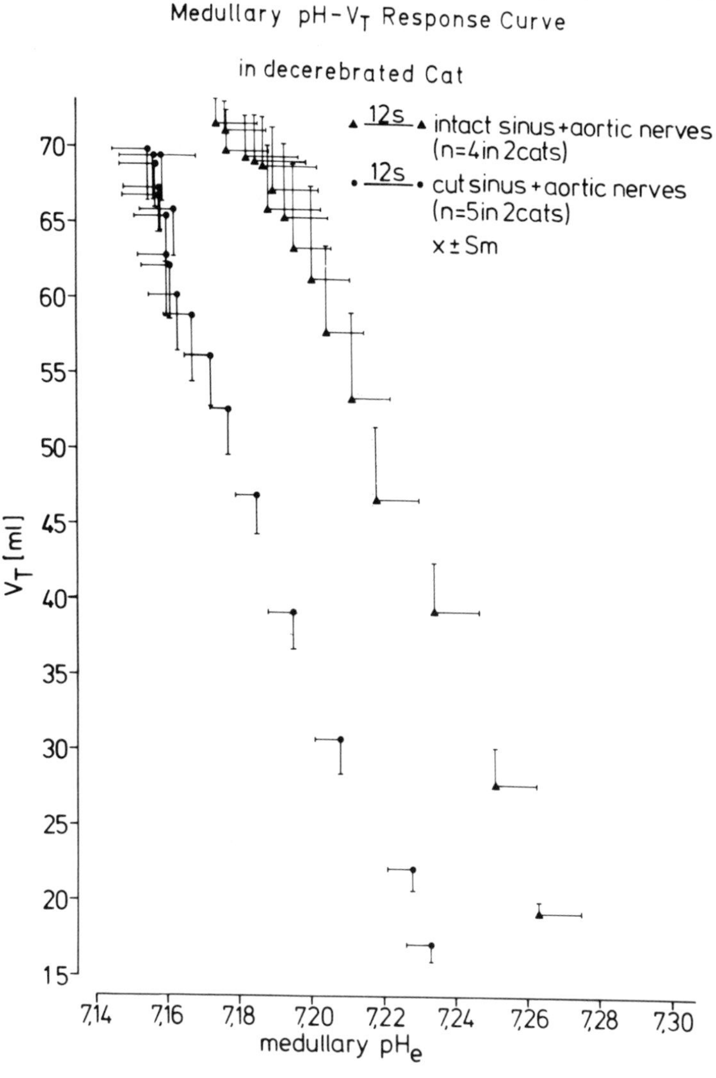

Medullary pH–V_T Response Curve

in decerebrated Cat

Source: Ahmad & Loeschcke (1983a); courtesy of Springer-Verlag.

explains the stability of extracellular pH under the experimental conditions, as observed, for example, by Loeschcke & Sugioka (1969) and by Leusen (1972).

As soon, however, as more acid was injected unexpectedly ventilation and tidal volume did not rise further and in most cases dropped back to values close to control (Figure 2.5). When in this stage, however, CO_2 was inhaled the response to pH$_e$ continued with approximately the same slope (Figure 2.6). Under degrees of metabolic acidosis the response curve in respiratory acidosis (produced by CO_2) to pH$_e$ remained unchanged in slope (Figure 2.7). The response curves were merely shifted to an increasingly more acid pH$_e$, suggesting a divergence of the responses to pH in metabolic and respiratory acidosis.

If the injection of acid is repeated whilst pH$_e$ is still diminished from previous injections the response of ventilation to acid occurs again; a further pH$_e$ decrease causes the reversal of the ventilatory response observed. The interpretation of these results is difficult. Tentatively it may be speculated that the H^+ receptor becomes inaccessible to further influx of H^+ after a given response. It may be assumed that the H^+ mechanism resides in a compartment which in the beginning is accessible to H^+ but under the action of acid the compartment is shut off to further entrance of H^+ while CO_2 continues to enter and act on the $[H^+]$ receptor. A synaptic gap, for example, would be a good model (Figure 2.8); this is mentioned because quite different experiments, discussed below, suggest an effect of H^+ on the synapse.

The drop of tidal volume at increasing $[H^+]$ after the initial response has been observed is unexplained, but Fukuda & Loeschcke (1977) showed a similar response in the neuronal activity. One of the possibilities to be discussed is a presynaptic inhibitory effect.

As a result of this discussion it may be stated that H^+ rather than CO_2 must be considered as the stimulus to central chemosensitivity. The system behaves as if stimulated by extracellular H^+ in metabolic as well as in respiratory acidosis. A modification of the assumption, however, seems to be necessary to explain the divergence of the metabolic and respiratory acidotic effects when higher doses of acid are injected. A location of the receptor for H^+ is suggested to which extracellular H^+ has free entrance up to a mild degree of acidosis while in severe metabolic acidosis an impediment for H^+ to reach the receptor is created while the passage of CO_2 is uninhibited. A synaptic gap may tentatively serve as a model.

Figure 2.5: Change of $\dot{V}$ (Upper Part) and of Endtidal PCO_2 (Lower Part) During an Infusion of H_2SO_4 Plotted Against the Decrease of pH Measured on the Medullary Surface. Ventilation increases in spite of a decrease of PCO_2 in endtidal air. The increase of $\dot{V}$, however, is limited and in higher degrees of acidosis is reversed. The plotted points are 1 min apart. Averages of seven cats.

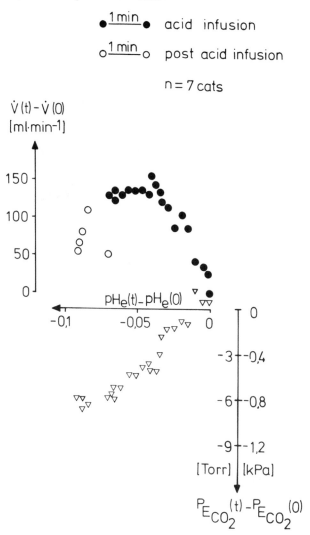

Source: Shams *et al.* (1981).

Figure 2.6: Ventilation in Respiratory (open circles) and Metabolic Acidosis (filled circles). As in Figure 2.5 metabolic acidosis is induced by acid infusion and $\dot{V}$ is plotted against medullary surface pH. Respiratory acidosis induced by inhaling CO_2. Up to a certain response there is a complete concordance of the responses to respiratory and to metabolic acidosis. After this point the effect of metabolic acidosis is reversed while the response to pH as induced by CO_2 inhalation continues with undiminished slope.

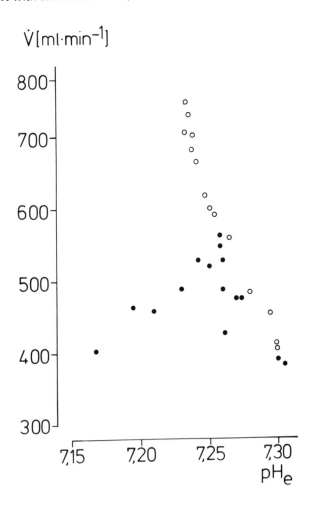

Source: Shams *et al.* (1981).

Figure 2.7: Responses of V_T to Medullary pH During Inhalation of CO_2. Extracellular medullary pH was additionally varied in several steps by i.v. injection of fixed acid. The slopes of the responses remain unchanged whereas the position is shifted to the acid side. Transient technique. The plotted points are 12s apart beginning with the lower point. Investigation in a single animal.

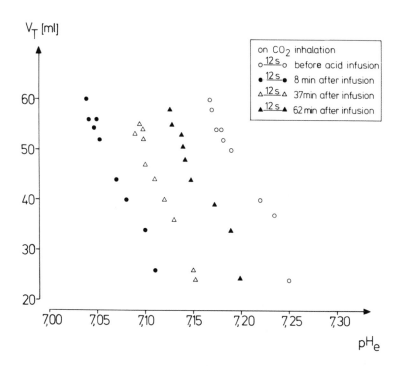

Localisation and Neurophysiology of Central Chemosensitivity

By applying pledglets of tissue paper soaked with acids or alkali (Mitchell *et al.*, 1963a, b) or by superfusion of small spots of the ventral medullary surface (Schläfke, See & Loeschcke, 1970) it was found that the chemosensitive responses were not uniformly distributed over the medulla. In two areas, one medial to the vagal root and the other medial to the hypoglossal root but lateral to the pyramids, increase of ventilation was observed when more acid buffer was applied. The two areas could also be identified by electrical stimulation (Loeschcke *et al.*, 1970).

Figure 2.8: Schematic Diagram Showing a Cholinergic Synapse Driven by Acetylcholine or Nicotine and Inhibited by Atropine or Hexamethonium. The hydrogen ion concentration modulates the cholinergic transmission as for example by an anticholinesterase effect (Gesell & Hansen, 1942, 1945). It may be speculated that the entrance of H^+ to the synaptic gap is prevented if $[H^+]$ increases.

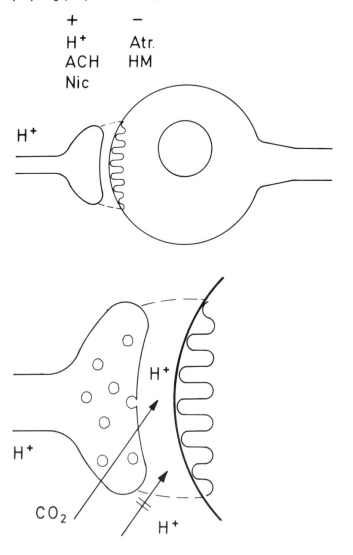

Figure 2.9: Elimination of Central Chemosensitivity by Local Cooling of the Intermediate Area. In this case respiration stops completely under cooling. Injection of NaCN is still effective indicating that the input from peripheral chemosensors at least partially bypasses the pathway from central chemosensitivity.

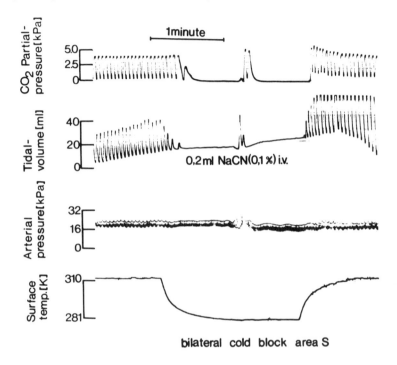

Source: Schlaefke *et al.* (1970); courtesy of Springer-Verlag.

In the area between these chemosensitive zones, called the intermediate area, no positive response to acid existed, a slight depression of ventilation occurred instead. This area became most interesting because either coagulation, or covering by procaine or cooling (to $12°C$) eliminated central chemosensitivity with the result that in previously chemodenervated animals breathing stopped completely even during inhalation of CO_2 (Schläfke *et al.*, 1970, 1979a, b; Cherniack *et al.*, 1979) (Figure 2.9). When in such an animal the central ends of the cut peripheral chemoreceptor nerves were stimulated electrically rhythmic breathing was taken up again (Loeschcke *et al.*, 1979) indicating that the respiratory centres were still functioning. But chemosensitivity to inhaled CO_2 was lost (Figure 2.10). Moreover in cats surviving elimination of both intermediate areas, where the peripheral chemoreceptors were kept intact, all or most of the steady-state response to inhaled CO_2 was lost indefinitely. A similar response was obtained in unanaesthesised cats (Schläfke *et al.*, 1979b).

Figure 2.10: Elimination of Central Chemosensitivity (Intact Sinus Nerve (A)) by Cooling the Intermediate Area or by Coagulation is Accompanied by a Loss of Response to Inhaled CO_2. This is tested by maintenance of respiration by stimulation of the central ends of the cut sinus nerves (B, C, E, D). If the test stimuli are applied rhythmically even an increased ventilation can be maintained (D). Also in this case respiration did not respond to increased PCO_2.

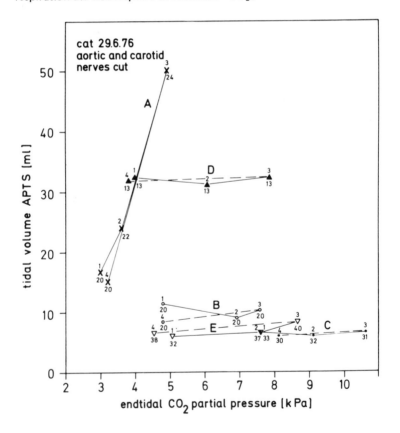

Source: Loeschcke *et al.* (1979); courtesy of Springer-Verlag.

These results suggest that the impulse traffic from the chemosensitive areas converge to the intermediate area where it could be interrupted by a small lesion or by cooling. Dev & Loeschcke (1979a) (Figure 2.11) clearly demonstrated this convergence by stimulating rostral and caudal chemosensitive areas with nicotine. Application of procaine to the intermediate areas on both sides reduced the nicotine effect to almost zero. Additional information is provided by recent experiments by Mikulski, Marek & Loeschcke (1981) who found that superficial cutting of the medulla close to the midline in peripherally chemo-denervated cats caused respiratory arrest (Figure 2.12). This means that the afferent impulse traffic decussates completely in a rather superficial layer of the medulla oblongata.

Evoked potentials have been observed in the paragigantocellular nucleus (Taber) and in the nucleus of the solitary tract after stimulating the caudal area by Davies & Loeschcke (1977) and Davies (1980). Inhibitory projections to the sympathetic chain (e.g. the splanchnic nerve) have been described (Schläfke, See & Burghardt, 1981) as well as increased impulse activity in several sympathetic nerves (Trzebski *et al.*, 1971, 1974) after increasing acidity on the ventral surface. Willshaw (1975, 1977) reported increase of efferent carotid nerve activity after alkalinisation of the medullary surface.

Stimulation of the carotid sinus nerve was still partially effective in driving respiration during cooling of the intermediate areas where the impulse traffic from central chemoreceptors decussate. Therefore the impulses from the carotid sinus nerve could bypass the site of the central chemosensitivity. Direct connections between carotid afferents and central chemoreceptors have been shown either by evoked potentials (Davies & Loeschcke, 1977; Davies, 1980) or by increased activity of H^+-activated neurones during sinus nerve stimulation (See & Schläfke, 1978). These connections may explain the reduced efficacy of sinus nerve stimulation during block of the intermediate areas.

Von Euler & Söderberg (1952) were the first to observe tonic discharges of neurones in an undefined location in the medulla oblongata which were strongly enforced (in frequency) by inhalation of CO_2. Shimada, Trouth & Loeschcke (1969) found H^+-activated neurones in the caudal area. Schläfke *et al.* (1975), Cakar & Terzioglu (1976), Prill (1977), Pokorski, Schlaefke & See (1975), and Pokorski (1976) (Figure 2.13) demonstrated that in all three medullary areas and also in the underlying paragigantocellular nucleus (Taber) action potentials could be obtained when CO_2 was inhaled, acid injected or acid superfused on the medullary surface (Figure 2.14). Most of the

Figure 2.11: Response of Ventilation to Nicotine Applied to the Rostral Area Before and After Application of Procaine to the Intermediate Area. Left column: control; middle column: mean of first 15 breaths after nicotine; right column: mean of 16th to 30th breaths after nicotine. This shows that the nicotine effect is a prolonged one. The effect is depressed by procaine on the intermediate area demonstrating the convergence of the impulse traffic from the rostral to the intermediate area.

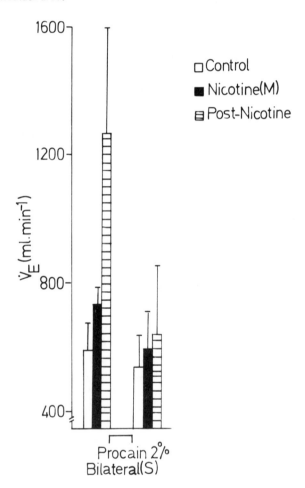

Source: Dev & Loeschcke (1979a, b); courtesy of Springer-Verlag.

Figure 2.12: Ventral Aspect of the Cat's Medulla. The hatched areas are
the rostral and the caudal chemosensitive areas. Cutting at 1 abolishes
most of central chemosensitivity, cuts 2 and 3 abolish the remainder.
This is interpreted to mean that the impulse traffic crosses to the
contralateral side completely where it reaches the contralateral intermediate area. There were relatively small diminutions of ventilation
after either cuts 4 or 5.

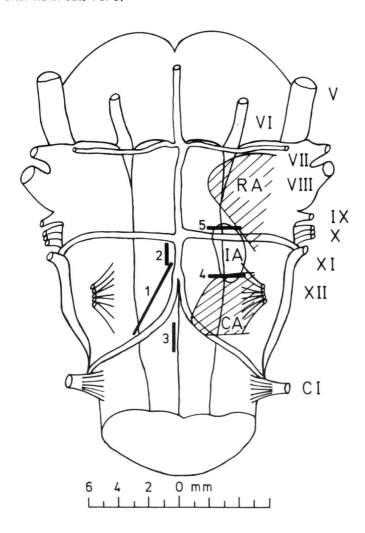

Source: Mikulski *et al.* (1981); courtesy of Drs Mikulski and Marek.

neurones observed fired in a slow rhythm but there were exemptions which discharged with higher frequencies; these were most suitable for the investigation of alkali injection. Only about 50 per cent of the spontaneously active neurones in this region reacted to pH changes. Others could either not be activated by identified stimuli, or some reacted to touch of the limbs.

Figure 2.13: Impulse Activity from a Neuron in the Intermediate Area. In the upper trace discharge before and after i.v. injection of 4 ml 0.1 N-HC1. In the middle and lower traces the continuous record counts of impulse frequency before and after the injection of acid (arrow).

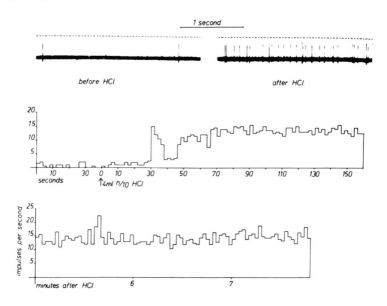

Source: Schlaefke *et al.* (1975); courtesy of Bull. Physio-Pathol. Resp.

To date most of the H^+-activated units investigated were of cellular origin. It has, so far, been impossible to keep intracellular potentials for any reasonable time. In a given neurone the correlation between discharge frequency and pH measured locally on the medullary surface layer was much better than to pH in any other location.

All hydrogen ion-activated neurones are also driven by electrophoretically applied acetylcholine (Schlaefke, personal communication). There are also neurones not reacting to H^+ but still responding to acetylcholine. Among these may be the candidates for the cardio-vascular effects which are seen if the ventral medullary surface is stimulated. Such effects have been amply observed by Loeschcke *et al.* (1958), Schläfke & Loeschcke (1967), by Feldberg (1976, 1980) and Guertzenstein (1979). As pH changes on the ventral surface of the medulla are without clear and reproducible effects on the arterial pressure and heart rate it must be doubted that the cardiovascular effects are transmitted by the same neurones (Loeschcke, 1980) as the chemosensitive respiratory ones. Furthermore, the reactions to some drugs are the same for respiratory and vascular reflexes while others produce opposite effects.

Figure 2.14: A Neurone in the Paragigantocellular Nucleus Marked by Procion Yellow and the Response (Impulses s^{-1}) of This Neurone Plotted Against Extracellular pH Measured on the Surface of the Medulla Oblongata during Superfusion with a Buffer of Varying pH.

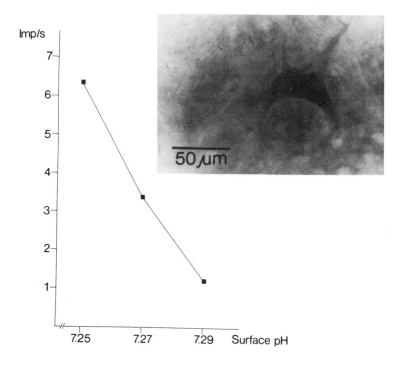

Source: Schlaefke (1981); courtesy of Dr M.E. Schlaefke.

Lipscomb & Boyarski (1972) failed to record action potentials activated by H^+ in the ventral superficial layer of the medulla oblongata. This is surprising in view of the body of evidence referred to above. It must, however, be remembered that the chemosensory mechanism is easily anaesthetised; furthermore in such superficial layers of the CNS the electrodes do not easily stay in place. The electrodes used by Schläfke *et al.* (1975) were miniaturised; this was necessary to make them 'floating'. It was also found that the electrodes were kept in place much better if the puncture occurred at a sharp angle to the surface thus providing support and friction in the relatively tough tissue of the marginal glia.

The hypothesis of Lipscomb & Boyarski (1972) that the respiratory effects of acid or temperature might be due to transport of acidity or heat should be taken more seriously especially since Trzebski *et al.* (1980) and Shams (1981) obtained the typical effects of acidification by injecting solutions into the vertebral arteries. However, a superfusion is not the same as an intra-arterial injection. It cannot be expected that the blood in the time of its passage equilibrates completely since it is buffered, while in an arterial bulk injection the blood is more or less replaced by the injected fluid. Nevertheless it cannot be denied that some acid or some heat might enter or leave the blood from a subarachnoid perfusion and exchange again in the tissue. The heat profile published by Schlaefke & Loeschcke (1967) does not support appreciable effects of such transported heat after spotwise cooling of the ventral surface. The observations by Mitchell & Herbert (1974) and Marino & Lamb (1975) are against an excitatory respiratory effect of transported acid because it has been shown by them that extracellular acid inhibits respiratory neurones rather than driving them.

The first cell type suspected to be chemosensitive was described by Trouth *et al.* (1973a, b). Schläfke *et al.* (1974) and Dermietzel (1976) gave detailed accounts of the different cell types occurring in the chemosensitive region. It is still not clear which cells are the chemosensitive ones. Several cell types have been visualised by electron microscopic technique (Ullah, 1973; Dermietzel, 1976; Leibstein, Willenberg & Dermitzel, 1981a). The cells possess many synapses on the somata and also on long dendrites. The distribution of cholinergic cells has been investigated by histochemical techniques for demonstrating acetylcholine-esterase. Nothing has been seen that does not also occur in other parts of the central nervous system except perhaps a denser packing of cholinergic cells in the chemosensitive region (Leibstein *et al.*, 1981). However, the topography of the intercellular spaces may be peculiar in so far as relatively wide spaces occur which are in open connection with basal membrane covered branched recesses opening to the subarachnoid space. This type of tissue has been called glia spongiosa (Dermietzel, 1976).

Fukuda & Honda (1975) and Fukuda *et al.* (1978) succeeded in keeping slices of about 0.4 mm thickness from the ventral surface of the rat brain viable in a perfusion chamber for several hours. They were able to record intracellular potentials of up to -90 mV. The membrane potentials decreased slightly if PCO_2 was elevated. These cells, however, were silent and could not be stimulated electrically. They were interpreted as glial cells. Later Fukuda & Loeschcke (1977) and Fukuda *et al.* (1979) also found spontaneously discharging cells in the same preparation. The action potentials of these cells were recorded extracellularly. Some of them were activated by H^+; a smaller number of cells were inhibited (Figure 2.15). The response in a single cell, however, under given condition was reproducible. Elevating H^+ in non-respiratory acidosis above a critical level caused the activity of the cell to decrease again. This is an observation which may be the cellular analogue of the similar reaction of ventilation in the injection experiments of Shams *et al.* (1981).

Figure 2.15: Discharge of Impulses Depending on Extracellular pH in an Excitatory E-neurone (A) and an Inhibitory I-neurone (B). pH was varied in the buffer bathing an isolated slice from the ventral medullary surface at constant PCO_2.

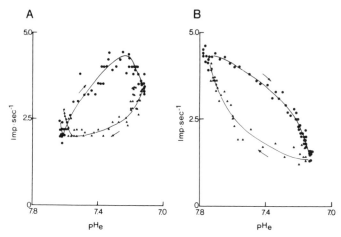

Source: Fukuda & Loeschcke (1977); courtesy of Springer-Verlag.

In the chemosensitive areas, the majority of cells were activated by H^+. In slices of the dorsal surface of the medulla the same types of cellular responses were observed; this time, however, the cells inhibited by H^+ were in the majority. A cell activated by H^+ if treated with a solution containing reduced calcium and increased magnesium concentrations could be reversed in its response, i.e. it became inhibited by H^+. Such behaviour is said to be indicative of a synaptic effect, and if correct will play an important role in the interpretation of the chemosensitive mechanism.

A Cholinergic Mechanism as a Link in the Chemosensitive Mechanism

Mitchell *et al.* (1960) described an inhibition of ventilation when acetylcholine was applied to the area postrema. There was, however, no effect of acid from this spot and hence the action of acetylcholine was no indication of chemosensitivity. However the same authors (Mitchell *et al.*, 1963a, b) applied acetylcholine to what later was called the rostral chemosensitive area; increased ventilation resulted. The location of the acetylcholine effect on ventilation and circulation was studied in more detail by Dev & Loeschcke (1979a, b). The topical distribution showed two peaks of maximal effect and these coincided with those of hydrogen ion action. Also nicotine could mimic acetylcholine and physostigmine enhanced the action of acetylcholine. Ventilation was much depressed by atropine and in later experiments by Tok & Loeschcke (1983) using higher concentrations, was in some cases completely arrested. Intravenous hexamethonium exhibited a depressing action on the acetylcholine effect.

Tok & Loeschcke (1983) reinvestigated the effect of progesterone on ventilation and it turned out that progesterone superfused on to the ventral medullary surface raised ventilation. This effect was counteracted by atropine.

This group of experiments suggests that the chemosensitive mechanism involves a cholinergic link which can be blocked by atropine.

At the same time Fukuda & Loeschcke (1979) studied the effects of cholinergic stimulating and inhibiting substances on the neurones activated by H^+ in superfused tissue slices of the ventral medullary surface of the rat. An almost complete mirror image was observed with the effects of drugs on ventilation if applied to the intact surface of the medulla. Acetylcholine like H^+ stimulated the neurones (Figure 2.16) and these effects were additive. Nicotine acted similarly as did eserine. Mecamylamine, an acetylcholine-inhibitor, stopped impulse generation (Figure 2.17). The most important observation was that mecamylamine did not only stop the effect of acetylcholine but also that of H^+. The atropine story was found to be slightly more complicated in so far as atropine did not stop the impulse generation and usually even increased it. Still under atropine the effect of H^+ was lost. This suggests that atropine competes with H^+ at the receptor blocking it for H^+.

Figure 2.16: A. Effects of pH Variation in an *in Vitro* Perfusion of a Slice of the Ventral Medullary Surface. From upward down pH, standard impulses, integrated impulses. B. Continuation of the recording. At the arrow ACh was injected into the perfusion fluid. The responses to increased acidity mimic the effect of acetylcholine.

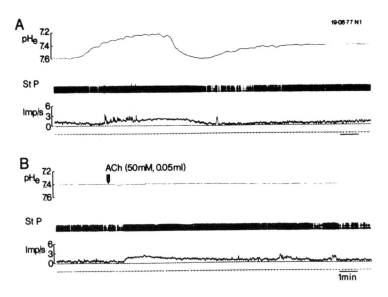

Source: Fukuda & Loeschcke (1979); courtesy of Springer-Verlag.

That decrease of calcium and increase of magnesium concentrations reverses H^+ sensitivity has been mentioned previously. If it is true that this is an effect characteristic of a synapse the overall conclusions of this chapter must be that the chemosensitive mechanism is an effect on a cholinergic synapse where H^+ has an action like that of acetylcholine and can be replaced by it. Blockade is possible with mecamylamine, atropine and partly by hexamethonium. This suggests muscarinic *and* nicotinic receptors.

Figure 2.17: A. Effects of Increased Acidity in a Tissue Slice of the Ventral Medullary Surface in an *in Vitro* Perfusion Chamber. From upward down: pH, standard impulses, integrated impulses. Strong increase of impulse generation after acidification. B. Under the influence of mecamylamine (an acetylcholine antagonist) the pH effect is abolished or even reversed. C. It recovers after washing out.

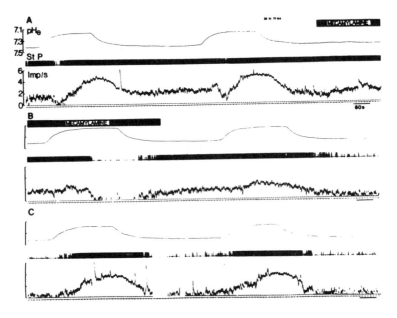

Source: Fukuda & Loeschcke (1979); courtesy of Springer-Verlag.

An Excusion into the Control of Breathing in Muscular Exercise

If we continue the argument developed in the previous section a cholinergic, chemosensitive synapse would suggest a synaptic neurogenic input. Whilst this suggestion is only speculative, there is an experiment, which looks even too good in this respect (Figure 2.18). Spode (1980) stimulated the anterior root in the lumbar part of the spinal medulla. In order to avoid any kind of adaptation to the stimulus this was given in trains alternating between the right and the left side. A marked

Figure 2.18: Rhythmic Stimulation, Alternating from Left to Right of the Lumbar Anterior Roots (L VII). From upward down: arterial pressure, stimulation left root VII, right root VII, expiratory PCO_2, tidal volume, temperature on both intermediate areas. Stimulation increases tidal volume. Cooling of the intermediate areas (s) causes respiratory arrest in spite of ongoing root stimulation. This indicates that the increase of ventilation in exercise cannot be maintained after elimination of central chemosensitivity.

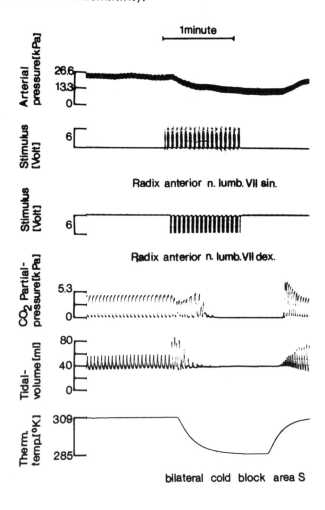

Source: Spode (1980); courtesy of Dr Spode.

increase of ventilation was observed and following the argument of Kao (1956) this should mainly be interpreted as a neuronal drive from the muscle. If central chemosensitivity was interrupted by local cooling of the intermediate areas (in a chemodenervated animal) breathing stopped completely during rest and also during exercise and the effect was reversible. It might then be assumed that neural signals from the muscle would reach the ventral chemosensitivity in the form of a synaptic input and the interruption of central chemosensitivity should also eliminate the propagation of the neural drive. This is what happened in this technically most difficult experiment. It could, of course, be argued that under this condition the neural drive might not really have been interrupted but might only not have sufficed to drive ventilation. It is known, however, that the respiratory drive from the hypothalamus in hyperthermia does survive the elimination of the central chemosensitivity and also that the drive from peripheral chemoreceptors can maintain ventilation after elimination of the central drive, making such an argument unlikely.

Schläfke *et al.* (1979) have shown in the nucleus paragigantocellularis that some neurones could be driven from the tibial nerve as well as by H^+. This corroborates the idea that the neuronal drive in exercise is relayed in otherwise chemosensitive neurones or in other words, that H^+ modulates the transmission of neural impulses related to exercise (Figure 2.19).

Generalisation Leading to a Tentative Model of Respiratory Chemosensitivity

The role of central chemosensitivity in the control system is schematically represented in Figure 2.19. Inputs from the muscle, the neurogenic drive and unspecific sensory inputs are thought to converge on the chemosensitive neurones driving them via cholinergic synapses. H^+ also acts to modulate the same synapses and there are also specific inputs from the peripheral chemosensors. Finally by whatever chain of neurones, the chemosensory excitation reaches the respiratory output neurones. This model reminds us of the classical ideas of Gesell & Hansen (1942, 1945) concerning the role of acetylcholine in respiratory control.

Figure 2.19: Schematic Diagram of Central Respiratory Chemosensitivity with its Principal Interconnections. Inhibitory and facilitatory unspecific afferents from the sympathetic nervous system and sensory fibres from the integument and the muscles and also the specific fibres from the peripheral chemoreceptors are supposed to feed synaptically to the central chemosensitive neurone. This synapse is supposed to be cholinergic and is also reached by acetylcholine from the outside. Also hydrogen ion acts in a excitatory way on the cholinergic synapse probably by a mechanism involving anticholinesterase. Afferent fibres from the peripheral chemoreceptor additionally bypass central chemoreceptor as do the fibres from the hypothalamic thermosensor. Finally they converge to an output neurone where the tonic discharge is commuted into a respiratory rhythm.

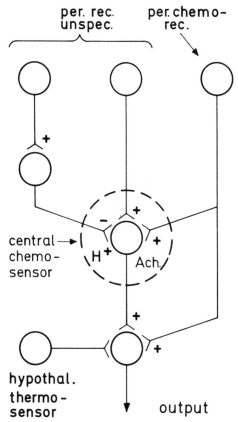

Source: Ahmad & Loeschcke (1983a); courtesy of Springer-Verlag.

Signal Transmission from Blood to Receptor

The receptor in the brain tissue responds to extracellular H^+ at least up to a certain level. The transmission from blood to the extracellular fluid is not the same for the three variables of the Henderson–Hasselbalch equation. There are two boundaries which have to be passed in parallel, the choroid plexus and the endothelium of the brain capillaries.

The blood flow in the choroid plexus is extremely high (Page *et al.*, 1980) and the tissue volume drained by it consists of the choroid plexus tissue only. It is to be expected that any a-v-difference in plexus blood chemistry is very small, and hence the average PCO_2 of the choroid plexus is low, close to arterial. However, in contact with the brain tissue the secreted plexus fluid takes up CO_2, equilibrating with the brain surface and making the fluid more acid; this partially explains the higher acidity of CSF in comparison with blood plasma. If is is assumed that in a steady state the composition of cerebrospinal fluid equals that of extracellular fluid this is obviously an approximation. The good exchange between the two compartments will serve to approach the two fluids to each other. Only in a most superficial layer, however, will the local extracellular pH be under the influence of the cerebrospinal fluid because the tissue is perfused by the blood circulation. On average in most regions of the tissues we may assume that the local PCO_2 is determined by the usual factors, local metabolism, CO_2 binding capacity of the blood, blood flow, and distance to the capillaries. Ponten & Siesjö (1966) proposed that tissue PCO_2 equalled the algebraic mean between venous and arterial PCO_2 of cerebral blood plus one Torr. This has been verified by Ahmad *et al.* (1976).

Several authors have determined the relation between HCO_3^- in blood and in CSF. In the dog and man it appears that HCO_3^- in CSF changes only about 40 per cent of any change in blood plasma (Pappenheimer *et al.*, 1965; Mitchell *et al.*, 1965; Kronenberg & Cain, 1968; Fencl, 1971). In cats it was even less; for example 20 per cent (Ahmad *et al.*, 1976). Meanwhile it was observed that though in chronic human cases this relation was maintained, in the cat at least this factor in extracellular fluid of the brain was time dependent and that the increase of bicarbonate concentration in ECF after an i.v. injection or infusion decreased with time in such a way that after an hour not much of the increase was left.

Even if the story about the bicarbonate exchange becomes more complicated it is well established that *bicarbonate does exchange* between blood plasma and extracellular fluid (ECF) (Leusen, 1972).

Figure 2.20: Passage of Bicarbonate from Blood Plasma to the Extracellular Fluid of the Brain After i.v. Injection of Bicarbonate. Transients of HCO_3^- exchanges against chloride in the superficial layer of the medulla oblongata. Averages of six experiments in four cats ($\pm$ SEM), points are 12 s apart starting in the upper left corner. There is an exchange from plasma to the endothelial cell and another exchange of the same kind from the interior of the endothelium to the extracellular fluid of the brain.

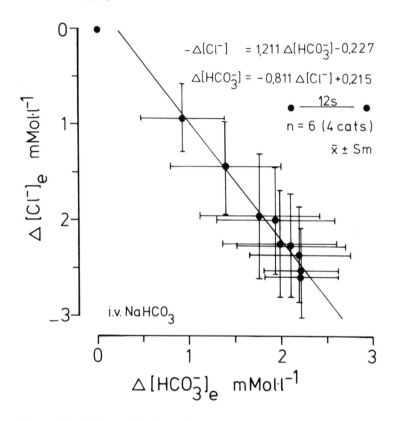

$$-\Delta[Cl^-] = 1{,}211\,\Delta[HCO_3^-] - 0{,}227$$
$$\Delta[HCO_3^-] = -0{,}811\,\Delta[Cl^-] + 0{,}215$$

12s

n = 6 (4 cats)

$\bar{x} \pm Sm$

i.v. $NaHCO_3$

Source: Ahmad & Loeschcke (1983a); courtesy of Springer-Verlag.

This is an exemption of HCO_3^- and Cl^- of the typical behaviour of the blood-brain barrier. It is also clear that the HCO_3^- exchange is a fast process occurring in minutes if not in seconds (Ahmad *et al.*, 1976, 1983a). Finally it is apparent that the exchange of HCO_3^- occurs simultaneously with an exchange of Cl^- in the opposite direction (Figure 2.20).

This exchange is now assumed to happen in a specific *anion exchange* channel most possibly by a protein carrier (Wieth, Brahm & Funder, 1980), and obviously occurs through the endothelium of the brain capillaries.

When CO_2 is inhaled the bicarbonate concentration in ECF increases. This was first shown by Pannier, Wayne & Leusen (1970). The increase of HCO_3^- in ECF was higher than that in blood plasma and even if by additional infusion of acid an increase of bicarbonate in plasma was avoided or even a decrease was obtained the $[HCO_3^-]$ in CSF rose. The source of HCO_3^- in this case could only be the cells of the brain tissue. Again by electrochemical methods Ahmad & Loeschcke (1983b) were able to show that there is another exchange of HCO_3^- against *chloride* between *extracellular fluid* and *cells* (Figure 2.21). This exchange was also an exchange against chloride, approximately one to one. It occurred in seconds and it could be inhibited by stilbene derivatives.

Figure 2.21: Transients of Bicarbonate and Chloride Concentrations in the Extracellular Fluid of the Brain During Inhalation of CO_2. Averages of nine experiments in four cats (± SEM), the points are 12 s apart. It must be assumed that there is an exchange of bicarbonate against chloride between the extracellular fluid of the brain and the brain cells (presumably astrocytes).

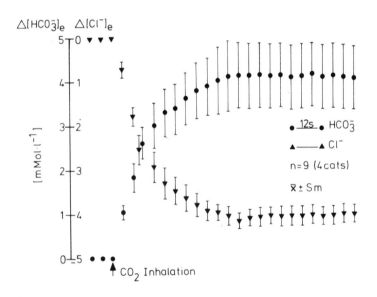

Source: Ahmad & Loeschcke (1983b); courtesy of Springer-Verlag.

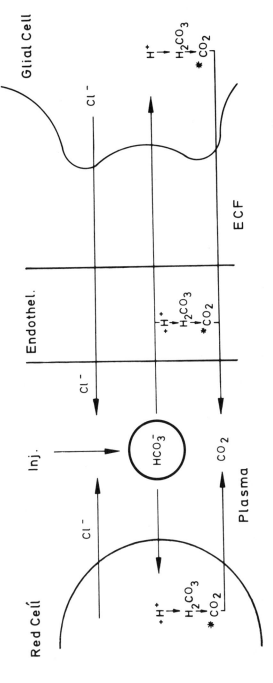

Figure 2.22: Schematic Diagram of the Exchange Processes Between the Five Compartments, Red Cells, Plasma, Endothelium, Extracellular Fluid and Brain Cells (Presumably Glial). Injection of HCO_3^-.

. Effect of bicarbonate injection on acid base parameters in plasma, ECF and cells.

✳ carbonic anhydrase

Source: Ahmad & Loeschcke (1983b); courtesy of Springer-Verlag.

Extracellular pH becomes a variable which depends on the two types of anion exchange between blood plasma and endothelium and through endothelium to the extracellular fluid and between brain cells and extracellular fluid. For any modelling of pH kinetics in ECF all five compartments (Ahmad & Loeschcke, 1981) namely red cells, plasma, endothelium, extracellular fluid and glial cells must be considered (Figure 2.22). This, of course, is true for metabolic as well as respiratory acidosis. The resultant mathematical model should give the quantitative answers about steady and transient states in all acid-base disturbances and especially about the extracellular pH under all these conditions.

The model of Middendorf & Loeschcke (1976a, b) of the regulation of breathing made use of the bicarbonate relation between plasma and CSF as determined by the authors. It approached reality reasonably well in respiratory and metabolic acidosis and among other features could also simulate variation of the total metabolism. It follows from the model that the response of ventilation to respiratory acidosis is much higher than that to metabolic acidosis. This response, however, is much more effective in compensating for a pH change in metabolic than in respiratory acidosis (Middendorf, 1974, 1976).

Summarising this chapter it may be stated that the extracellular pH in four ventrolateral areas of the medulla oblongata is the main chemical signal determining the response of ventilation. There are also neuronal signals which are modulated by the extracellular pH (in the peripheral chemoreceptors). This pH depends on the tissue PCO_2 and the bicarbonate concentration in the tissue. The distribution factor of bicarbonate between plasma and CSF under given conditions can be experimentally determined. There are processes counteracting a full equilibrium. Any prediction of the extracellular pH should be based on the exchange processes between the five compartments red cells, plasma, endothelium of the brain capillaries, extracellular fluid and brain cells.

References

Ahmad, H.R., Berndt, J. & Loeschcke, H.H. (1976) 'Bicarbonate Exchange between Blood, Brain Extracellular Fluid and Brain Cells at Maintained PCO_2', in H.H. Loeschcke (ed.) *Acid Base Homeostasis of the Brain Extracellular Fluid and the Respiratory Control System*, Thieme Verlag, Stuugart, pp. 19-27

Ahmad, H.R. & Loeschcke, H.H. (1983a) 'Transient and Steady State Response of Pulmonary Ventilation to the Medullary Extracellular pH', *Pflügers Arch.*, *395*, 285-92

Ahmad, H.R. & Loeschcke, H.H. (1983b) 'Fast Bicarbonate-Chloride Exchange between Brain Cells and Brain Extracellular Fluid in Respiratory Acidosis', *Pflügers Arch.*, *395*, 293-304

Ahmad, H.R. & Loeschcke, H.H. (1983c) 'Fast HCO_3^--Cl Exchange between Plasma and Brain Extracellular Fluid at Maintained PCO_2', *Pflügers Arch.*, *395*, 305-25

Åström, A. (1952) 'On the Action of Combined Carbon Dioxide Excess and Oxygen Deficiency in the Regulation of Breathing', *Acta Physiol. Scand.*, *27*, Suppl. 98

Benzinger, T., Opitz, E. & Schoedel, W. (1938) 'Atmungserregung durch Sauerstoffmangel', *Pflügers Arch.*, *241*, 71-7

Çakar, L. & Terzioğlu, M. (1976) 'The Response of the Chemosensitive Areas of the Cat to the Breathing of Hypercapnia Gas Mixtures', *Bull. Physio-Pathol. Resp.*, *12*, 224 P

Cherniack, N.S., v. Euler, C., Homma, I. & Kao, F.F. (1979) 'Graded Changes in Central Chemoreceptor Input by Local Temperature Changes on the Ventral Surface of the Medulla', *J. Physiol.*, *287*, 191-211

Cragg, P., Patterson, L. & Purves, M.J. (1977) 'The pH of Brain Extracellular Fluid in the Cat', *J. Physiol.*, *272*, 137-66

Davies, R.O. (1980) 'Evidence that Neural Paths from the Caudal and Cranial Chemoreceptor Zones of the Ventral Medulla Converge in the Intermediate Zone', *Proc. Int. Un. Physiol. Sci.*, *XIV*, 371

Davies, R.O. & Loeschcke, H.H. (1977) 'Neural Activity Evoked by Electrical Stimulation on the Chemosensitive Areas on the Ventral Medullary Surface', *Proc. Int. Un. Physiol. Sci.*, *XIII*, 164

Dermietzel, R. (1976) 'Central Chemosensitivity, Morphological Studies', in H.H. Loeschcke (ed.) *Acid-Base Homeostasis of the Brain Extracellular Fluid and the Respiratory Control System*, Thieme Verlag, Stuttgart, pp. 52-65

Dev, N.B. & Loeschcke, H.H. (1979a) 'Topography of the Respiratory and Circulatory Responses to Acetylcholine and Nicotine on the Ventral Surface of the Medulla Oblongata', *Pflügers Arch.*, *379*, 19-27

Dev, N.G. & Loeschcke, H.H. (1979b) 'A Cholinergic Mechanism Involved in the Respiratory Chemosensitivity', *Pflügers Arch.*, *379*, 29-36

v. Euler, C. & Söderberg, U. (1952) 'Medullary Chemosensitive Receptors', *J. Physiol.*, *118*, 545-59

Feldberg, W. (1976) 'The Ventral Surface of the Brainstem: a Scarcely Explored Region of Pharmacological Sensitivity', *Neuroscience*, *1*, 427-41

Feldberg, W. (1980) 'Cardiovascular Effects of Drug Acting on the Ventral Surface of the Brain Stem', in H.P. Koepchen, S.M. Hilton & A. Trzebski (eds) *Central Interaction between Respiratory and Cardiovascular Control Systems*, Springer, Berlin, Heidelberg, New York, pp. 45-55

Fencl, V. (1971) 'Distribution of H^+ and HCO_3^- in Cerebral Fluids', in B.K. Siesjö & S.C. Sørensen (eds), *Ion Homeostasis of the Brain. Alfred Benzon Symposium III*, Munksgaard, Copenhagen, pp. 175-85

Fencl, V., Miller, T.B. & Pappenheimer, J.R. (1966) 'Studies on the Respiratory Response to Disturbances of Acid-Base Balance, with Deductions Concerning the Ionic Composition of Cerebral Interstitial Fluid', *Am. J. Physiol.*, *210*, 459-72

Fukuda, Y. & Honda, Y. (1975) 'pH Sensitive Cells at Ventrolateral Surface of the Rat Medulla Oblongata', *Nat. New. Biol.*, *256*, 317-8

Fukuda, Y., Honda, Y., Schlaefke, M.E. & Loeschcke, H.H. (1978) 'Effect of H^+ on the Membrane Potential of Silent Cells in the Ventral and Dorsal Surface Layer of the Rat Medulla *in Vitro*', *Pflügers Arch.*, *376*, 229-35

Fukuda, Y. & Loeschcke, H.H. (1977) 'Effect of H^+ on Spontaneous Neuronal

Activity in the Surface Layer of the Rat Medulla Oblongata', *Pflügers Arch.*, *371*, 125-34

Fukuda, Y. & Loeschcke, H.H. (1979) 'A Cholinergic Mechanism Involved in the Neuronal Excitation by H^+ in the Respiratory Chemosensitive Structures of the Ventral Medulla Oblongata of Rats *in Vitro*', *Pflügers Arch.*, *379*, 125-35

Fukuda, Y., See, W.R., Schlaefke, M.E. & Loeschcke, H.H. (1979) 'Chemosensitivity and Rhythmic Activity of Neurons in the Ventral Surface Layer of the Rat Medulla Oblongata *in Vitro* and *in Vivo*', *Pflügers Arch.*, *379*, R 50

Gesell, R. & Hansen, E.T. (1942) 'Eserine, Acetylcholine, Atropine and Nervous Integration', *Am. J. Physiol.*, *139*, 371-85

Gesell, R. & Hansen, E.T. (1945) 'Anticholinesterase Activity of Acid as a Biological Instrument of Nervous Integration', *Am. J. Physiol.*, *144*, 126-63

Gray, J.S. (1950) *Pulmonary Ventilation and its Physiological Regulation*, Thomas, Springfield, Ill.

Guertzenstein, P.G. (1979) 'Blood Pressure Effects Obtained by Drugs Applied to the Ventral Surface of the Brain', *J. Physiol.*, *229*, 395-408

Hori, T., Roth, G.I. & Yamamoto, W.S. (1970) 'Respiratory Sensitivity of Rat Brain-stem to Chemical Stimuli', *J. Appl. Physiol.*, *28*, 721-4

Hugelin, A. & Cohen, M.I. (1963) 'The Reticular Activating System and Respiratory Regulation in the Cat', *Ann. N.Y. Acad. Sci.*, *109*, 568-603

Kao, F.F. (1956) 'Regulation of Respiration During Muscular Activity', *Am. J. Physiol.*, *185*, 145

Kronenberg, R.S. & Cain, S.M. (1968) 'Effects of Acetazolamide and Hypoxia on Cerebrospinal Fluid Bicarbonate', *J. Appl. Physiol.*, *24*, 17-20

Lambertsen, C.J., Semple, S.J.G., Smith, M.G. & Gelfand, R. (1961) 'H^+ and PCO_2 as Chemical Factors in Respiratory and Cerebral Circulatory Control', *J. Appl. Physiol.*, *16*, 473-84

Leibstein, A.G., Willenberg, I. & Dermitzel, R. (1981) 'Untersuchung zur Transmittercharakteristik von Neuronen an der ventralen Oberfläche der Medulla oblongata', *Pflügers Arch.*, *391*, 226-30

Leusen, I. (1954a) 'Chemosensitivity of the Respiratory Center. Influence of CO_2 in the Cerebral Ventricles on Respiration', *Am. J. Physiol.*, *176*, 39-44

Leusen, I. (1954b) 'Chemosensitivity of the Respiratory Center. Influence of Changes of H^+ and Total Buffer Concentrations in the Cerebral Ventricles on Respiration', *Am. J. Physiol.*, *176*, 45-51

Leusen, I. (1972) 'Regulation of CSF Composition with Reference to Breathing', *Physiol. Rev.*, *52*, 1-56

Lipscomb, W.T. & Boyarski, L.L. (1972) 'Neurophysiological Investigations of Medullary Chemosensitive Areas of Respiration', *Resp. Physiol.*, *16*, 362-76

Loeschcke, H.H. (1969) 'On Specificity of CO_2 as a Respiratory Stimulus', *Bull. Physio-Pathol. Resp.*, *5*, 13-25

Loeschcke, H.H. (1973) 'The Respiratory Control System: Analysis of Steady State Solutions for Metabolic and Respiratory Acidosis-Alkalosis and Increased Metabolism', *Pflügers Arch.*, *341*, 23-42

Loeschcke, H.H. (1980) 'Introduction', in H.P. Koepchen, S.M. Hilton & A. Trzebski (eds) *Central Interaction between Respiratory and Cardiovascular Control Systems*. Springer, Berlin, Heidelberg, New York, pp. 45-6

Loeschcke, H.H., Koepchen, H.P. & Gertz, K.H. (1958) 'Über den Einfluβ von Wasserstoffionenkonzentration und CO_2-Druck im Liquor cerebrospinalis auf die Atmung', *Pflügers Arch.*, *266*, 565-85

Loeschcke, H.H., de Lattre, J., Schlaefke, M.E. & Trouth, C.O. (1970) 'Effects on Respiration and Circulation of Electrically Stimulating the Ventral Surface of the Medulla Oblongata', *Resp. Physiol.*, *10*, 184-97

Loeschcke, H.H., Schlaefke, M.E., See, W.R. & Herker-See, A. (1979) 'Does CO_2

Act on the Respiratory Centers?' *Pflügers Arch.*, *381*, 249-54

Loeschcke, H.H. & Sugioka, K (1969) 'pH of Cerebrospinal Fluid in the Cisterna Magna and on the Surface of the Choroid Plexus of the 4th Ventricle and its Effect on Ventilation in Experimental Disturbances of Acid-Base Balance', *Pflügers Arch.*, *312*, 161-88

Marino, P.L. & Lamb, T.W. (1975) 'Effects of CO_2 and Extracellular H^+ Iontophoresis on Single Cell Activity in the Cat Brainstem', *J. Appl. Physiol.*, *38*, 688-95

Middendorf, T. (1974) 'Analysis of the Efficiency of the Respiratory Control System', in W. Umbach & H.P. Koepchen (eds), *Central Rhythmic and Regulation*, Hippokrates, Stuttgart, pp. 117-20

Middendorf, T. & Loeschcke, H.H. (1976a) 'Mathematische Simulation des Respirationssystems', *J. Mathem. Biol.*, *3*, 149-77

Middendorf, T. & Loeschcke, H.H. (1976b) 'Analysis of the Respiratory Control System', in H. Loeschcke (ed.) *Acid-Base Homeostasis in the Brain Extracellular Fluid and the Respiratory Control System*, Thieme, Stuttgart, pp. 190-201

Mikulski, A., Marek, W. & Loeschcke, H.H. (1983) 'Interconnections between Central Chemosensitive Areas, 'Respiratory Centres' and Efferent Output', *Pflügers Arch.* (in press)

Mitchell, R.A., Carman, C.J., Severinghaus, J.W., Richardson, B.W., Singer, M.M. & Schnider, S. (1965) 'Stability of Cerebrospinal Fluid pH in Chronic Acid-Base Disturbances in Blood', *J. Appl. Physiol.*, *20*, 443-52

Mitchell, R.A. & Herbert, D.A. (1974) 'Synchronized High Frequency Synaptic Potentials in Medullary Respiratory Neurons', *Brain Res.*, *75*, 350-5

Mitchell, R.A., Loeschcke, H.H., Massion, W.H. & Severinghaus, J.W. (1963a) 'Respiratory Responses Mediated through Superficial Chemosensitive Areas on the Medulla', *J. Appl. Physiol.*, *18*, 523-33

Mitchell, R.A., Loeschcke, H.H., Severinghaus, J.W., Richardson, B.W. & Massion, W.H. (1963b) 'Regions of Respiratory Chemosensitivity on the Surface of the Medulla', *Ann. N.Y. Acad. Sci.*, *109*, 661-81

Mitchell, R.A., Massion, W., Carman, C.T. & Severinghaus, J.W. (1960) '4th Ventricle Respiratory Chemosensitivity and the Area Postrema', *Fed. Proc.*, *19*, 374

Nielsen, M. (1936) 'Untersuchungen über die Atmungsregulation beim Menschen', *Skand. Arch. Physiol.*, *74*, Suppl. 10, 83-208

Page, R.B., Funsch, D.I., Brennan, R.W. & Hernandez, M.J. (1980) 'Choroid Plexus Blood Flow in the Sheep', *Brain Res.*, *197*, 532-7

Pannier, J.C., Weyne, J. & Leusen, I. (1970) 'The CSF/Blood Potential and the Regulation of the Bicarbonate Concentration of CSF During Acidosis in the Cat', *Life Sci.*, 287-300

Pappenheimer, J.R. (1967) 'The Ionic Composition of Cerebral Extracellular Fluid and its Relation to Control of Breathing', *Harvey Lectures*, *61*, 71-94

Pappenheimer, J.R., Fencl, V., Heisey, S.R. & Held, D. (1965) 'Role of Cerebral Fluids in Control of Respiration as Studied in Unanaesthetized Goats', *Am. J. Physiol.*, *208*, 436-50

Pokorski, M. (1976) 'Neurophysiological Studies on Central Chemosensor in Medullary Ventrolateral Areas', *Am. J. Physiol.*, *230*, 1288-95

Pokorski, M., Schlaefke, M.E. & See, W.R. (1975) 'Neurophysiological Studies on the Central Chemosensitive Mechanism', *Pflügers Arch.*, *355*, R 33

Pontén, U. & Siesjö, B.K. (1966) 'Gradients of CO_2 Tension in the Brain', *Acta Physiol. Scand.*, *67*, 129-40

Prill, R.K. (1977) 'Das Verhalten von Neuronen des caudalen chemosensiblen Feldes in der Medulla oblongata der Katze gegenüber intravenösen Injektionen

76   Central Chemoreceptors

von NaHCO$_3$ und HCl', Thesis. *Abt. Naturwiss. Med.*, Ruhr-Universität Bochum

Schlaefke, M.E., Kille, J., Folgering, H., Herker, A. & See, W.R. (1974) 'Breathing Without Central Chemosensitivity', in W. Umbach & H.P. Koepchen (eds) *Central Rhythmic and Regulation*, Hippokrates, Stuttgart, pp. 97-104

Schlaefke, M.E., Kille, J.F. & Loeschcke, H.H. (1979b) 'Elimination of Central Chemosensitivity by Coagulation of a Bilateral Area on the Ventral Medullary Surface in Awake Cats', *Pflügers Arch.*, *379*, 231-41

Schlaefke, M.E. & Loeschcke, H.H. (1967) 'Lokalisation eines an der Regulation von Atmung und Kreislauf beteiligten Gebietes an der ventralen Oberfläche der Medulla oblongata durch Kälteblockade', *Pflügers Arch.*, *297*, 201-20

Schlaefke, M.E., Pokorski, M., See, W.R., Prill, R.K., Kille, J.F. & Loeschcke, H.H. (1975) 'Chemosensitive Neurons on the Ventral Medullary Surface', *Bull. Physio-Pathol. Resp.*, *11*, 277-84

Schlaefke, M.E., See, W.R., Herker-See, A. & Loeschcke, H.H. (1979a) 'Respiratory Response to Hypoxia and Hypercapnia after Elimination of Central Chemosensitivity', *Pflügers Arch.*, *381*, 241-8

Schlaefke, M.E., See, W.R. & Kille, J.F. (1979c) 'Origin and Afferent Modification of Respiratory Drive from Ventral Medullary Areas', in C. von Euler & H. Lagercrantz (eds), *Central Nervous Control Mechanisms in Breathing*, Pergamon Press, Oxford, New York, Toronto, Sidney, Paris, Frankfurt, pp. 25-34

Schlaefke, M.E., See, W.R. & Loeschcke, H.H. (1970) 'Ventilatory Response to Alterations of H$^+$-ion Concentration in Small Areas of the Ventral Medullary Surface', *Resp. Physiol.*, *10*, 198-212

Schlaefke, M.E., See, W.R., Massion, W.H. & Loeschcke, H.H. (1969) 'Die Rolle 'spezifischer' und 'unspezifischer' Afferenzen für den Antrieb der Atmung, untersucht durch Reizung und Blockade von Afferenzen an der decerebrierten Katze', *Pflügers Arch.*, *312*, 189-205

See, W.R. & Schlaefke, M.E. (1978) 'The Influence of Sinus Nerve Stimulation on Neuronal Activity of Ventral Medullary Neurones', *Neurosci. Lett.*, Suppl. *1*, 519

Shams, H., Ahmad, H.R. & Loeschcke, H.H. (1982) 'The Dependence of Ventilation on H$^+$-Activity in the Extracellular Fluid of the Brain in Respiratory or Metabolic Acidosis', *Pflügers Arch.*, *212*, R 53

Shimada, K., Trouth, C.O. & Loeschcke, H.H. (1969) 'Von der H$^+$-Ionenkonzentration des Liquors abhängige Aktivität von Neuronen im Gebiet der chemosensiblen Zonen der Medulla oblongata', *Pflügers Arch.*, *312*, R 61

Spode, R. (1980) 'Ausschaltung der zentralen und peripheren chemosensiblen Atemantriebe bei der anaesthesierten Katze während Muskelarbeit', Thesis. *Abt. Nat. Med.*, Ruhr-Universität Bochum

Tok, T. & Loeschcke, H.H. (1983) 'Untersuchung über die zentrale Wirkung von Progesteron auf die Atmung und Vasomotorik bei Katzen', *Z. Atemwegs- und Lungenkrankheiten,* Dustri-Verlag, München-Deisenhofen (in press)

Trouth, C.O., Loeschcke, H.H. & Berndt, J. (1973a) 'A Superficial Substrate on the Ventral Surface of the Medulla Oblongata Influencing Respiration', *Pflügers Arch.*, *339*, 135-52

Trouth, C.O., Loeschcke, H.H. & Berndt, J. (1973b) 'Histological Structures in the Chemosensitive Regions on the Ventral Surface of the Cat's Medulla Oblongata', *Pflügers Arch.*, *339*, 171-83

Trzebski, A., Mikulski, A. & Przybyszewski, A. (1980) 'Effects of Stimulation of Chemosensitive Areas by Superfusion on Ventral Medulla and by Infusion into Vertebral Artery of Chemical Stimuli in Non-anaesthetized "Encephale Isole" Preparation in Cats', in H.P. Koepchen, S. Hilton & A. Trzebski (eds), *Central*

Interaction Between Respiratory and Cardiovascular Control Systems, Springer, Berlin, Heidelberg, New York

Trzebski, A., Zielinski, A., Lipski, J. & Majcherczyk, S. (1971) 'Increase of Sympathetic Preganglionic Discharges and of the Peripheral Resistance Following Stimulation by H^+ Ions of the Superficial Chemosensitive Areas in the Medulla Oblongata in Cats', *Proc. Int. Un. Physiol. Sci.*, *9*, 571

Trzebski, A., Zielinksi, A., Majcherczyk, S., Lipski, J. & Szulczyk, P. (1974) 'Effect of Chemical Stimulation and Depression of the Medullary Superficial Areas on the Respiratory Motoneurone Discharges, Sympathetic Activity and Efferent Control of Carotid Area Receptors', in W. Umbach & H.P. Koepchen (eds) *Central Rhythmic and Regulation*, Hippokrates, Stuttgart, pp. 170-7

Ullah, Z. (1973) 'Elektronenmikroskopische Untersuchungen über ein chemosensibles Areal für die Atmungsregulation im Bereich der Medulla oblongata der Hauskatze', Thesis. *Abt. Theoretische Medizin*, G.H. Essen

Wieth, J.O., Brahm, J. & Funder, J. (1980) 'Transport and Interactions of Anions and Protons in the Red Blood Cell Membrane', *Ann. N.Y. Acad. Sci.*, *341*, 394-418

Willshaw, P. (1975) 'Sinus Nerve Efferents as a Link Between Central and Peripheral Chemoreceptors', in M.T. Purves (ed.), *The Peripheral Arterial Chemoreceptors*, Cambridge University Press, Cambridge, pp. 253-68

Willshaw, P. (1977) 'Mechanism of Inhibition of Chemoreceptor Activity by Sinus Nerve Efferents', in H. Acker, S. Fidone, D. Pallot, C. Eyzaguirre & D.W. Lübbers (eds), *Chemoreception in the Carotid Body*, Springer, Berlin, Heidelberg, New York, pp. 168-72

Winterstein, H. (1911) 'Die Regulierung der Atmung durch das Blut', *Pflügers Arch.*, *138*, 167-84

Winterstein, H. (1956) 'Chemical Control of Pulmonary Ventilation III. The Reaction Theory of Respiratory Control', *N. Engl. J. Med.*, *255*, 331-7

3 LUNG AND AIRWAY RECEPTORS

A.S. Paintal

There exist some sensory receptors, with fibres running in the branches of the sympathetic system, located near the roots of the lungs but these are more related to the mediastinal structures rather than the lungs. Apart from the occasional impulses generated in these receptors by lung movements apparently no other information from the lungs reaches the central nervous system via the sympathetic routes (Holmes & Torrance, 1959). This is in contrast to the relatively vast amount of information that travels to the brain through the vagus nerves during rest and exercise.

The activity during rest consists mainly of impulses from the pulmonary stretch receptors — both slowly and rapidly adapting ones. During exercise the activity in these receptors is increased greatly owing to the increase in the tidal volume and frequency of inspiration but to this traffic is added the impulses from the type J receptors which reflexly tend to terminate exercise (Paintal, 1969, 1970). However, none of the reflex effects of the receptors will be dealt with in this chapter; it will be limited to a description of the responses of the two main types of receptors (stretch and J) in addition to the bronchial receptors recently described by Coleridge & Coleridge (1975, 1977b).

The slowly and rapidly adapting pulmonary stretch receptors will be regarded as a functionally homogenous group in keeping with the earlier conclusion based on available evidence (Paintal, 1977b). (See also note below on rapidly adapting pulmonary stretch receptors.)

Pulmonary Stretch Receptors

Much is known about the pulmonary stretch receptors. Their responses first described precisely by Adrian in his classical paper (Adrian, 1933) have been reviewed in the past. These reviews should be consulted for the well-established knowledge about them (Paintal, 1963, 1973, 1977b; Fillenz & Widdicombe, 1972). Further information about the responses of the receptors to various stimuli is summarised in Table 3.1.

Location and Blood Supply

There has been an impression that the pulmonary stretch receptors are largely located in the extrapulmonary airways and the larger intra-pulmonary airways near the hilum of the lungs as indicated in the recent review by Pack (1981). This impression is the result of observa-tions in dogs by Sant'Ambrogio (1973), Miserocchi & Sant'Ambrogio (1974) who used the technique of occluding the airways with a balloon for locating the pulmonary stretch receptors. Similar observations were made by them in the cat (Sant'Ambrogio & Miserocchi, 1973). The latter was unexpected in view of certain casual observations made by the present author from time to time and so the problem was reinvestigated in cats.

In the first study on 20 receptors in 12 cats it was found using punctate local stimulation for localising the receptor, that 85 per cent of the receptors were located in the lung parenchyma away from the hilum. This result was reported in a review (Paintal, 1977a). Thereafter further observations were made on a larger number of receptors and these results have been reported in a preliminary communication recently (Paintal & Ravi, 1980). Once again the majority (83 per cent of the 233 receptors examined) were found to be located in the intra-pulmonary airways.

The same apparently holds true in the dog because using the same experimental approach (local probing) Sant'Ambrogio & Sant'Ambrogio found more recently (1981) a similar distribution although it differed from their earlier observations (Sant'Ambrogio & Sant'Ambrogio, 1980). They found that out of a total of 57 slowly adapting tracheo-bronchial receptors 74 per cent were located in the intrapulmonary airways (Sant'Ambrogio & Sant'Ambrogio, 1981).

An interesting point that has emerged is that there are differences between the location of the low-threshold receptors (i.e. receptors firing not only during inspiration but also during expiration) and the higher-threshold ones (i.e. receptors firing only during inspiration). It was found that 89 per cent of the latter were located in the intra-pulmonary airways whereas only 71 per cent of the low-threshold ones were located in this region (Paintal & Ravi, 1980). It needs to be clarified that a distinction between these two groups of receptors was first made because of the fact that the low-threshold receptors (of which there are 50 per cent) seemed to be responsible for producing Head's paradoxical reflex (Paintal, 1966).

A second important difference between the two is that whereas

83 per cent of the higher-threshold ones are accessible to chemical substances such as veratrine through the pulmonary circulation (i.e. they are stimulated by injection of veratrine into the right atrium) only 25 per cent of the low-threshold ones are accessible in this way (Paintal & Ravi, 1980). A possible explanation for this difference could be that the higher-threshold receptors are (as their location would suggest) located in the more distal airways which, as pointed out by Staub (1975) are closely related to the alveoli and are therefore close to the capillaries. It is therefore not altogether surprising that substances reach the receptors located in these airways via the pulmonary circulation. However, initially these results came as a surprise for although Armstrong & Luck (1974b) had shown that several receptors were accessible via the pulmonary circulation one did not expect that such a large proportion (89 per cent) of one group (the higher-threshold) would be accessible only via the pulmonary circulation.

The location of the higher-threshold receptors in the more peripheral parts of the lung is consistent with the observations of Fahim & Jain (1979) who found that the higher-threshold receptors need a longer exposure to be blocked by bipuvacaine aerosol than the low threshold ones (see below).

Differential Block of Pulmonary Receptors

For certain investigations on respiratory reflexes in both man and animals it is useful to be able to block receptors differentially. In the case of the pulmonary stretch receptors (both slowly adapting and rapidly adapting) this has been achieved in the cat (see Table 3.1) by using an aerosol of a local anaesthetic — bipuvacaine which blocks the pulmonary stretch receptors without blocking the reflex effects of the J receptors (Fahim & Jain, 1979). It can be assumed that these results apply also to the dog, rabbit and man in view of the observations on the reflex changes in these animals using such anaesthetic aerosols (Jain, Trenchard, Reynolds, Noble & Guz, 1973; Dain, Boushy & Gold, 1975; Cross, Guz, Jain, Archer, Stevens & Reynolds, 1976).

A noteworthy observation by Fahim & Jain (1979) was that the aerosol did not block all the higher-threshold receptors. On the other hand not only were all the low-threshold ones blocked but they were blocked sooner than the higher-threshold ones which led them to suggest that those receptors that proved less susceptible to the local anaesthetic were located more deeply in the lungs and so the anaesthetic did not reach them in adequate concentrations. This conclusion is consistent with the observations of Paintal & Ravi (1980) described above.

Another method of blocking the stretch receptors in the rabbit is by the administration of sulphur dioxide. In these animals Davies, Dixon, Callanan, Huszczuk, Widdicombe & Wise (1978) found that when 200 ppm of sulphur dioxide was administered, 23 out of 26 pulmonary stretch receptors were abruptly silenced and the inflation reflex abolished simultaneously, leaving the reflex effects of J receptors intact. This method of block is of limited value as it apparently does not work in cats and dogs (see Davies *et al.*, 1978).

Effect of Carbon Dioxide

It is a physiological curiosity that reduction of the alveolar PCO_2 to values of the order of 3 mm Hg (i.e. levels that cannot possibly occur in the living animal) increases the activity of pulmonary stretch receptors; the reverse is also true (Table 3.1). However, raising the alveolar PCO_2 from normal levels to higher ones (i.e. hypercapnia) has hardly any effect – reducing the activity of slowly adapting receptors by only 4.5 per cent and without any change in the rapidly adapting receptors (Table 3.1). Like earlier investigators (Sant'Ambrogio, Miserocchi & Mortola, 1974) Coleridge *et al.* (1978) believe that the effects of CO_2 are due to a direct action on the sensory ending itself. However, Mitchell, Cross, Hiramoto & Scheid (1980) have found evidence to show that part of the effect is due to changes in bronchomotor tone.

Role in Pulmonary Congestion, Oedema and Microembolism

There is no noteworthy effect of pulmonary congestion or oedema on the responses of pulmonary stretch receptors. In fact there is either no effect as shown by Bülbring & Whitteridge (1945) or there is a small increase of 10 per cent in the frequency of discharge for the same volume during congestion (Table 3.1). However Marshall & Widdicombe (1958) showed that there is a reduction in compliance of the lungs and this could explain the increase in the frequency of discharge. Costantin (1959) also observed an increase in activity.

The increase in activity during pulmonary oedema is also small and variable (Widdicombe, 1961; Paintal, 1969).

Injection of microemboli does not have any direct stimulant or sensitising action on the receptors. Walsh (1946) observed no change after injecting starch microemboli and the small increase of about 10 per cent observed by Armstrong, Luck & Martin (1976) in both the slowly and rapidly adapting receptors (Table 3.1) is wholly explained by the reduction in the compliance of the lungs with the consequent rise in intratracheal pressure.

Conclusion

Thus it is clear that unlike the type J receptors (see below) pulmonary congestion, pulmonary oedema and multiple pulmonary embolism do not have any significant excitatory effect on pulmonary stretch receptors.

Note on Rapidly Adapting Pulmonary Stretch Receptors

The rapidly adapting group of pulmonary stretch receptors were put in a separate category by Larabbee & Knowlton (1946 in Knowlton & Larabbee, 1946), apparently in order to identify a group of receptors that would explain the excitatory effects of inflating the lung (with about 40-60 ml or more of air) on some single phrenic motor neurones of the cat. From the data obtained by them (see Table 3.1 and Figures 5 and 11 in Knowlton & Larabbee, 1946) there was evidence for them to conclude that these receptors constituted the more rapidly adapting section of one functionally homogenous group of pulmonary stretch receptors. However, they were apparently searching for receptors that stimulated inspiration – a search that stemmed from an assumption that inflation of the lungs should excite inspiration at higher volumes of inflation. This assumption arose from an extended interpretation of Head's paradoxical reflex (Larabbee & Knowlton, 1946). It was assumed by them that Head's paradoxical reflex which appears when the vagi are cooled to suitable temperatures (Head, 1889) also occurs when the vagi are not cooled (Larabbee & Knowlton, 1946). They used this assumption in order to explain some observations on certain single phrenic motor neurones which showed an increase in activity when the lungs were inflated to volumes of the order of 40-60 ml in cats in which the chest wall had been removed and a continuous discharge in single phrenic motor neurones was produced after stopping the respiratory pump. They concluded that this excitation was produced reflexly by the rapidly adapting group of pulmonary stretch receptors. It is possible that the effects of inflation on the phrenic motor neurones described by them exist under the unusual conditions of their experiments. These excitatory effects could be due to both slowly as well as rapidly adapting receptors. However, they are contrary to the accepted facts that inflation of the lungs at various volumes inhibits the mass activity of the phrenic discharge leading to the termination of inspiration (Clark & von Euler, 1972).

Thus there is no firm evidence for an excitatory effect under ordinary conditions on inspiration by inflation of the lungs. Moreover as pointed out earlier (Paintal, 1977b) it is certain that neither Head's paradoxical reflex nor the deflation reflex (both exciting inspiration) can be due to the activity of the rapidly adapting pulmonary stretch receptors since they adapt rapidly.

One rather disconcerting aspect relating to past studies on the rapidly adapting stretch receptors is that one cannot be certain about what fraction of the rapidly adapting units that were isolated were actually located in the lungs since Sampson & Vidruck (1975) found that 38 per cent of receptors behaving like typical rapidly adapting pulmonary receptors were in fact located in the oesophagus. If one assumes that a similar percentage also applies to cats then the relative number of rapidly adapting receptors becomes still smaller. For example since Knowlton & Larabbee (1946) found that 15 per cent of their receptors had an adaptation index of 100 per cent it follows that perhaps only 9 per cent of such receptors were actually located in the lungs. In the case of Widdicombe's results (Widdicombe, 1954a) this figure would amount to only 4 per cent.

Figure 3.1: Frequency Distribution of Pulmonary Stretch Receptors at Various Adaptation Indices. Note, the receptors with adaptation index greater than 70 do not form a separate group.

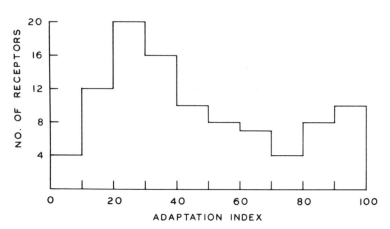

Source: Redrawn from Widdicombe, 1954a.

From data relating to the frequency distribution of all pulmonary stretch receptors (Figure 3.1) Widdicombe (1954a) could not support the conclusion of Knowlton & Larabbee (1946) that the slowly adapting and rapidly adapting receptors fell into separate groups. This conclusion of Widdicombe (1954a) is justified from Figure 3.1 and now if one says that the rapidly adapting group with adaptation index greater than 70 per cent should be reduced by 38 per cent it will further support Widdicombe's conclusion relating to Figure 3.1.

A comparison of the responses of the rapidly adapting receptors with the slowly adapting ones is presented in Table 3.1. From this Table it is now clear that except for some differences in the inflation threshold of the receptors, the location, blood supply and the response of the rapidly adapting receptors to about 13 kinds of stimuli is similar to that of the slowly adapting receptors. Moreover if the observations of Sant'Ambrogio & Sant'Ambrogio (1981) showing that 100 per cent of the rapidly adapting receptors of dogs are accessible through the pulmonary circulation (Table 3.1) are also applicable in the cat then these receptors must justifiably form part of the higher threshold group of pulmonary stretch receptors since they also have this property in common with them (Table 3.1 item 5).

Two additional features deserve attention. First that animals do not hold their breath and so rapid adaptation has little role to play from the point of view of the responses of the receptors to maintained inflation of the lungs. Second, the normal tidal volume of cats is about 40 ml (Nadeau & Colebatch, 1965), a volume that will stimulate about 70 per cent of the rapidly adapting receptors along with all the slowly adapting ones. During exercise the tidal volume will more than double bringing into activity according to the same pattern, all the pulmonary stretch receptors both slowly adapting and rapidly adapting. Further with inspiratory time less than 0.5 s (at a respiratory frequency of 60/min) even the more rapid adaptation rate of the rapidly adapting receptors will have little effect on the activity of these receptors. Thus the temporal pattern of discharge of the rapidly adapting receptors will, for all practical purposes, be just like that in the slowly adapting ones and one would therefore expect them to have the same reflex effects as the slowly adapting receptors.

Thus at present it is best to regard the rapidly adapting receptors as forming one functionally homogenous group along with the higher-threshold slowly adapting stretch receptors. Functionally the low-threshold stretch receptors are different as they produce the deflation reflex and Head's paradoxical reflex (Paintal, 1977b) which the higher-threshold ones and the rapidly adapting ones do not do.

Table 3.1: Comparison of the characteristics and responses of slowly and rapidly adapting pulmonary stretch receptors of cats; data from other animals is specified. Numbers in parenthesis relate to source of information given in the list of references at the bottom

Sl. No.	Stimulus or characteristic of receptors	Slowly adapting stretch receptor	Rapidly adapting stretch receptor	Inference
1.	Conduction velocity of fibres (mean)	36 m/sec (9) (dogs) 31.6 (11)	25 m/sec (9) (dogs) 28.8 (11)	Different Similar
2.	Inflation threshold	Below 40 ml in 91-100% of receptors (7, 16)	Below 40 ml in 50-68% of receptors (7, 16)	Different
3.	Effect of deflation	33% stimulated (7)	66% stimulated (7)	Different
4.	Location	83% in lung parenchyma (10) (dogs) 74% intra-pulmonary (13)	— (dogs) 86% intra-pulmonary (13)	Similar
5.	Accessibility to chemical substances through pulmonary circulation	83% of higher-threshold and 25% of low-threshold accessible (10) (dogs) 90% of intra-pulmonary receptors accessible (13)	— (dogs) 100% of intra-pulmonary receptors accessible (13)	Similar
6.	Effect of CO_2			
	(a) Effect of lowering CO_2 from 32 to 3mm Hg	(dogs) activity increased (4)	(dogs) activity increased (4)	Similar
	(b) Increasing CO_2 from low levels	(dogs) activity increased (4)	(dogs) activity increased (4)	Similar
	(c) Raising CO_2 from 32 to 50mm Hg	(dogs) reduction of activity by 4.5% (4)	(dogs) no change (4)	Similar
7.	Histamine			
	(a) intravenous	Sensitized ? (17)	Sensitized (1)	Similar
	(b) aerosol	Stimulated (18)	(dogs) stimulated (12)	Similar
8.	Pulmonary congestion	Stimulated (8)	(Rabbits) stimulated (14)	Similar
9.	Pulmonary micro-embolism	Stimulated (2)	Stimulated (2)	Similar
10.	Sulphur dioxide	15% sensitized (16)	11% sensitized (16)	Similar
11.	Insufflation of powders	Sensitization present (16)	60% sensitized (16)	Similar
12.	Ammonia vapour	66% stimulated (16)	82% stimulated (16)	Similar
13.	Ether vapour	60% partially depressed (16)	75% depressed (16)	Similar
14.	Halothane	Sensitized (15, 3)	Sensitized (1)	Similar
15.	Veratridine into right atrium	(dogs) stimulated (13)	(dogs) stimulated (13)	Similar
16.	Bradykinin	(dogs) variable stimulation/sensitization (6)	(dogs) stimulation/sensitisation in 33% (6)	Similar
17.	Benzonatate (right atrium)	(dogs) blocked (13)	(dogs) blocked (13)	Similar
18.	Bipuvacaine aerosol	Depressed-blocked (5)	Depressed-blocked (5)	Similar

1. Armstrong & Luck (1975); 2. Armstrong *et al.* (1976); 3. Coleridge *et al.* (1968); 4. Coleridge *et al.* (1978); 5. Fahim & Jain (1979); 6. Kaufmann *et al.* (1980); 7. Knowlton & Larabbee (1946); 8. Marshall & Widdicombe (1958); 9. Paintal (1953b); 10. Paintal & Ravi (1980); 11. Sampson (1977); 12. Sampson & Vidruck (1975); 13. Sant'Ambrogio & Sant'Ambrogio (1981); 14. Sellick & Widdicombe (1969); 15. Whitteridge (1958); 16. Widdicombe (1954a); 17. Widdicombe (1954b); 18. Widdicombe (1961).

Coleridges' Bronchial Receptors

In 1975 Coleridge & Coleridge reported in a preliminary communication the existence of a group of C-fibre endings in the bronchi of the lungs that differed from the J receptors because they were accessible to chemical substances (i.e. capsaicin) through the bronchial circulation and were also stimulated by phenyl diguanide (Coleridge & Coleridge, 1975). So far only endings with non-medullated fibres have been reported but as in the case of the J receptors and aortic chemoreceptors (Paintal & Riley, 1966) it is possible that a fraction of them may also have small medullated fibres. The responses of these receptors in dogs have been described in several papers by the Coleridges and their co-workers (Coleridge & Coleridge, 1977a, b, c; Coleridge, 1980; Coleridge *et al.*, 1976, 1978; Kaufmann *et al.*, 1980).

In the cat Delpierre *et al.* (1981) found no difference between the responses of the J receptors and the bronchial receptors to CO_2, effect of deflation of the lungs, effect of hyperinflation and to changes in stroke volume on the spontaneous discharge and they therefore lumped their J receptors and bronchial ones into one group. It should be noted that even though their responses to various stimuli actually turns out to be the same as that of the J receptors, it is certain, assuming that there are a significant number of bronchial receptors, that they do not produce respiratory or somatic reflex effects typical of the J receptors since no such reflex effects have been observed following injection of phenyl diguanide after blocking the cardiac receptors by intrapericardial injection of xylocaine (Anand & Paintal, 1980) – a technique of block that has been validated by the recent report by Arndt *et al.* (1981). Obvious reflex effects are also absent in dogs and rabbits (see Paintal, 1982).

In the dog the bronchial receptors yield a lower frequency of discharge following injection of capsaicin and a greater frequency following injection of phenyl diguanide (Coleridge & Coleridge, 1977b) and bradykinin (Kaufmann *et al.*, 1980) than the J receptors. On the other hand the responses of both groups to the prostaglandins are similar. This applies to both the bronchoconstrictor one, i.e. PgF_2a as well as the bronchodilator prostaglandins of the E series E_1 and E_2, the latter having a greater excitant action (Coleridge *et al.*, 1976, 1978).

The bronchial ones are also stimulated by certain prostaglandin analogues, i.e. cyclic ethers I and II, two stable analogues of cyclic endoperoxide PGH_2, an intermediate substance in the biosynthesis of prostaglandins. Indeed injection of as little as 0.2 μg/kg into the blood stream produced a vigorous response (Coleridge *et al.*, 1978).

Type J Receptors

Introduction

The juxta-pulmonary capillary receptors (i.e. type J receptors initially called deflation receptors (Paintal, 1955) are mechanoreceptors located in the interstium close to the pulmonary capillaries (Figure 3.2).

Figure 3.2: Schematic Representation of Type J Receptor Lying in Interstitial Tissue, between Capillary and Alveolus, Containing Collagen Fibres (vertical lines). Ending is stimulated by an increase in interstitial volume when inflow of fluid into interstitial tissue (which acts like a sponge) exceeds removal of fluid. This excitation is affected through the generator region (F) of the endings. Excitation by chemical substances from outside (through the alveolar air) or inside through the blood occurs by an action on the regenerative region (R).

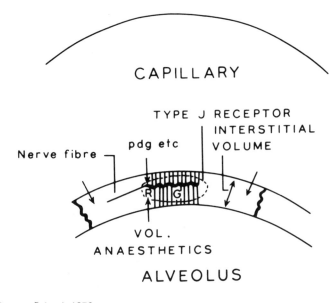

Source: Paintal, 1970.

Although earlier observations (Paintal, 1957) provided evidence for this location it was only later (Paintal, 1969) from additional observations on the effects of rise in pulmonary capillary pressure and pulmonary oedema that it was possible to conclude that they were a kind of stretch receptor whose natural stimulus was an increase in interstitial volume (Paintal, 1969, 1970). The gradual rise and fall in discharge following a much sharper rise and fall in pulmonary vascular pressures (Figure 3.9) was good evidence showing that the endings did not respond to changes in pulmonary capillary pressure *per se* but that their stimulation was linked to slower events associated with the increases and decreases in pulmonary capillary pressure. However, it is clear that their response to local increases in interstitial volume is not slow because many of them respond within five seconds to the rise in pulmonary capillary pressure that must occur as a result of increased blood flow (Figures 3.5 and 3.7).

Impulses from these receptors have been recorded in cats, and dogs (see below) as well as rabbits (Sellick & Widdicombe, 1970; Russell & Trenchard, 1980, 1981) and rats (Sapru, Willettee & Krieger, 1981). Impulses have also been recorded from similar receptors in the gills of fish (Satchell, 1978; Poole & Satchell, 1979).

Identification

Although the natural stimulus for these receptors is an increase in interstitial volume it is not convenient experimentally to use this stimulus for identifying the receptors. At present they are identified, as being J receptors, if they are stimulated within 2.5 s following the injection of an excitatory substance into the right atrium (e.g. phenyl diguanide in the cat) and they are not stimulated by a similar injection into the aorta or left atrium (Paintal, 1969; Coleridge & Coleridge, 1977b). Additionally in order to establish firmly that the endings are located in the lungs it is necessary to confirm that the endings are stimulated within 0.3 s following insufflation of a volatile anaesthetic, into the lungs, such as halothane (Paintal, 1969) or that they are stimulated by localised pressure in the lung (Coleridge & Coleridge, 1977b). Either of these two procedures is essential for excluding cardiac receptors that might be stimulated owing to phenyl diguanide reaching them with an unusually short latency either through the coronary circulation or directly through the injured walls of the heart which are easily injured during manipulation of the catheter. One such receptor was stimulated within a short latency following injection of phenyl diguanide into the right ventricle but since it was also stimulated by injecting phenyl diguanide at the root of the aorta, doubt about its location arose. The uncertainty was removed after it was found that the receptor could be stimulated by squeezing the right ventricle locally. This therefore illustrates the value of local mechanical probing for the localisation of receptors particularly in doubtful situations, a lesson that was first learnt in connection with the location of the type B atrial receptors (Paintal, 1953a). However, as pointed out by Delpierre *et al.* (1981) direct mechanical localisation of the receptor certainly does produce local disturbances. Indeed local pressure on the lungs can set off a prolonged discharge in certain J receptors (Paintal, 1969).

The J receptors should not be identified merely by the fact that the conduction velocities of their afferent fibres is within the non-medullated range of conduction velocities (i.e. below 2.5 m/s) because although the majority of the J receptors have non-medullated fibres there are several whose conduction velocity is well above the non-medullated range not only in the cat (Paintal, 1969) but also in the rabbit (see Russell & Trenchard, 1980) and the dog. Indeed in the dog Coleridge, Coleridge & Luck (1965) found that the conduction velocity of three of their 16 fibres (i.e. 20 per cent) were between 3.4 and 3.7 s. This puts these three fibres in the medullated class because following Gasser's observations, it is generally assumed that the maximum conduction velocity of non-medullated fibres is 2.5 m/s (Gasser, 1950). It must be kept in mind that certain properties of the smallest medullated fibres overlap those of non-medullated fibres (Paintal, 1967b). Thus it is clear that about 20 per cent of the fibres of Coleridge *et al.* (1965) were well above the non-medullated range. The same is true of the sample obtained by Dixon, Jackson & Richards (1980) who reported a mean conduction velocity of 3.1 m/s with a standard error of 1.5 for their seven fibres. It must follow that about half their fibres were in the medullated range of conduction velocities — some according to calculation being as high as 8 m/s. Apparently they must have known this since they referred to their fibres as 'C' fibres (Dixon *et al.*, 1980). Thus it is certain that injection of phenyl diguanide in the cat (or capsaicin in the dog) will excite J receptors with both medullated and non-medullated fibres and adoption of procedures by which selection is limited to the non-medullated group would exclude the study of the variation in the responses to chemical substances (as in the case of aortic chemoreceptors; Paintal, 1967a) that depend on whether the regenerative region is medullated or non-medullated (Paintal, 1964, 1977c).

Histological Features

Although the histological characteristics of the actual sensory endings is

Figure 3.3: Electronmicrograph from Hung *et al.* (1972) showing a likely J receptor ending (NE) in the wall of the alveolar duct. The ending contains numerous small mitochondria. It has a bare surface facing the process of a type I pneumocyte (I) and a basal lamina (BL) separates the ending and the cell. The opposite surface of the axon is capped by a Schwann cell sheath (S). A nerve bundle consisting of a single axon (arrow) is in the vicinity of the ending (NE). Alv., lumen of the alveolar duct; II, type II pneumocyte. $\times$ 12,600.

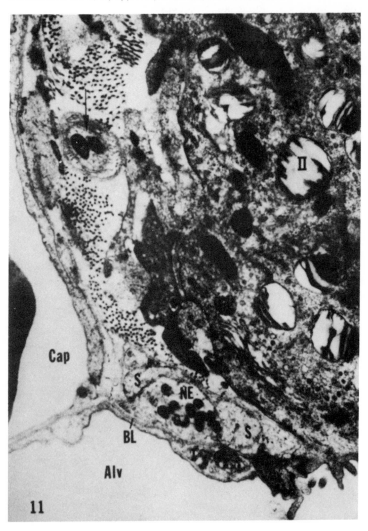

not known there is a fair amount of information regarding the general layout of the fibres of the receptors and what the sensory endings might look like.

Initially from the responses of the receptors to chemical substances and particularly their excitation during pulmonary congestion and pulmonary oedema it was predicted that the endings must lie in the collagen tissue close to the membranes of the pulmonary capillaries and alveoli without linkages to smooth muscle (Paintal, 1970). Based on this information a search for these receptors was made in the lungs of the rat and it was found that non-medullated sensory fibres were present in the predicted location but they were rare (Meyrick & Reid, 1971a, b). Subsequently Hung, Hertweck, Hardy & Loosli (1972) found that there were many in the mouse lung and they also found nerve terminals possessing the characteristics of sensory endings e.g. terminals with mitochondria etc. (Figure 3.3). Similar nerve terminals have been found in the lungs of man by Hertweck & Hung (1980) and by Fox, Bull & Guz (1980) (Figure 3.4). Although there is no proof that that these (e.g. Figure 3.3) are the final terminations, they do possess several characteristics of final sensory terminations (see Andres, 1974). It is worth noting that the J receptors are the only ones in the lung of which at least something is known about their structure with some degree of certainty.

Natural Stimulus

It has been known since 1955 that the J receptors are stimulated by pulmonary congestion (Paintal, 1955). However, it was not until 1969 that it came to be realised that this was their natural stimulus and that these receptors were in fact interstitial stretch receptors that were stimulated when the volume of fluid in the interstitium was increased due to an increase in pulmonary capillary pressure and permeability of the capillaries e.g. after administration of alloxan or 0.5 per cent chlorine in cats (Paintal, 1969, 1970). This was confirmed by Coleridge & Coleridge (1977a) in dogs who while agreeing that the J receptors could act as interstitial stretch receptors at left atrial pressures such as 25-30 mm Hg felt, keeping in view some observations of Guyton & Lindsey (1959), that at lower levels such as 3-5 mm Hg, fluid would not flow out into the interstitium. However as pointed out earlier (Paintal, 1970) it must follow that for every increment in pulmonary capillary pressures there must occur (other factors remaining unchanged) a corresponding increase in fluid in the interstitium.

Here the reader's attention is drawn to another point, i.e. any increase

Figure 3.4: Axons in interstitium of alveolar wall, probably from a J receptor in the lung of a man. There is air space, (A) contaminated with red blood cells, on three sides; lower side, border of capillary. The axons contain mitochondria, neurofilaments, agranular and dense-cored vesicles and are surrounded by Schwann cell cytoplam. Bar, 1 μm.

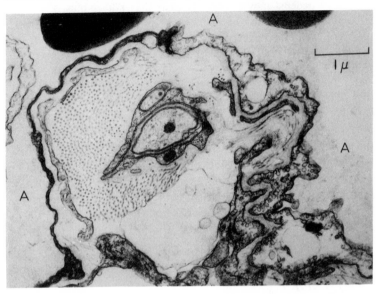

in blood flow through the capillaries must be accompanied by an increase in the pressure gradient across the capillary. Thus when blood flow increases through any particular section of the pulmonary vascular bed, there will be an increase in the pulmonary capillary pressure in that bed and therefore a corresponding increase, albeit transient, in the interstitial volume. Thus it is now easy to understand why an increase in pulmonary blood flow causes an increase in the activity of J receptors described below.

Effect of Blood Flow

One of the significant recent findings is that the J receptors are stimulated by an increase in blood flow (Anand & Paintal, 1980). Figure 3.5 shows that the latency for this increase in less than 5 to 10 s and the fall in activity on reducing blood flow is also as rapid. These latencies are smaller than the latencies for increase in activity produced by increase in left atrial pressure in some receptors (Figure 3.9). Not all the receptors examined were stimulated by a particular increase in blood flow. For example two receptors were not stimulated by doubling blood flow (Table 1 in Anand & Paintal, 1980); it is possible that these may have been located in regions of the lung in which the pulmonary circulation was not affected by the rise in pulmonary artery pressure that is the cause of the increase in blood flow. On the other hand there are others that are stimulated by increases of only 27 per cent (Figure 3.5).

Figure 3.5: Effect of Increase in Blood Flow in Right Lung Lobes on Three J Receptors of Three Cats by Occluding the Left Pulmonary Artery at Upward Arrow. In A, the blood flow increased by 147%; in B by 113% in the case of receptor shown by filled circles and by 27% in the one shown by open circles. Downward arrows mark the release of the occlusion. The ordinate shows the frequency of discharge averaged over 10 s. Note that maximum activity occurred at different times in the three receptors, but it started within 10 s in all three of them

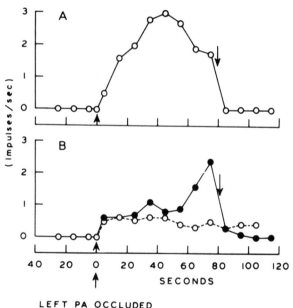

Source: Anand & Paintal, 1980.

The variation in the increase in activity of six receptors is shown in Figure 3.6. When the whole population of J receptors is taken into account it is found that the average activity when blood flow is increased by an average of 133 per cent is 0.75 impulses/s (Figure 3.7) (Anand & Paintal, 1980). The important finding is that the increase in blood flow shown in Figures 3.5, 3.6 and 3.7 are those that would occur during ordinary exercise. Thus it is certain that the J receptors are stimulated during ordinary exercise at sea level and not only at high altitude when even pulmonary oedema may occur during exercise (see Paintal, 1969, 1970).

Figure 3.6: Effect of Increase in Blood Flow on the Activity of Six J Receptors in the Right Lung Showing Variation in Responses. There are two points for each receptor, the lower one shows activity at rest and the upper one the activity during increased blood flow produced by occluding the left pulmonary artery. Abscissa, blood flow, ml/min/g of lung tissue. Ordinate, maximum frequency of discharge averaged over 10 s

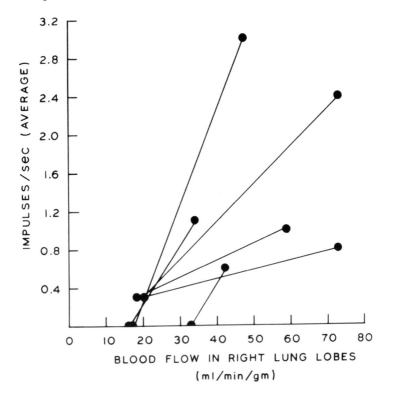

Source: Paintal, 1982

Figure 3.7: Effect of Increasing Blood Flow by an Average of 133% on the Average Activity of Ten J Receptors in the Right Lung. Blood flow was increased by occluding the left pulmonary artery at arrow. Vertical bars represent standard error.

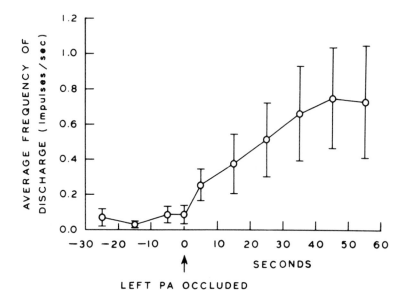

Source: Anand & Paintal, 1980.

Effect of Pulmonary Oedema

The J receptors are markedly stimulated during conditions leading to pulmonary oedema whether this is produced by injecting alloxan or inhalation of chlorine (Figure 3.8). The activity averaged in all the receptors during pulmonary oedema is about 7.5 impulses/s which is very large when it is compared with the activity of 0.75 impulses/s that produces marked visceral and somatic reflex effects (Anand & Paintal, 1980). This is not surprising when it is noted that pulmonary oedema is the most potent form of natural stimulation for the J receptors as it represents the extreme manifestation of pulmonary congestion — a commonly used term for raised pulmonary capillary pressure. The observations in cats following injection of alloxan have been confirmed by Coleridge & Coleridge (1977a) in dogs and in corresponding receptors in the gills of fish by Satchell (1978) and Poole & Satchell (1979).

Figure 3.8: Excitation of a Type J Receptor Following Injection of Alloxan 150 mg µg/kg into the Right Atrium at Zero Time. The ordinate on the left represents impulses/s averaged over a 20 s period. Oedema fluid appeared in the tracheal cannula after the 7th minute at arrow. Cat on artificial ventilation. ITP, intratracheal pressure. Note the gradual onset of excitation, after the initial short excitation, due to the sudden rise in the pulmonary vascular pressures, at about 1 min after injection of alloxan. The maximum activity is about 6.5 impulses/s which approximates to the average excitation (7.5 impulses/s) in J receptors during pulmonary oedema.

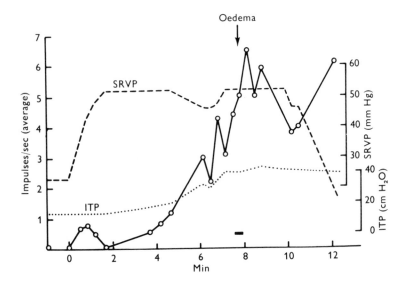

Source: Paintal, 1969.

Effect of Carbon Dioxide

It is known that injection of 0.5–3 ml of CO_2 into the right ventricle stimulates the J receptors of cats and produces reflex effects identical to those produced by injections of suitable doses of phenyl diguanide; the receptors are also stimulated by inflating the lungs with 20–100 per cent CO_2 (Dickinson & Paintal, 1968, 1970). Because of the high concentrations of CO_2 used for inflating the lungs, the physiological significance of the excitatory effect of CO_2 during hypercapnia (say less than 10 per cent CO_2) was open to question. However, Delpierre *et al.* (1981) have shown recently that the activity of the receptors increased when the end-tidal CO_2 was raised from 2 to 10 per cent but the endings adapted to this stimulus. Curiously in some receptors the activity increased on *reducing* the end-tidal CO_2 back to initial levels. Unfortunately it is not easy to assess their results quantitatively as they lumped their data from the J receptors along with those of the bronchial endings because according to them the effect of CO_2 was similar on both types of endings. They concluded that the changes in alveolar CO_2 concentration constituted the usual stimulus for them. One should keep in mind that some of the receptors in their sample were stimulated after relatively long latencies following injection of relatively large doses of phenyl diguanide (100 µg/kg) into the right atrium and so it is possible that aortic chemoreceptors, which are also stimulated by phenyl diguanide (Paintal, 1967a) may have been included in their sample of J receptors.

In contrast to the J receptors of the cat those of the dog show a significant reduction of the 'spontaneous' activity when CO_2 is raised from 2 to 19 mm Hg in the left lung while the left pulmonary artery is occluded. However the reverse is not true because reducing the end-tidal CO_2 from 35 to 2 mm Hg produced no change in their activity (Coleridge *et al.*, 1978).

Effects of Chemical Substances

Like several other groups of sensory receptors with non-medullated nerve fibres, the J receptors can be stimulated in a non-specific manner by a variety of chemical substances that do not stimulate receptors with medullated fibres (Paintal, 1964, 1977). Interest in the effects of chemical substances arises from two other sources. First, the use of phenyl diguanide and capsaicin as a means of isolating the J receptors (since they are normally either silent or show low activity and the use of natural stimulation (pulmonary congestion) as a means of isolating the receptors is impractical) and studying their reflex effects. Secondly, interest arises from the possibility that various metabolites such as 5-hydroxytryptamine, histamine and the prostaglandins may, through their excitatory effects, be involved in the process of natural stimulation of the receptors.

Effect of Phenyl Diguanide on the J Receptors of Cats

The effects of phenyl diguanide on the J receptors is well known (see

Paintal, 1973). New information of special interest is that it has been possible to equate known levels of activity produced in the receptors by phenyl diguanide with levels of activity generated by natural stimulation of the J receptors. Specifically it has been found that 0.75 impulses/s (average) are generated in J receptors by 12–18 µg of phenyl diguanide. Equivalent activity is generated by increasing pulmonary blood flow by 133 per cent (Figure 3.7) (Anand & Paintal, 1980). Thus by injecting 12–18 µg/kg of phenyl diguanide into the right atrium, one can, as far as the activity of the J receptors is concerned, mimic the effects of doubling blood flow on the receptors for studying the reflex effects of the receptors in a meaningful manner.

Effect of Phenyl Diguanide on the J Receptors of Dogs

At the Krogh centenary symposium held in 1974 Dr John Coleridge reported briefly that phenyl diguanide stimulated the J receptors of dogs (see p. 406 in Paintal & Gill-Kumar, 1977) and subsequently published the data in detail (Coleridge & Coleridge, 1977a). They observed that 40 per cent of their J receptors (10 out of 25) were stimulated by phenyl diguanide to yield an average frequency of discharge of 3.3 impulses/s following injection of 10 µg/kg phenyl diguanide into the right atrium; for the entire population of 25 receptors the average would be 1.3 impulses/s. Thus at *this* dose the J receptors of dogs yield more activity than those of the cat which produce less than 0.5 impulses/s (see Figure 5 in Anand & Paintal, 1980). Moreover they observed that there were occasional dogs in which the J receptors yielded marked activity of the order of 17 impulses/s (average) at this dose and in these dogs there followed the typical effects of J receptors, i.e. bradycardia and hypotension (Coleridge & Coleridge, 1977b). However, Coleridge & Coleridge emphasise the fact that compared to capsaicin, the excitatory effects of phenyl diguanide is much smaller (see Figure 3 in Coleridge & Coleridge, 1977) and apparently for this reason the fact that phenyl diguanide does stimulate the J receptors of dogs is often ignored by them (e.g. see Coleridge *et al.*, 1978; Kaufman *et al.*, 1980).

Effects of Metabolites on J Receptors

From time to time various substances normally produced in the body (metabolites) have been implicated in the excitation of J receptors. The first of these was 5-hydroxytryptamine which was found to stimulate the J receptors in the same way as phenyl diguanide: indeed

6 μg/kg of the base was effective in producing a pronounced discharge (Paintal, 1955). Since it is now known that 0.75 impulses/s can produce marked reflex effects (Anand & Paintal, 1980) one may estimate from the published records (i.e. Figure 10 in Paintal, 1955) that 1-2 μg/kg would stimulate the J receptors effectively. This might be of significance in situations in which the concentration of 5-hydroxytryptamine in the blood rises *suddenly* to such levels.

Histamine

In doses of 150 μg (about 50 μg/kg) histamine stimulates about half the receptors (Paintal, 1974). The excitatory effect is quite unlike phenyl diguanide as it develops gradually and after a significant delay (Paintal, 1974, 1977b). Moreover the excitatory effect is characterised by intermittent bursts and in view of the fact that this effect is greatly enhanced during pulmonary congestion it was concluded that it was due to some intermittent event such as the movement of fluid (Paintal, 1974). Armstrong & Luck (1974) also found that histamine stimulated the J receptors of cats. So far the effects of histamine on the J receptors of dogs has not been reported. However, in view of the fact that in three out of eight dogs injection of 10 μg/kg histamine into the right atrium produced apnoea of 7-22 s duration as an immediate response (Singh, Jain & Kumar, 1982) it is likely the effects of histamine on the J receptors of dogs might be even greater than those of the cat.

Effects of Prostaglandins

See under 'Coleridges' Bronchial Receptors' above.

Stimulation of J Receptors under Pathophysiological Conditions

Since the J receptors are stimulated by increase in interstitial volume it follows that all pathophysiological conditions that lead to increase in interstitial volume will lead to stimulation of these receptors. These conditions are the following:

1. Conditions causing increase in regional blood flow due to block of other parts of the pulmonary vascular bed.
2. Conditions causing rise in left atrial pressure.
3. Conditions causing increased permeability of the capillaries.
4. Inflammatory and other conditions affecting the alveolar interstitium.

Conditions Causing Increase in Regional Blood Flow

In all conditions where there is partial blockage of the pulmonary vascular bed there must occur some increase in pulmonary artery pressure which will increase the blood flow in the unoccluded parts of the lungs. This occurs in multiple pulmonary embolism, hypoxic vasoconstriction and occlusion of major branches of the pulmonary artery.

Effect of Pulmonary Microembolism

The J receptors are stimulated following injection of starch (Paintal, 1955) or plastic microemboli (Paintal, Damodaran & Guz, 1973). The pattern of activity that develops is similar to that during pulmonary congestion/oedema following injection of alloxan i.e. the discharge develops gradually reaching its peak several minutes after the rise in pulmonary artery pressure. The evidence indicated that the receptors were not stimulated directly by the mechanical irritation by the emboli but rather by the rise in pulmonary artery pressure that follows the injection of the microemboli. These observations have been confirmed by Armstrong, Luck & Martin (1976) who have considered the possibility of biologically active amines being involved in the process of excitation. However, the simplest explanation for the excitation is that the increased blood flow that occurs in the unembolised vessels leads to the stimulation of the J receptors near the capillaries of those vessels i.e. this is an example of stimulation of J receptors owing to increase in regional blood flow.

Effect of Hypoxia

It is well known that pulmonary vasoconstriction occurs during hypoxia, the best manifestation of this being the rise in pulmonary artery pressure at high altitudes. Indeed the pressure rises to remarkable levels in man at high altitudes and this can lead to pulmonary oedema particularly as a result of exercise (see Paintal, 1969, 1970). One would therefore expect that the effect of this rise in pressure would be similar to that which occurs during microembolism, i.e. the J receptors would be stimulated in regions where the vascular bed is open and is subject to the rise in vascular pressures.

Occlusion of Branches of Pulmonary Arteries

Since it has been shown that occluding the left pulmonary artery leads to stimulation of the J receptors in the right lung owing to increased blood flow in the latter (Figures 3.5-3.7) it follows that in clinical situations where similar blocks occur will also result in an increase in the activity of the J receptors in the unoccluded vascular bed during exercise or when cardiac output increases owing to other causes. This is presumably what happened in the patient with gross pulmonary vascular obstruction described by Guz, Noble, Eisele & Trenchard (1970) so that sectioning of the vagus of the normal lung to which the blood flow was diverted led to the marked reduction in dyspnoea during mild exercise.

Figure 3.9: Responses of a Type J Receptor to Occlusion of the Aorta at Zero Time. Note the lag between the rise in systolic right ventricular pressure (SRVP) (which is nearly the same as systolic pulmonary artery pressure) and the excitation of the endings. There is a similar lag at the end of occlusion at arrow. The ordinate (impulses/s) represents the reciprocal of the interval between individual impulses.

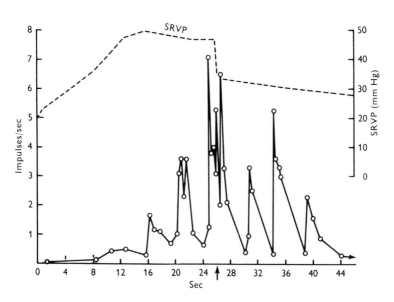

Source: Paintal, 1969.

Conditions Causing Rise in Left Atrial Pressure

Any condition that causes a rise in left atrial pressure will lead to stimulation of the J receptors owing to consequent rise in pulmonary capillary pressure. Thus obstruction to blood in the aorta causes a marked increase in activity as shown in Figure 3.9. The greater the increase in left atrial pressure the greater the activity in the J receptors as shown in Figure 3.10 (Coleridge & Coleridge, 1977a). Thus the receptors will be stimulated in all those clinical conditions where similar increase in left atrial pressure occurs. The most dramatic of these is failure of the left ventricle which in severe cases leads to pulmonary oedema resulting in the well-known dyspnoea and the other reflex effects of J receptors (see Paintal, 1973, 1977a, b). In chronic conditions, e.g. mitral stenosis, the effects would be similar though less dramatic and would be particularly manifested during exercise (see Reed, Ablett & Cotes, 1978).

Figure 3.10: Stimulation of Three J Receptors of Dogs by Congestion of the Lungs; All Three Fibres were Non-medullated. The ordinate represents the frequency of discharge (impulses/s) averaged over 20 s periods. The abscissa represents left atrial pressure which was raised by distending a balloon in the left atrium. Note different thresholds and sensitivities of the three receptors.

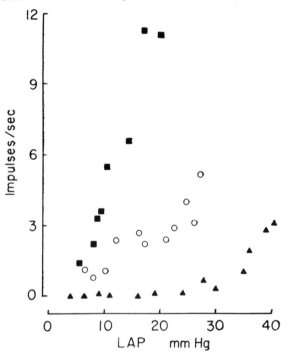

Source: Coleridge & Coleridge, 1977a.

Conditions Causing Increased Permeability of the Capillaries

Inhalation of chlorine causes pulmonary oedema owing to increased permeability of the pulmonary capillaries and since this leads to marked stimulation of the J receptors (Paintal, 1969) it follows that the same should occur when damage of the pulmonary capillaries occurs following other external agents or endogenously produced toxins.

Other Conditions Affecting the Alveolar Interstitium

There is evidence that inflamatory conditions affecting the alveolar interstitium such as those produced by instillation of carageenin cause tachypnoea which can be conveniently explained as being due to stimulation of J receptors (Trenchard, Gardner & Guz, 1972). Thus the same must be taking place in pneumonia. One can speculate that in other conditions such as allergic alveolitis similar stimulation of J receptors occurs.

Acknowledgements

Financial support was provided by a grant from SERC, Department of Science & Technology, New Delhi.

References

Adrian, E.D. (1933) 'Afferent Impulses in the Vagus and their Effect on Respiration', *J. Physiol.*, *79*, 332-58

Anand, A. & Paintal, A.S. (1980) 'Reflex Effects Following Selective Stimulation of J Receptors in the Cat', *J. Physiol.*, *299*, 553-72

Andres, K.H. (1974) 'Morphological Criteria for the Differentiation of Mechanoreceptors in Vertebrates', *Rheinisch Westfalische Akademie der Wissenschaften*, *53*, 135-51

Armstrong, D.J. & Luck, J.C. (1974a) 'A Comparative Study of Irritant and Type J Receptors in the Cat', *Respiration Physiol.*, *21*, 47-60

Armstrong, D.J. & Luck, J.C. (1974b) 'Accessibility of Pulmonary Stretch Receptors from the Pulmonary and Bronchial Circulations', *J. Appl. Physiol.*, *36*, 706-10

Armstrong, D.J., Luck, J.C. & Martin, V.M. (1976) 'The Effect of Emboli upon Intrapulmonary Receptors in the Cat', *Respiration Physiol.*, *26*, 41-54

Arndt, J.O., Pasch, U., Samodelov, F. & Weibe, H. (1981) 'Reversible Blockage of Myelinated and Non-myelinated Cardiac Afferents in Cats by Instillation of Procaine into the Pericardium', *Cardiovascular Res.*, *15*, 61-7

Bülbring, E. & Whitteridge, D. (1945) 'The Activity of Vagal Stretch Endings during Congestion in Perfused Lungs', *J. Physiol.*, *103*, 477-87

Clark, F.J. & Euler, C.V. (1972) 'On the Regulation of Depth and Rate of Breathing', *J. Physiol.*, *222*, 267-95

Coleridge, H.M. (1980) 'Afferent Fibers Involved in Defence Reflexes from Respiratory Tract', *Proc. Int. Union Physiol. Sci.*, *14*, 85

Coleridge, H.M. & Coleridge, J.C.G. (1975) 'Two Types of Afferent Vagal C-fiber in the Dog Lung: Their Stimulation by Pulmonary Congestion', *Fed. Proc.*, *34*, 372

Coleridge, H.M. & Coleridge, J.C.G. (1977a) 'Afferent Vagal C-fibers in the Dog Lung: Their Discharge During Spontaneous Breathing, and Their Stimulation by Alloxan and Pulmonary Congestion', in A.S. Paintal & P. Gill-Kumar (eds) *Respiratory Adaptations, Capillary Exchange and Reflex Mechanisms*, Vallabhbhai Patel Chest Institute, Delhi, pp. 396-405

Coleridge, H.M. & Coleridge, J.C.G. (1977b) 'Impulse Activity in Afferent Vagal C-fibres with Endings in the Intrapulmonary Airways of Dogs', *Respir. Physiol.*, *29*, 125-42

Coleridge, J.C.G. & Coleridge, H.M. (1977c) 'Afferent C-fibers and Cardiorespiratory Chemoreflexes', *Am. Rev. Resp. Dis.*, *115*, (Suppl), 251-60

Coleridge, H.M., Coleridge, J.C.G., Baker, D.G., Ginzel, K. H. & Morrison, M.A. (1978) 'Comparison of the Effects of Histamine and Prostaglandin on Afferent C-fiber Endings and Irritant Receptors in the Intrapulmonary Airways', in R.S. Fitzgerald, H. Gautier & S. Lahiri (eds), *The Regulation of Respiration during Sleep and Anesthesia*, Plenun Press, New York, pp. 291-305

Coleridge, H.M., Coleridge, J.C.G. & Banzett, R.B. (1978) "II Effect of CO_2 on Afferent Vagal Endings in the Canine Lung', *Respir. Physiol.*, *34*, 135-51

Coleridge, H.M., Coleridge, J.C.G., Ginzel, K.H., Baker, D.G., Banzett, R.B. & Morrison, M.A. (1976) 'Stimulation of 'Irritant' Receptors and Afferent C-fibres in the Lungs by Prostaglandins', *Nature*, *264*, 451-3

Coleridge, H.M., Coleridge, J.C.G. & Luck, J.C. (1965) 'Pulmonary Afferent Fibres of Small Diameter Stimulated by Capsaicin and by Hyperinflation of the Lungs', *J. Physiol.*, *179*, 248-62

Coleridge, H.M., Coleridge, J.C.G., Luck, J.C. & Norman, J. (1968) 'The Effect of Four Volatile Anaesthetic Agents on the Impulse Activity of Two Types of Pulmonary Receptor', *Br. J. Anesth.*, *40*, 484-92

Costantin, L.L. (1959) 'Effect of Pulmonary Congestion on Vagal Afferent Activity', *Am. J. Physiol.*, *196*, 49-53

Cross, B.A., Guz, A., Jain, S.K., Archer, S., Stevens, J. & Reynolds, F. (1976) 'The Effect of Anaesthesia of the Airway in Dog and Man: a Study of Respiratory Reflexes, Sensations and Lung Mechanics', *Clin. Sci.*, *50*, 439-54

Dain, D.S., Boushey, H.A. & Gold, W.M. (1975) 'Inhibition of Respiratory Reflexes by Local Anesthetic Aerosols in Dogs and Rabbits', *J. Appl. Physiol.*, *38*, 1045-50

Davies, A., Dixon, M., Callanan, D., Huszczuk, A., Widdicombe, J.C. & Wise, J.C.M. (1978) 'Lung Reflexes in Rabbits During Pulmonary Stretch Receptor Block by Sulphur Dioxide', *Respir. Physiol.*, *34*, 83-101

Dawes, G.S. & Mott, J.C. (1950) 'Circulatory and Respiratory Reflexes Caused by Aromatic Guanidines', *Br. J. Pharmac. Chemother.*, *5*, 65-76

Delpierre, S., Crimaud, Ch., Jammes, Y. & Mei, N. (1981) 'Changes in Activity of Vagal Bronchopulmonary C Fibres by Chemical Stimuli in the Cat', *J. Physiol.*, (in press)

Dickinson, C.J. & Paintal, A.S. (1968) 'Stimulation of Lung Deflation Receptors by the Injection of Carbon Dioxide Gas into the Right Ventricle', *J. Physiol.*, *196*, 70-1P

Dickinson, C.J. & Paintal, A.S. (1970) 'Stimulation of Type-J Pulmonary Receptors in the Cat by Carbon Dioxide', *Clin. Sci.*, *38*, 33P

Dixon, M., Jackson, D.M. & Richards, I.M. (1980) 'The Action of Sodium Cromoglycate on 'C' Fibre Endings in the Dog Lung', *Br. J. Pharmac.*, *70*, 11-13

Fahim, M. & Jain, S.K. (1979) 'The Effect of Bupivacaine Aerosol on the Activity of Pulmonary Stretch and 'Irritant' Receptors', *J. Physiol.*, *288*, 367-78

Fillenz, M. & Widdicombe, J.G. (1972) 'Receptors of the Lungs and Airways', in E. Neil (ed.), *Enteroceptors, Handbook of Sensory Physiology*, vol. III/1, Springer-Verlag, Berlin, pp. 81-112

Fox, B., Bull, T.B. & Guz, A. (1980) 'Innervation of Alveolar Walls in the Human Lung: an Electron Microscopic Study', *J. Anat.*, *131*, 683-92

Gasser, H.S. (1950) 'Unmedullated Fibers Originating in Dorsal Root Ganglia', *J. Gen. Physiol.*, *33*, 651-90

Guyton, A.C. & Lindsey, A.W. (1959) 'Effect of Elevated Left Atrial Pressure and Decreased Plasma Protein Concentration on the Development of Pulmonary Edema', *Circulation Res.*, *7*, 649-57

Guz, A., Noble, M.I.M., Eisele, J.H. & Trenchard, D. (1970) 'Experimental Results of Vagal Block in Cardiopulmonary Disease', in R. Porter (ed.) *Breathing, Hering-Breuer Centenary Symposium*, Churchill, London, pp. 315-29

Head, H. (1889) 'On the Regulation of Respiration', Part I, Experimental, *J. Physiol.*, *10*, 1-70

Hertweck, M.S. & Hung, K.S. (1980) 'Ultrastructural Evidence for the Innervation of Pulmonary Alveoli', *Experientia*, *36*, 112

Holmes, R. & Torrance, R.W. (1959) 'Afferent Fibres of the Stellate Ganglion', *Q. J. Exp. Physiol.*, *44*, 271-81

Hung, K-S., Hertweck, M.S., Hardy, J.D. & Loosli, C.G. (1972) 'Innervation of Pulmonary Alveoli of the Mouse Lung: An Electron Microscopic Study', *Am. J. Anat.*, *135*, 477-95

Jain, S.K., Trenchard, D., Reynolds, F., Noble, M.I.M. & Guz, A. (1973) 'The Effect of Local Anaesthesia of the Airway on Respiratory Reflexes in the Rabbit', *Clin. Sci.*, *44*, 519-38

Kaufmann, M.P., Coleridge, H.M., Coleridge, J.C.G. & Baker, D.G. (1980) 'Bradykinin Stimulates Afferent Vagal C-fibers in Intrapulmonary Airways of Dogs', *J. Appl. Physiol.*, *48*, 511-17

Knowlton, G.C. & Larabbee, M.G. (1946) 'A Unitary Analysis of Pulmonary Volume Receptors', *Am J. Physiol.*, *147*, 100-14

Larabbee, M.G. & Knowlton, G.C. (1946) 'Excitation and Inhibition of Phrenic Motoneurones by Inflation of the Lungs', *Am. J. Physiol.*, *147*, 90-9

Marshall, R. & Widdicombe, J.G. (1958) 'The Activity of Pulmonary Stretch Receptors During Congestion of the Lungs', *Q. J. Exp. Physiol.*, *43*, 320-30

Meyrick, B. & Reid, L. (1971a) 'Intra-alveolar Wall Nerve in Rat Lung: an Electron-Microscopic Study', *J. Physiol.*, *214*, 6-7P

Meyrick, B. & Reid, L. (1971b) 'Nerves in Rat Intra-acinar Alveoli: an Electron Microscopic Study', *Respir. Physiol.*, *11*, 367-77

Miserocchi, G., Mortola, J. & Sant'Ambrogio, G. (1973) 'Localization of Pulmonary Stretch Receptors in the Airways of the Dog', *J. Physiol.*, *235*, 775-82

Miserocchi, G. & Sant'Ambrogio, G. (1974) 'Distribution of Pulmonary Stretch Receptors in the Intrapulmonary Airways of the Dog', *Respir. Physiol.*, *21*, 71-5

Mitchell, G.S., Cross, B.A., Hiramoto, T. & Scheid, P. (1980) 'Effects of Intrapulmonary CO_2 and Airway Pressure on Phrenic Activity and Pulmonary Stretch Receptor Discharge in Dogs', *Respir. Physiol.*, *40*, 29-48

Nadeau, R.A. & Colebatch, H.J.H. (1965) 'Normal Respiratory and Circulatory Values in the Cat', *J. Appl. Physiol.*, *20*, 836-8

Pack, A.I. (1981) 'Sensory Inputs to the Medulla', *Ann. Rev. Physiol.*, *43*, 73-90

Paintal, A.S. (1953a) 'A Study of Right and Left Atrial Receptors', *J. Physiol.*, *120*, 596-610

Paintal, A.S. (1953b) 'The Conduction Velocities of Respiratory and Cardiovascular

Afferent Fibres in the Vagus Nerve', *J. Physiol.*, *121*, 341-59

Paintal, A.S. (1955) 'Impulses in Vagal Afferent Fibres from Specific Pulmonary Deflation Receptors. The Response of These Receptors to Phenyl Diguanide, Potato Starch, 5-hydroxytryptamine and Nicotine, and Their Role in Respiratory and Cardiovascular Reflexes', *Q. J. Exp. Physiol.*, *40*, 89-111

Paintal, A.S. (1957) 'The Location and Excitation of Pulmonary Deflation Receptors by Chemical Substances', *Q. J. Exp. Physiol.*, *42*, 56-71

Paintal, A.S. (1964) 'Effects of Drugs on Vertebrate Mechanoreceptors', *Pharmac. Rev.*, *16*, 341-80

Paintal, A.S. (1966) 'Re-evaluation of Respiratory Reflexes', *Q.J. Exp. Physiol.*, *51*, 151-63

Paintal, A.S. (1967a) 'Mechanism of Stimulation of Aortic Chemoreceptors by Natural Stimuli and Chemical Substances', *J. Physiol.*, *189*, 63-84

Paintal, A.S. (1967b) 'A Comparison of the Nerve Impulses of Mammalian Non-medullated Nerve Fibres with those of the Smallest Diameter Medullated Fibres', *J. Physiol.*, *193*, 523-33

Paintal, A.S. (1969) 'Mechanism of Stimulation of Type J Pulmonary Receptors', *J. Physiol.*, *203*, 511-32

Paintal, A.S. (1970) 'The Mechanism of Excitation of Type J Receptors, and the J Reflex', in R. Porter (ed.), *Breathing, Hering-Breuer Centenary Symposium*, Churchill, London, pp. 59-71

Paintal, A.S. (1973) 'Vagal Sensory Receptors and their Reflex Effects', *Physiol. Rev.*, *53*, 159-227

Paintal, A.S. (1974) 'Fluid Pump of Type J Receptors of the Cat', *J. Physiol.*, *238*, 53-4P

Paintal, A.S. (1977a) 'Thoracic Receptors Connected with Sensation', *Br. Med. Bull.*, *33*, 169-74

Paintal, A.S. (1977b) 'The Nature and Effects of Sensory Inputs into the Respiratory Centres', *Fed. Proc.*, *30*, 2428-32

Paintal, A.S. (1977c) 'Effects of Drugs on Chemoreceptors, Pulmonary and Cardiovascular Receptors', *Pharmac. Ther. B.*, *3*, 41-63

Paintal, A.S. (1982) 'Reflex Effects of J Receptors', in M.E. Schlaefke (ed.), *Central Neurons, Environment and the Control Systems of Breathing and Circulation*, Springer-Verlag, Berlin (in press)

Paintal, A.S., Damodaran, V.N. & Gus, A. (1973) 'Mechanism of Excitation of Type J Receptors', *Acta Neurobiol. Exp.*, *33*, 15-19

Paintal, A.S. & Gill-Kumar, P. (1977) 'Respiratory Adaptations, Capillary Exchange and Reflex Mechanisms', Vallabhbhai Patel Chest Institute, Delhi

Paintal, A.S. & Ravi, K. (1980) 'The Relative Location of Low- and Higher-threshold Pulmonary Stretch Receptors' *J. Physiol.*, *307*, 50-1P

Paintal, A.S. & Riley, R.L. (1966) 'Responses of Aortic Chemoreceptors, *J. Appl. Physiol.*, *21*, 543-8

Poole, C.A. & Satchell, G.H. (1979) 'Nociceptors in the Gills of the Dogfish *Squalus acanthias*', *J. Comp. Physiol.*, *130*, 1-7

Reed, J.W., Ablett, M. & Cotes, J.E. (1978) 'Ventilatory Responses to Exercise and to Carbon Dioxide in Mitral Stenosis Before and After Valvulotomy: Causes of Tachypnoea', *Clin. Sci.*, *54*, 9-16

Russell, N.J.W. & Trenchard, D. (1980) 'Non-myelinated Vagal Lung Receptors in the Rabbit', *J. Physiol.*, *300*, 31 P

Russell, N.J.W. & Trenchard, D. (1981) 'Chemoreflexes of Pulmonary Origin Elicited by Sodium Dithionite in the Anaesthetized Rabbit', *J. Physiol.*, *310*, 63-4 P

Sampson, S.R. (1977) 'Sensory Neurophysiology of Airways', *Am. Rev. Resp. Dis.*, *115* (suppl), 107-15

Sampson, S.R. & Vidruk, E.H. (1975) 'Properties of 'Irritant' Receptors in Canine Lung', *Respir. Physiol.*, 25, 9-22

Sant'Ambrogio, G. & Miserocchi, G. (1973) 'Functional Localization of Pulmonary Stretch Receptors in the Airways of the Cat', *Arch. Fisiol*, 70, 1-7

Sant'Ambrogio, G., Miserocchi, G. & Mortola, J. (1974) 'Transient Responses of Pulmonary Stretch Receptors in the Dog to Inhalation of Carbon Dioxide', *Respir. Physiol.*, 22, 191-7

Sant'Ambrogio, F.B. & Sant'Ambrogio, G. (1980) 'Circulatory Accessibility of Localized Airways Receptors', *Proc. Int. Union Physiol. Sci.*, 14, 680

Sant'Ambrogio, F.B. & Sant'Ambrogio, G. (1981) 'Circulatory Accessibility of Afferent Endings in the Tracheo-bronchial Tree of the Dog', *J. Physiol.*, (in press)

Sapru, H.N., Willette, R.N. & Krieger, A.J. (1981) 'Stimulation of Pulmonary J Receptors by an Enkephalin-analog', *J. Pharmac. Exp. Ther.*, 217, 228-34

Satchell, G.H. (1978) 'Type J Receptors in the Gills of Fish', in R. Porter (ed.), *Studies in Neurophysiology*, Cambridge University Press, London, pp. 131-42

Sellick, H. & Widdicombe, J.G. (1969) 'The Activity of Lung Irritant Receptors During Pneumothorax, Hyperpnoea and Pulmonary Vascular Congestion', *J. Physiol.*, 203, 359-81

Sellick, H. & Widdicombe, J.G. (1970) 'Vagal Deflation and Inflation Reflexes Mediated by Lung Irritant Receptors', *Q.J. Exp. Physiol.*, 55, 153-63

Singh, S., Jain, S.K. & Kumar, A. (1982) 'Vagal Sensory Mechanisms Involved in Reflex Bronchoconstriction in Dogs', *Ind. J. Med. Res.*, (in press)

Staub, N.C. (1975) 'Some Aspects of Airways Structure and Function', *Post-grad. Med. J.*, 51, (Suppl. 7), 21-34

Trenchard, D., Gardner, D. & Guz, A. (1972) 'Role of Pulmonary Vagal Afferent Nerve Fibres in the Development of Rapid Shallow Breathing in Lung Inflammation', *Clin. Sci.*, 42, 251-63

Walsh, E.G. (1946) 'Vagal Nerve Fibre Activity Following Multiple Pulmonary Embolism', *J. Physiol.*, 106, 466-70

Whitteridge, D. (1958) 'Effects of Anaesthetics on Mechanical Receptors', *Br. Med. Bull.*, 14, 5-7

Widdicombe, J.G. (1954a) 'Receptors in the Trachea and Bronchi of the Cat', *J. Physiol.*, 123, 71-104

Widdicombe, J.G. (1954b) 'The Site of Pulmonary Stretch Receptors in the Cat', *J. Physiol.*, 125, 336-51

Widdicombe, J.G. (1961) 'The Activity of Pulmonary Stretch Receptors during Bronchoconstriction, Pulmonary Oedema, Atelectasis and Breathing Against a Resistance', *J. Physiol.*, 159, 436-50

4 RESPIRATORY REFLEXES

Andrzej Trzebski

Introduction

Reflexes originating from the receptors situated in the tracheobronchial tree and within the lung parenchyma (for review see Paintal, 1973, 1977a, b; Widdicombe, 1974; Sant'Ambrogio, 1982) are an important component of the respiratory feedback loops and constitute therefore the proprioceptive reflexes of the control system of breathing (see von Euler, 1977; Cohen, 1979). Some of these reflexes are involved in the defence responses from the respiratory tract (see Coleridge & Coleridge, 1981) and recent data support the concept (Daly, 1972) on their role in the cardiovascular-respiratory interactions and in the tonic control of the circulation (for review see Shepherd, 1981).

Four main types of receptors innervated by the afferent vagal fibres in the lungs and airways are distinguished; they have been discussed in detail by Paintal in Chapter 3.

Although the reflex responses produced by the stimulation or inactivation of the different types of receptors differ, a distinction between them appears often difficult. The reason is a species-dependent overlap in their sensitivity to mechanical or chemical stimuli. There is also a marked overlap in fibre size and conduction velocities between individual SARs and RARs myelinated afferents (see for review Paintal, 1973; Sant'Ambrogio, 1982). Thus the usual techniques of selective blockade, or selective stimulation, of afferents are of limited value in providing specific reflex responses confined to the stimulation of only one type of receptor, although SARs are apparently selectively blocked by sulpur dioxide inhalation (Callannan, Dixon & Widdicombe, 1975; Davies *et al.*, 1978).

One cannot overlook the fact that most of the physiological and pathophysiological changes in the lungs stimulate or inactivate more than one type of receptor. Afferent input from the lungs and tracheo-bronchial tree is heterogeneous as it originates from different receptors either co-activated or inactivated simultaneously. Thus the study of the interaction between reflexes rather than the analysis of an artificially isolated reflex may provide information on the natural means whereby the reflex control operates. Such an approach remains largely unexplored as it requires a complex neurophysiological analysis of the interactions between the inputs from individual types of receptor upon central neurones.

Vagal sensory endings are the main source of the reflexes originating from the lungs and tracheobronchial tree. There are, however, some indications that also non-vagal afferent pathways may convey information from the respiratory system, since some reflex changes in breathing could be evoked by the inhalation of chemical irritants in vagotomised cats and these effects were abolished by the removal of the stellate ganglia (Widdicombe, 1954a); in addition the vasodepressor reflex response to lung inflation is in part mediated by fibres passing through the stellate ganglion (Daly & Robinson, 1968). Sympathetic afferent fibres from the lungs and mediastinum have been identified by Holmes & Torrance (1959). In dogs a respiratory modulation and non-adapting characteristics of the discharge related to transpulmonary pressure has been observed in afferent sympathetic myelinated fibres of the thoracic white rami (Kostreva *et al.*, 1975); furthermore stimulation of the sympathetic afferents in the ansa subclavia, containing fibres from the lungs, produces inhibition of the phrenic nerve and external intercostal muscle activity (Kostreva *et al.*, 1978). The localisation of the receptors innervated by sympathetic afferents, whether in the lung parenchyma, pulmonary vessels or in the mediastinum, and their functional role is still obscure. The reflex effects mediated by the sympathetic afferents appear, however, of small magnitude and thus less significant than those involving the vagal pathway.

Respiratory Reflexes

Reflexes Originating from the Slow Adapting Receptors (SARs)

Control of Inspiration by the SAR Reflex

Since the classical discovery of the reflex inhibition of inspiratory activity provoked by stimulation of the slowly adapting receptors by Hering & Breuer (1868) and Breuer (1968), the Hering-Breuer reflex has been recognised as the important negative feed-back system in the control of breathing (for review see Bradley, 1977). The volume, or rather transpulmonary pressure, related SARs response (Widdicombe, 1954b; Davis, Fowler & Lambert, 1956) causes a reflex inhibition of the inspiratory motor output to the diaphragm, the inspiratory intercostal nucleus (Sant'Ambrogio & Widdicombe, 1965) and to the inspiratory laryngeal muscle fibres (Bartlett, Remmers & Ganthier, 1973; Glogowska, Stransky & Widdicombe, 1974; Cohen, 1975). Volume thresholds are different for the various motor outputs being higher for the diaphragm response than for the external intercostal muscles (Bruce, Euler & Yamashira, 1979) or for the inspiratory laryngeal muscle responses (Cohen, 1975).

The lung volume necessary to terminate inspiration falls during the progress of inspiration, and thus the volume threshold of the reflex constitutes a hyperbolic function of the time during inspiration (Clark & Euler, 1972); this decline of the reflex threshold has been explained as progressive lowering of the threshold of the central inspiratory 'off-switch' mechanism (Euler & Trippenbach, 1976). In the course of a normal breath lung volume rises until it reaches the 'off-switch' threshold, overriding the timing of the central pattern generator. Clark & Euler (1972) suggested that the termination of the inspiratory phrenic burst by lung volume information occurred in a trigger-like or 'all-or-none' fashion. According to this concept, the vagal inspiration-inhibiting reflex would not exert any inhibitory effect until shortly before the peak of inspiration, once above a critical threshold. These results were incorporated in von Euler's model (Bradley *et al.*, 1975; Euler, 1977). The central pattern generator for respiratory activity consisted of two mechanisms: a ramp-like generator of central inspiratory activity (CIA) and the inspiratory 'off-switch' mechanism. The 'off-switch' mechanism would be supplied by additive excitatory inputs from SARs, from growing CIA activity, from the pneumotaxic centre and other sources. CIA activity could be stopped only when the 'off-switch' mechanism threshold had been reached. However, more recent data have introduced important modifications to this original model. SARs induced inhibition of the CIA appears to occur during the progress of inspiration in an additive, graded and reversible fashion as the first stage of the inspiratory 'off-switching' followed by an irreversible cessation of inspiration (Younes *et al.*, 1978; Remmers, Baker & Younes, 1979; Cross, Jones & Guz, 1980).

Karczewski *et al.* (1976, 1980a) and Gromysz *et al.* (1980) stimulated afferent vagal fibres and showed that there was a non-linear temporal and spatial summation by the respiratory controller so that the rise of inspiratory activity could be controlled 'on-line' by the continuous vagal input. The time constant of this integrative process was long enough to influence not only actual but also subsequent respiratory cycles in a way resembling a 'short-term' memory (Karczewski *et al.*, 1978; Budzinska *et al.*, 1979). This last result does not support the original von Euler concept (Clark & von Euler, 1972) of the lack of any interaction between the duration of the expiratory phase and the subsequent inspiratory period. Similar findings of a graded and reversible inhibition of the phrenic nerve activity by shortlasting moderate lung inflations was reported by Kubin & Lipski (1979).

Cohen (1975) used the technique of withholding SAR input over a single inspiratory period by tracheal occlusion at the end of expiration and observed a graded activation of the inspiratory laryngeal nerve activity and, in some experiments, also a graded activation of the phrenic nerve activity. More recently, using a no inflation test for one respiratory cycle (by withholding inflation delivered by a respiratory cycle-triggered pump) Cohen *et al.* (1982a) observed an increase in the peak frequency of discharge both in the early-onset and in the late-onset phrenic motoneurones. The latter responded, however, significantly more which is consistent with their greater excitability to the pulmonary vagal inhibitory input (Karczewski *et al.*, 1980a).

Together, these findings modify the original von Euler model. SAR dependent, graded, inhibition of the inspiratory activity is presumably due to a more general non-linear property of the central 'off-switch' mechanism (Euler, 1980) which appears to continuously inhibit the inspiratory activity, even at the subthreshold level.

The inhibitory input from SARs in each lung is transmitted ipsilaterally by the vagal afferents in dogs and centrally summated with a mutual facilitation of the responses both in the reduced T_I and peak amplitude of integrated phrenic nerve discharge. If opposite V_T changes are applied in each lung, the phrenic nerve and T_I reflex responses are significantly reduced and the summation of both responses differs quantitatively; yet, the effects are the same on both sides, in each phrenic nerve. This result indicates that either there is some common 'off-switch mechanism' for the two halves of the brain stem or that a strong synchronisation of inspiratory timing between the two halves of the respiratory control system over-rides the pattern of T_I and the magnitude of phrenic discharge imposed by the SAR input from each lung individually (Cross, Guz & Jones, 1981).

Under appropriate experimental conditions, under light anaesthesia, a positive-feedback component of the SAR reflex, consisting of an increase in the rate of rise of the inspiratory discharge, has been demonstrated in dogs (Bartoli *et al.*, 1975; Cross *et al.*, 1980; Pack, Delaney & Fishman, 1971) and in pigs (Huszczuk *et al.*, 1977), but not in rabbits (Karczewski *et al.*, 1980b). In cats, DiMarco *et al.* (1981) applied very small static lung inflations, which were apparently subthreshold for the high-threshold rapidly adapting receptors (RARs), and observed a higher rate of rise of inspiratory nervous activity. This increase was more pronounced in the external intercostal than in the phrenic nerve and was abolished by bilateral vagotomy. This facilitatory component of the inspiratory terminating Hering–Breuer reflex may have an obvious function in the precise 'on-line' control of breathing by compensating shortened inspiratory duration by the increased rate of inspiratory discharge. Besides, as suggested by DiMarco *et al.* (1981), it may increase the relative contribution of the external intercostal muscles with increasing respiratory drive thus explaining progressive increments in the relative contribution of the rib cage movement to the tidal volume with hyperpnea.

The mechanism of the inspiration-augmenting component of the SAR reflex is obscure, but it presumably depends on the central organisation of the reflex. It has been shown that the input from SARs brings about a subthreshold excitation of the type I-alpha dorsal group of inspiratory neurones (Lipski *et al.*, 1979), which are known to have excitatory bulbo-spinal projections to the phrenic motoneurones (Graham & Duffin, 1982). On the other hand, I-beta inspiratory neurones, excited by the lung inflation and until recently thought of as the neurones involved in the inhibition of the I-alpha neurones and to be facilitatory to the 'off-switch' mechanism, have been shown to be a heterogeneous population (Marino, Davies & Pack, 1981; Cohen, 1982a). Some of them have a high dynamic sensitivity and, possibly, receive input from the inspiration-facilitatory RARs, a suggestion originally made by Berger (1977). Convergence of the inputs from the two types of pulmonary receptors, SARs and RARs, on the inspiratory I-beta neurones could explain a facilitatory component in the SAR reflex. In agreement with this hypothesis Lipski, Kubin & Jodkowski, (1983) have found recently an excitatory bulbo-spinal projection to the phrenic motoneurones from the I-beta neurones which is excited by the SAR input.

Yet, some other possibilities should be kept in mind. It has been found that the SAR discharge from the extrathoracic airways behaves differently to the SAR activity from intrathoracic sites (i.e. it decreases during inspiration and increases during expiration) (Sant'Ambrogio & Mortola, 1977). More than 60 per cent of the SARs are situated in the extrapulmonary airways (Miserocchi *et al.*, 1973, 1974; Sant'Ambrogio & Miserocchi, 1973) and about 17 per cent in the extrathoracic trachea (Bartlett *et al.*, 1976). Thus, during spontaneous breathing, the inhibitory input from the SARS situated within the extrathoracic trachea appears to be paradoxically reduced in inspiration especially with progressing hyperpnea, and hence the inspiratory activity may be disinhibited. The functional role of the SARs in the extrathoracic trachea appears to be related more to the dynamic sensitivity of airflow, and airflow direction, than to static volume changes (Sant'Ambrogio & Mortola, 1977; for discussion see Sant'Ambrogio, 1982). At high levels of lung volume SARs respond even more to increase in flow-rate than to increase in volume (Lloyd, 1979) so they become more like rate receptors (Pack, 1982). How the information related to flow-rate is processed by the central respiratory neurones and how it influences the overall reflex response from the SARs is not known.

SARs located in the extrapulmonary airways are possible involved in the mechanism of the apnoea produced by high frequency (15 Hz), small volume oscillations (HFO) of positive pressure ventilation. Apnoea during HFO was observed both in anaesthetised animals (Bohn *et al.*, 1981) and in humans (Butler *et al.*, 1980), and is of reflex origin as it is reversed by bilateral vagotomy (Thompson *et al.*, 1981). This type of ventilation is able to maintain adequate gas exchange and blood gas homeostasis. Inflation volumes applied by HFO ventilation are considerably less than normal tidal volume, thus the volume sensitive intrapulmonary SARs are not likely to be involved. High frequency pressure amplitudes (13 cmH$_2$O, Thompson *et al.*, 1981) are presumably significantly damped in the small intrapulmonary bronchi so that the extrapulmonary and extrathoracic airway receptors, which are known to be sensitive to the change in flow volume acceleration (Bartlett *et al.*, 1976; Mortola & Sant'Ambrogio, 1979), appear to be more significant for the reflex inhibition of breathing during HFO.

Neuromuscular blockade, like vagotomy, initiates rhythmic phrenic activity in cats made apnoeic by HFO (England, Onayemi & Bryan, 1982). Therefore, some extravagal factor, possibly an inhibitory reflex from the primary spindle afferents of the respiratory muscles which is known to inhibit inspiratory activity (Remmers, 1970), contributes to reflex apnoea induced by HFO. Sternal and chest wall vibration at 100Hz, a stimulus to spindle receptors, reflexly decreases tidal volume in conscious human subjects if applied in inspiration (Gandevia & McCloskey, 1976; Homma, 1980). Sternal vibration in cats does not mimic the Hering-Breuer inhibitory reflex as T_I increases; the T_T/T_I curve remains unchanged and decrease of the tidal volume is due to the reduced rate at which inspiration proceeds (Colebath, Gandevia & McCloskey, 1977).

Reflex Control of Expiration by the SAR Reflex

Increase in the expiratory phase duration produced by lung inflation, and facilitation of inspiration by lung deflation, (pulmonary deflation reflex) have been observed since the discovery of the Hering–Breuer reflex in 1868. Clark & Euler (1972) showed that short-lasting lung inflation, applied during expiration, delayed the subsequent inspiration. The magnitude of the reflex is time dependent over the expiratory phase and is greater with increasing time of expiratory duration. Since low-threshold SARs are active over the whole respiratory cycle, including expiration at FRC volume (for review see Sant'Ambrogio, 1982), a lung deflation will reduce total SAR firing frequency. Progressive decline of the SAR expiration facilitatory input, as lung volume returns to FRC, tends to shorten the expiratory phase and to facilitate the subsequent inspiration (Knox, 1979). Such disfacilitation of the central expiratory mechanism as a cause of the inspiratory facilitatory deflation reflex was proposed by Adrian (1933) and is consistent with the original concept of the discoverers of the reflex (Hering & Breuer, 1868; Breuer, 1868). Some contribution of RARs to the initiation of the pulmonary deflation reflex during eupnoeic breathing has been postulated chiefly because RAR activity occurs during moderate lung deflations (e.g. Sellick & Widdicombe, 1970; Luck, 1970; Bartole *et al.*, 1973); yet, this mechanism has not been proven until recently. Davies *et al.* (1981) demonstrated that in rabbits deprived of SAR activity by SO_2 inhalation and artificially ventilated, phrenic inspiratory bursts were linked to lung deflations. Bilateral vagotomy abolished this relation. The tidal volume in most of their experiments had to be slightly increased, about 30 per cent above the values similar to those during quiet breathing. Yet, a possible contribution to the deflation reflex of the sensory endings of non-myelinated vagal fibres could not be excluded (see below).

It has been shown that lung inflation excites expiratory neurones situated rostral to the obex (Feldman & Cohen, 1978) and in the retroambigual area of the medulla in a graded and sensitive fashion over the volume range below the FRC (Koepchen *et al.*, 1979). The final 30 per cent of the expiratory period duration is insensitive to SAR facilitatory input (Knox, 1973), a finding consistent with the results of Feldman & Cohen (1978) who showed that only 'early onset'-expiratory neurones are excited while 'late onset'-expiratory neurones are inhibited by lung inflation. About 50 per cent of the 'late onset'-expiratory neurones, situated caudal to the obex, are upper motor neurones exciting thoracic and abdominal expiratory motor neurones.

Their inhibition corresponded to the inhibition of the expiratory intercostal nerve activity by lung inflation observed by Cohen *et al.* (1982).
However, the problem of whether lung deflation reduces or augments
the expiratory output, in addition to increasing expiratory duration,
is more complex. Maintained lung inflation has been shown to recruit
vagally dependent expiratory abdominal muscle activity in cats (Russel
& Bishop, 1976). A biphasic inhibitory–excitatory response to lung
inflation in the expiratory, internal intercostal muscles has also been
reported, the excitatory phase being facilitated by hypercapnia (Arita
& Bishop, 1982). Different volume stimuli and different experimental
procedures may explain the discrepancies in the results. Apparently,
the SAR reflex influences the two separate central mechanisms in a
different fashion: one, expiratory–inspiratory 'off-switch' mechanism
determining the expiratory duration and the other responsible for the
build-up of the expiratory discharge. The former mechanism is inhibited
by the SAR reflex, thus expiratory duration is increased. The latter one
seems to be influenced in a more variable fashion being recruited when
some expiratory load is imposed, e.g. during breathing against positive
airway pressure (Farber, 1982). This reflexly augmented expiratory
discharge during loading is entirely due to the SAR reflex as the blockade of SARs by SO_2 inhalation abolishes forced expiratory abdominal
activity (Davies *et al.*, 1980).

Functional Significance of the SAR Respiratory Reflex

Role of the SAR reflex in the control of the pattern of breathing. Clark
& von Euler (1972) found an inverse hyperbolic relation between
tidal volume (V_T) and inspiratory duration (T_I) and also subsequent
expiratory duration (T_E) suggesting a positive correlation between the
magnitude of the SAR reflex and frequency of breathing. In vagotomised
animals an increase in the respiratory drive, induced by CO_2 inhalation,
augmented the build-up of inspiration and the tidal volume, but not
the frequency of breathing. Similar results were obtained in conscious
human subjects following local vagal anaesthesia (Guz *et al.*, 1966a,
b). As the selective blockade of the SARs by SO_2 inhalation brings
about the same changes as vagotomy (Davies *et al.*, 1978), it is reasonable to suggest that the vagally dependent control of the pattern of
breathing has to be attributed mainly to the SAR reflex.

These results provided the framework of the current concepts on the
primary role of the SAR reflex in the control of timing the respiratory
cycle. A decrease in lung compliance (Marshall & Widdicombe, 1958)
and other factors known to increase the activity of SARs and the

responsiveness of the reflex (for references see Sant'Ambrogio, 1982) tend to switch the pattern to more frequent and shallow breathing.

The role of the SAR reflex in setting the pattern of breathing is not, however, exclusive. T_{TOT}, T_I and T_E changes, although less regular, occur in vagotomised animals during increased respiratory drive induced by the stimulation of either central or peripheral chemoreceptors (e.g. Miserocchi, 1976; Feldman & Gauthier, 1976; St. John, 1979; Eldridge & Gill-Kumar, 1980). Polypnoeic breathing in moderate hyperthermia and thermal panting have a central origin and are independent of the SAR reflex (Euler *et al.*, 1970; Karczewski *et al.*, 1972; Widdicombe & Winning, 1976); furthermore hypocapnic polypnoea is vagally independent (Cohen, 1964). The role of the level of alertness on V_T-T_I relationship in humans (Newsom Davis & Stagg, 1975) and of the reticular activating system in the control of breathing pattern is well known and has been frequently reported since the paper by Hugelin & Cohen (1963). Thus rostral brain areas and the pneumotaxic centre, the vagally independent brain stem respiratory frequency controller, also influence the central 'off-switch' mechanism in the absence of the pulmonary vagal input. It may be assumed that the central respiratory controller is disinhibited, and thus more effective, following the disruption of the vagal input, as lung inflation has been found tonically to inhibit the respiratory modulation of neurones in the pontine pneumotaxic centre (Feldman, Cohen & Wolotsky, 1976; Cohen, 1979). The contribution to the inspiratory and expiratory 'off-switch' of either mechanism − a peripheral one from SARs and a central one − appears to be species and anaesthesia dependent thus varying under different experimental conditions.

The SAR reflex is resistant to anaesthesia and, over some range, even facilitated in anaesthetised animals (Sant'Ambrogio & Widdicombe, 1965; Bouverot, Crance & Dejours, 1970; Bystrzycka *et al.*, 1972). A negative hyperbolic relation between V_T and T_I (Clark & Euler, 1972) has not been shown in conscious man (Newsom Davis & Stagg, 1975) and in non-anaesthetised dog (Iscoe, Young & Jennings, 1982). On the contrary, a positive correlation between V_T and T_I was reported as a possible consequence of the changes in the theoretical 'off-switch' threshold curve (Iscoe *et al.*, 1982) or explained in accordance with v. Euler's model by suggesting that mean inspiratory flow (rate of inspiration), and not tidal volume, is the controlled variable (Newsom Davis & Stagg, 1975).

The sensitivity of the SAR reflex is high in rabbits and low in humans (Widdicombe, 1961) and in seals (Angell-James, Elsner & de Durgh

Daly, 1981). Adult man shows no volume-dependent reflex effects on the pattern of breathing over a range exceeding twice that of the normal eupnoeic tidal volume (Clark & Euler, 1970, 1972); furthermore local vagal anaesthesia in conscious subjects does not influence pattern of breathing, minute ventilation, or end-tidal CO_2 concentration (Guz *et al.*, 1964, 1966a). Thus, the functional role of the SAR reflex in humans is not significant in eupnoeic breathing and only becomes apparent during increased respiratory drive and hyperpnoea (Guz *et al.*, 1966b); the low responsiveness of the SAR reflex under eupnoeic conditions is not caused by low sensitivity of SARs themselves as they discharge during these conditions (Langrehr, 1964; Guz & Trenchard, 1971a). In contrast, newborn babies exhibit the SAR reflex readily (Cross *et al.*, 1960; Bodegard *et al.*, 1969), and hence its sensitivity decreases with maturation. Similar decreases in the responsiveness of SAR reflex with maturation have also been reported in other species (cats – Trippenbach *et al.*, 1979; and rabbits – Schwieler, 1968). The exact mechanism of this age-dependent attenuation of the reflex, presumably linked to the postnatal development of a high-threshold central mechanism (Euler *et al.*, 1970), is still obscure.

Role of the SAR reflex in the control of ventilation. The role of SAR reflex in the control of ventilation is more controversial as several variable factors interact to determine either an inhibitory or facilitatory influence of the total vagal input. Vagally mediated depressions of the ventilatory response to moderate hypoxia (Kashani & Haig, 1975, Kiwull-Schöne & Kiwull, 1979), to stimulation of the carotid sinus nerve (Kiwull & Wiemer, 1971) and to hyperoxic hypercapnia (Richardson & Widdicombe, 1969; Wiemer & Kiwull, 1972) have all been demonstrated. Conversely, with increasing hypercapnia and ventilation a vagally mediated ventilatory facilitation has been demonstrated as inactivation of vagal traffic by total cold block or bilateral vagotomy decreases ventilatory response (Richardson & Widdicombe, 1969; Phillipson *et al.*, 1970; Wiemer & Kiwull, 1972). Local vagal anaesthesia in conscious human subjects depresses the slope of the $\dot{V}/CO_2$ curve (Guz *et al.*, 1966b) The fact that vagal block under normocapnic and mild hypocapnic conditions enhances the sensitivity to CO_2 (Kiwull-Schöne *et al.*, 1981) fits well the suggestion that the SARs exert a primarily inhibitory influence upon ventilatory responses (Richardson & Widdicombe, 1969; Wiemer & Kiwull, 1972; Kiwull-Schöne & Kiwull, 1979) under these conditions. It has been postulated that increasing ventilation recruits positive feedback reflexes mediated by RARs (Phillipson *et al.*, 1973; Davies *et al.*, 1978) which facilitate expiratory termination (Kiwull-Schöne, Ward & Kiwull, 1981). Alternatively the positive feedback may arise from non-myelinated vagal afferents (Anand *et al.*, 1982). Facilitatory, vagally mediated, mechanisms seem to oppose and entirely override the inhibitory influence of the SAR reflex upon ventilation since a selective elimination of the SAR reflex by differential vagal cold blockade (Phillipson *et al.*, 1973; Kiwull-Schöne *et al.*, 1981) or by sulphur dioxide inhalation (Davies *et al.*, 1978) does not result in diminished ventilatory responsiveness to hypercapnia or hypoxia.

Another factor complicating the role of the SAR reflex in the ventilatory response is a direct local inhibitory effect of intrapulmonary CO_2 upon SAR discharge (Sant'Ambrogio *et al.*, 1974; Bartlett & Sant'-Ambrogio, 1976; Bystrzycka & Nail, 1980) which reduces the sensitivity of the SAR reflex and, by disinhibition, facilitates ventilatory CO_2 responsiveness (Bradley *et al.*, 1976; and see below). The effect is exerted, however, only over the hypocapnic range of intrapulmonary CO_2 (Coleridge *et al.*, 1978a; Mitchell *et al.*, 1980) and thus apparently constitutes a negative feedback mechanism which prevents excessive hyperventilation by increasing the responsiveness (disinhibiting) of the SAR reflex with progressive hypocapnia (Banzett, Coleridge & Coleridge, 1978).

SAR reflex and the control of the airway resistance. The SAR reflex decreases airway resistance by the reduction of bronchomotor tone, relaxation of the tracheobronchial smooth muscle and increase of the airway calibre (Looffbourrow *et al.*, 1952; Widdicombe & Nadel; 1973; Widdicombe, 1963). This influence is maintained in a tonic fashion during eupnoeic breathing since a differential cold block of the vagal traffic in thick myelinated fibres brings about a persistent bronchoconstriction (Karczewski & Widdicombe, 1969c). The bronchodilator effect is due to the reflex inhibition, late in inspiration of the spontaneous activity in efferent vagal fibres, to the airway smooth muscle (Widdicombe, 1966). The possibility of a reflex activation of, recently discovered, non-cholinergic and non-adrenergic bronchodilator fibres in the vagal trunk (Irvin *et al.*, 1980) has not been, so far, checked.

The SAR reflex control of the laryngeal muscles merits particular interest as the upper airway resistance retards expiratory airflow and lung deflation, thus, via volume-related SAR reflex feedback, controls T_E and frequency of breathing (Gauthier, Remmers & Bartlett, 1973). The effect of SARs stimulation on the larynx is consistent with the action on other respiratory muscles resulting in inhibition of the phasic inspiratory motor discharge to the posterior cricoarytenoid muscle (Cohen, 1975), which abducts the vocal cords and hence decreases laryngeal resistance. The SAR reflex control of overall laryngeal resistance is complex (see Dixon *et al.*, 1974; Bartlett, 1980). One reason for the discrepancies is uncertainty about the correlation between inspiratory activity in the recurrent laryngeal nerve, and motor unit activity in the posterior cricoarytenoid muscle; what is certain is that motor fibre activity does not correlate well with the decrease in the laryngeal resistance (Glogowska *et al.*, 1974). Rhythmical inspiratory discharge to the posterior cricoarytenoid muscle tends to decrease laryngeal resistance in inspiration and may account for the phasic variations in laryngeal resistance over the respiratory cycle. The SAR reflex, by inhibiting this inspiratory activity, attenuates phasic variations of laryngeal resistance in dogs (McCaffrey & Kern, 1980). Prolonged lung inflation delays the onset of the rhythmical inspiratory burst to abductor muscles but paradoxically also brings about a sustained tonic activation of the abductor muscles (Bartlett *et al.*, 1973). Apparently, the abductor motor neurone pool is heterogeneous; some inspiratory abductor motor units are activated and others inhibited by lung inflation (Barillot & Bianchi, 1971), and the abductor motor neurones have connections with both inspiratory and expiratory central neurones (Bartlett *et al.*, 1973). Activity of the motor fibres firing irregularly in expiration and supplying abductor (constrictor) laryngeal muscles is inhibited by the SAR reflex (see Szereda-Przestaszewska & Widdicombe, 1973).

The SAR reflex, in addition to its inhibitory influence upon phasic variations in laryngeal resistance, exerts a powerful tonic effect in decreasing laryngeal resistance (Dixon *et al.*, 1974; McCaffrey & Kern, 1980) mainly in expiration (Bartlett *et al.*, 1973) both in animals and man (Spann & Hyatt, 1971). This tonic influence is of particular significance during hypoxia or peripheral chemoreceptor stimulation. During hyperventilation in intact animals the vocal cords are more widely abducted, particularly during expiration, expiratory laryngeal resistance is decreased and hence adjusted to permit augmented airflow (Dixon *et al.*, 1974; Bartlett, 1980). However, following bilateral vagotomy hypoxia and chemoreceptor stimulation brings about a narrowing of the glottic airway (Dixon *et al.*, 1974) chiefly due to a decrease in the abductor muscle activity (Bartlett, 1980). The blockade of SARs by SO_2 inhalation in rabbits reverses the abductor muscle expiratory response to both hypoxia and hypercapnia and produces expiratory narrowing of the glottic airway similar to the reversal following bilateral vagotomy (Bartlett, Knuth & Knuth, 1981). Thus the vagal SAR reflex serves to maintain the patency of the laryngeal airway during hypoxia by opposing the effect of chemoreceptor stimulation which by itself exerts a narrowing effect upon the laryngeal calibre (Dixon *et al.*, 1974).

SAR reflex and the control of work of breathing. Slowly adapting pulmonary receptors provide information on the mechanical conditions of breathing: airway flow, airway calibre and lung volume, their activity being related to lung compliance and airway calibre (see other chapters of this volume and Sant'Ambrogio, 1982 for review).

SARs situated in the extrapulmonary tracheobronchial tree appear the most suitable for the reflex adjustment of the airway resistance to the optimal values for different flow rates as the rate and the direction of flow is signalled by the SARs in large airways (for review see Sant'-Ambrogio, 1982). There is a continuous processing of this sensory information, together with information from the chest wall and respiratory muscles, on the multiple level of the central nervous system (Remmers & Bartlett, 1977). Reflex respiratory responses adjust the pattern of breathing and the airway calibre to the ventilatory demand in such a way that changes in airway flow rates, airway resistance and lung volume minimise the mechanical work of the respiratory muscles (Otis, Fenn & Rahn, 1950; Goldman, Grimby & Head, 1976; Grimby, 1977) and the force of the muscular contraction (Mead, 1960) thus optimising the relation between the respiratory neural and mechanical outputs (Younes & Riddle, 1981). Vagal input appears to be an important component of this whole control system as bilateral vagotomy (Lim, Luft & Grodins, 1958) or cold block of vagal traffic (Zechman, Salzano & Hall, 1958) significantly augment the mechanical work of breathing. Interactions between the SAR reflex and defence reflexes from the respiratory tract have also been shown (e.g. the magnitude of the cough reflex is significantly enhanced by the accompanying SAR reflex provoked by the lung inflation in the first, inspiratory phase of coughing — for reference see Korpas & Tömori, 1979).

Reflexes Originating from the Rapidly Adapting Receptors (RARs)

Rapidly adapting receptors were identified originally as the high-threshold mechanoreceptors excited by lung hyperinflation (Larabbee & Knowlton, 1946); the newer term irritant receptors, introduced by Widdicome and his associates, is now broadly accepted. This term is based on the findings that, in addition to mechanical stimulation of the trachea and bronchi, rapid lung deflation, pneumothorax or lung hyperinflation, a multitude of airway irritants, aerosols, carbon dust, histamine, as well as pulmonary anaphylaxis, lung congestion and oedema exert a strong stimulatory effect upon the rapidly adapting receptors (Widdicombe, 1954b; Mills *et al.*, 1969, 1970; Sellick & Widdicombe, 1969; Widdicombe & Glogowska, 1973). However, at least a part of the excitatory effect of such chemical stimuli is indirect and due to secondary mechanical deformation (see other chapters of this volume, also for review, Sant'Ambrogio, 1982). Chemosensitivity of RARs is also species dependent; less pronounced in dogs (Sampson & Vidruk, 1975) than in rabbits, cats or guinea pigs (Bergren & Sampson, 1982). The reflex effects of RAR stimulation consist of an increased frequency of breathing, both in hyperpnoea and hyperventilation (Mills *et al.*, 1969, 1970; Sellick and Widdicombe, 1970; Glogowska *et al.*, 1972, for review see Widdicombe, 1974); both T_I and T_E are reflexly reduced (Widdicombe & Winning, 1976). These conclusions were chiefly based on the experiments in which histamine, thought to be a specific stimulus for RARs, was used. However, it has been shown subsequently that histamine also excites pulmonary endings of non-myelinated vagal afferents — J-receptors in rabbits (Stransky *et al.*, 1973) and in cats (Armstrong & Luck, 1974; Paintal, 1977b) and bronchial non-myelinated afferents in dogs (Coleridge *et al.*, 1978a). The technique of differential vagal blockade, frequently used in these experiments, has also been the subject of criticism as it does not exclude some impairment of the traffic in the non-myelinated vagal afferents (see for discussion Coleridge & Coleridge, 1981). Caution must be exercised therefore in attributing all effects of the stimuli, thought to be specific for RARs, to the RAR reflex as reflexes originating from the sensory endings of non-myelinated vagal afferents may also be involved (for discussion see Paintal, 1977b; Coleridge & Coleridge, 1981).

Brief and intense stimulation of RARs by pulses of lung deflation or inflation in rabbits, whose SARs had been previously blocked by SO_2 inhalation, provokes an augmented inspiration of fixed duration in an all-or-none fashion; the T_I increase is due to the overlap of the spontaneous and provoked phrenic nerve activity (Davies & Roumy, 1982). The effect is timed over the inspiratory phase, pulses delivered late in inspiration are less effective. Pulses of either deflation or inflation delivered in expiration shorten that expiration and trigger an all-or-none excitatory phrenic nerve response (Davies & Roumy, 1982). A characteristic feature of the reflex, except in paralysed and artificially ventilated animals, is an unusually long refractory period, (more than 1 min), following each augmented breath. The mechanism of this phenomenon is obscure. Another feature of the reflex is the capacity of a single pulse of inflation or deflation to affect the duration of the subsequent respiratory cycles (Davies & Roumy, 1982), a finding consistent with the hypothesis that there is some central integrative processing of the RAR input which has a time constant exceeding the duration of one respiratory cycle (Karczewski *et al.*, 1976, 1980a, b).

Even before the excitatory reflex effect of large pulmonary inflations had been assigned to the rapidly adapting pulmonary receptors by Larabbee & Knowlton (1946), it was suggested (Hammouda & Wilson, 1935, 1939) that breathing was the subject of two opposing, continuous reflex vagal influences, inhibitory and facilitatory. This conclusion was derived from the finding that cooling the vagal trunks to 8–10 °C in dogs blocked the inhibitory Hering-Breuer reflex and unmasked a facilitatory, continuous, vagal influence, which accelerated respiration.

Until recently the problem of the importance of the RARs facilitatory vagal feedback in the control of breathing, especially in eupnoea was analysed by methods involving differential blockade of vagal traffic such as cold or d.c. anodal block of the nerve trunks (Guz & Trenchard, 1971b; Fishman, Phillipson & Nadel, 1973; D'Angelo, 1978 and others). Fishman *et al.* (1973) concluded that both vagal mechanisms, the respiratory-inhibitory Hering-Breuer reflex and the respiratory-stimulatory RAR reflex, contributed to the control of eupnoeic breathing and that interaction between them constituted a mechanism by which optimal respiratory frequency was achieved. However, vagal cold block does not differentiate myelinated afferents of different diameters adequately. The maximum frequency of discharge which the nerve may transmit varies with the local temperature and conduction velocities of the fibres, a factor which may influence the effect of cold block on breathing and make interpretation equivocal (see for discussion Paintal, 1973; Coleridge & Coleridge, 1981). Recently, however, Davies *et al.* (1981) have shown in rabbits whose SARs were inactivated by SO_2 inhalation, that lung inflation and deflation, in the range only slightly (c. 30 per cent) above eupnoeic tidal volume, were associated with phrenic inspiratory bursts. Bilateral vagotomy disrupted this effect. These important results supported the earlier concept that a drive from RARs during expiration triggered the next inspiration and contributed to the termination of expiration along with the Hering-Breuer deflation reflex. The role of the RAR reflex in the control of inspiration seems less certain as the unmasking of the RAR reflex by SO_2 inhalation, has been shown to have no significant influence on the rate of increase in integrated phrenic nerve activity (Davies *et al.*, 1978).

Although the use of SO_2 blockade of SARs enabled a more selective analysis of the RAR reflex, the sensory endings on non-myelinated vagal fibres remained active and their contribution to the reflex responses cannot be ignored (see below).

The cough reflex originates from RARs situated in the trachea and main bronchi as the local mechanical irritation of bronchi produce only weak expiratory effects (Widdicombe, 1954a; Widdicombe *et al.*, 1962, for discussion see Widdicombe, 1974). Local anaesthesia of the airways by inhalation of lignocaine aerosol attenuates intractable cough in patients, possibly by desensitisation of irritant receptors (Howard *et al.*, 1977). The depressive effect of general anaesthesia on the cough reflex makes acute experiments less suitable for the study of the cough reflex, which appears to be related also to other areas of the respiratory system and possibly mediated also by non-myelinated vagal afferents (see, for review, Korpas & Tömöri, 1979 and below).

RAR Reflex Effects upon the Bronchomotor Tone and Laryngeal Airflow Resistance

Reflex bronchoconstriction due to the stimulation of RARs has been reported frequently (Mills *et al.*, 1969; Armstrong & Luck, 1974; Sampson & Vidruk, 1975). Bronchoconstrictor drugs, histamine or 5-hydroxytryptamine act by a combination of both reflex vagally mediated effects and by direct contraction of airway smooth muscle; vagotomy, cooling of the vagal trunks or atropine significantly reduce their bronchoconstrictor effects (Mills & Widdicombe, 1970; Dixon, Jackson & Richards, 1979). Similar reflex mechanisms of bronchoconstriction have been identified in experimental, allergen-induced, asthma in rabbits (Karczewski & Widdicombe, 1969a) and in dogs (Gold, Kessler & Yu, 1972) and is strongly suggested as a mechanism by which asthmatic attack in humans is provoked (Widdicombe & Sterling, 1970; Yu, Galant & Gold, 1972; Nadel, 1973, 1977; Gold, 1975). A reflex bronchoconstrictor response, once triggered, initiates a dangerous positive-feedback mechanism as bronchoconstriction facilitates a further RAR reflex.

However, neither the magnitude nor the time course of the excitatory effects upon RAR activity of histamine or 5-hydroxytryptamine match their reflex bronchoconstrictor activity (Dixon *et al.*, 1979).

This dissociation may be partly due to variable effects of both drugs upon the airway responsiveness to efferent vagal stimulation (Dixon *et al.*, 1980), but, apparently, depends also on the involvement of the bronchoconstrictor reflex from the pulmonary J receptors and/or bronchial C-fibre afferents (see below) as the temperature of the vagal block (0.5 °C) used by Dixon *et al.* (1979) was sufficient to block non-myelinated vagal afferents.

Histamine, thought of as a specific RAR stimulant, brings about a reflex laryngeal constriction in expiration due to the increased activity in expiratory (adductor) laryngeal motor neurones and also an increase in laryngeal resistance during inspiration as the number of spikes per inspiratory phase in the inspiratory (abductor) laryngeal motor neurones is reflexly reduced (Stransky *et al.*, 1973). Similar expiratory reflex constriction of the larynx has been shown in pneumothorax (Dixon *et al.*, 1974) and following other procedures known to excite RARs like pulmonary oedema, which increases activity in motor neurones supplying the laryngeal abductor muscles (Glogowska *et al.*, 1974). The possibility that non-myelinated afferents may be involved in the reflex cannot be overlooked for all the experimental procedures are not specific for RAR activation and may excite also pulmonary J type receptors and bronchial receptors (see below and Chapter 3).

Functional Significance of the RAR Respiratory Reflexes
The RAR reflex is active in near eupnoeic breathing, at least in rabbits (Davies *et al.*, 1981), and appears to contribute, along with the SAR reflex, to the control of the pattern of breathing by initiating inspiration during deflation. This initiating inspiratory drive may be of particular importance during the hyperpnoea associated with high frequency breathing when T_E is short and each inspiration starts immediately after reaching the end-expiratory level (Davies *et al.*, 1981). Whether a similar role of the RAR reflex exists in other species awaits confirmation.

The RAR reflex constitutes a positive vagal feedback mechanism which augments ventilation with increasing respiratory drive and minute ventilation and hence opposes the SAR reflex (see above). A variable balance between these two antagonistic inputs may optimise the pattern of breathing and magnitude of ventilation to the prevailing conditions of the respiratory system.

The RAR reflex has usually been ascribed to the category of respiratory defence reflexes (Coleridge & Coleridge, 1981). This role is obvious in cough. The laryngeal and bronchoconstrictor reflex response to irritant gases, aerosols, and cigarette smoke shown by animals and humans (Robertson *et al.*, 1969; Rees *et al.*, 1982), may have a protective role. More equivocal in this regard is the role of the ventilatory response. A decrease in tidal volume and increase in respiratory frequency, a breathing pattern produced by the introduction of such irritant substances into the respiratory tract could be considered protective by limiting the extent of lung tissue exposed to irritants, and by facilitating their deposit in the airways.

The RAR reflex is presumably involved in the inspiratory augmenting reflex or gasping inspirations of newborn babies (Cross, 1961) which are facilitated by lung collapse and reducing lung compliance (Reynolds, 1962); RARs like SARs exhibit a lowered threshold when lung compliance is decreased (Sellick & Widdicombe, 1970). In newborn, preterm, infants the RAR reflex appears immature as direct bronchial stimulation does not provoke the typical reflex inspiratory effort. In preterm infants such stimulation often has the opposite effect; that is apnoea (Fleming *et al.*, 1982).

The significance of the RAR reflex in pathological events such as asthma attack (see above) or the pattern of breathing in pulmonary congestion, oedema, microembolism or anaphylaxis (Mills *et al.*, 1969, 1970) is less certain; some maintain that reflexes originating from the sensory endings of non-myelinated vagal afferents are predominant under these conditions (Karczewski & Widdicombe, 1969a; Koller & Ferrer, 1973; Bleecker *et al.*, 1976, for discussion see Coleridge & Coleridge, 1981).

A characteristic feature of the respiratory defence reflexes is an increase in mucus secretion due to the reflex stimulation of the submucosal glands in the airways. In the dog at least vagal secretomotor fibres supplying the upper trachea run in the superior laryngeal nerve and those supplying the lower trachea in the recurrent laryngeal nerves (Davis *et al.*, 1976). It has been shown in cats that mechanical irritation of the trachea and bronchi reflexly stimulates vagally innervated submucosal glands, while irritation of the upper airways reflexly excites submucosal glands via sympathetic secretomotor fibres (Phipps & Richardson, 1976). On the other hand, German, Ueki & Nadel (1980) found that mechanical stimulation of the larynx increased submucosal gland secretion via a secretory reflex whose efferent pathways were vagal and which seem to be a part of cough reflex. Transient lung inflation reflexly reduces submucosal gland secretion apparently via a SAR reflex which consists of an inhibition of the tonically active vagal secromotor fibres (Chinn *et al.*, 1981).

Reflexes Originating from the Receptors Innervated by the Non-myelinated Vagal Afferents

In recent years there has been a growing interest in the reflexes mediated via the non-myelinated vagal afferents; indeed, effects previously attributed to RAR stimulation have been re-examined, and partly assigned to the stimulation of the sensory endings of non-myelinated vagal afferents (see for discussion, Paintal, 1977b; Coleridge & Coleridge, 1981). When considering the functional significance of these reflexes it must be remembered that in cats at least, non-myelinated vagal afferents outnumber myelinated ones by 10:1 as they represent about 90 per cent of the total population of vagal afferent fibres from the bronchi and lungs (Jammes *et al.*, 1982). This figure is based on electron-microscopic analysis, and exceeds the 80 per cent figure based on light-microscopy studies (Evans & Murray 1954; Agostoni *et al.*, 1957). In spontaneously breathing dogs about 40 per cent of non-myelinated afferent pulmonary vagal fibres are tonically active, their activity (mean firing rate two impulses per second) being slightly modulated over the respiratory cycle (Coleridge & Coleridge, 1977a, b). The aggregated input from so many non-myelinated vagal afferents constitutes an important, continuous signal to the central nervous system.

Reflexes from the Pulmonary Receptors (J Type Receptors)

Receptors within the lung parenchyma have been called juxtapulmonary capillary receptors or J type receptors as they have been localised close to the pulmonary capillaries in the interstitial tissue between capillary and alveolus (Paintal, 1969, 1970 and Chapter 3). A natural stimulus for them has been shown to be the mechanical distortion produced by an increase in the interstitial volume or pressure when pulmonary capillary pressure rises and when flow of fluid into the pulmonary interstitial space exceeds its outflow (Paintal, 1969). J type receptors are, therefore, mechanoreceptors excited by strong lung deflation (Paintal, 1955) or by hyperinflation (Coleridge *et al.*, 1965; Armstrong & Luck, 1974). The problem is, however, still unsolved (see Paintal, 1973, 1977b) as in intact chest preparation of the cat a moderate lung inflation inhibits the spontaneous activity of the non-myelinated vagal afferents recorded in the nodose ganglion (Delpierre *et al.*, 1981).

Due to the technical difficulty of applying specific, graded stimulation by increasing the pulmonary interstitial pressure, phenyl diguanide, a chemical stimulus, thought to be a specific stimulus for J type receptors (Paintal, 1955, 1969) has been widely used for studying J-receptor reflex (for review see Paintal, 1973, 1977b). Chemosensitivity of a mechanoreceptor is a common feature in many receptors (see Chapter 1 of this volume, also for review Paintal, 1973; Sant'Ambrogio, 1982). J type receptors are excited by several chemicals (see for review, Paintal, 1973) among them by the opiate receptor agonists, the enkephalins (Willette & Sapru, 1982), a finding which may explain the clinical observations on the precipitation of asthmatic attacks in susceptible individuals by morphine (Jaffe & Martin, 1980). The excitatory properties of phenyl diguanide have been found, however, to be species dependent. Phenyl diguanide is a strong stimulant of J receptors in cats and in rabbits but not in dogs, in which capsaicine, active also in cats, exerts an excitatory effect (Coleridge *et al.*, 1965; Coleridge & Coleridge, 1977b). Phenyl diguanide excites arterial chemoreceptors and does not provoke any significant pulmonary reflex in humans whereas lobeline is a more specific stimulant there (Jain *et al.*, 1972). Phenyl diguanide is not a specific stimulus for J type receptors as it has been shown to stimulate also cardiac receptors and to produce reflex effects which mimic those elicited from J type pulmonary receptors (Anand & Paintal, 1980).

In an extensive study Anand & Paintal (1980) correlated the magnitude of the natural stimulus for J type receptors, increased pulmonary blood flow, with the effect of different doses of phenyl diguanide. They looked for a low, 'physiological' dose of the drug, producing J receptor discharge comparable with that induced by increased pulmonary blood flow. They re-examined the reflex effects of J type receptor stimulation, known before largely from the results provided by the high doses of phenyl diguanide. In carefully controlled experimental conditions, after cardiac receptors had been previously blocked by intrapericardial injections of xylocaine, they confirmed earlier results (for review see Paintal, 1973) and found that stimulation of J type receptors caused reflex apnoea followed by rapid shallow breathing and inhibition of the somatic, monosynaptic knee jerk reflex (Deshpande & Devanandan, 1970). Respiratory inhibition was expressed by the reduced total number of phrenic impulses in each breath and decreased mean frequency of discharge per each phrenic burst. Both T_I and T_E were reduced in most of the experiments, a finding in agreement with earlier results by Winning & Widdicombe (1976) and Miserocchi *et al.* (1978). Sometimes reduction in T_E preceded a reduction in T_I. Such dissociation between these two variables would not occur if the recruitment of the SAR reflex was responsible for the induced pattern of breathing. Thus J type receptors appear to have a direct inhibitory input to both central mechanisms controlling the magnitude and duration of inspiratory discharge and the duration of expiratory discharge. The pontine pneumotaxic centre has been shown to be involved in the excitatory frequency response of the J receptor reflex but not in the mechanism of the magnitude of respiratory output (Miserocchi & Trippenbach, 1981). Pronounced respiratory effects are produced by levels of activity in J receptors corresponding to only 133 per cent increase in pulmonary blood flow, in other words the range of cardiac outputs which occur in moderate exercise (Anand & Paintal, 1980). This result supports the original concept of Paintal (1969) that the J reflex plays a role in exercise, not only at high altitudes, but also at sea level.

Intrapulmonary CO_2 stimulates J type receptors in cats (Dickinson & Paintal, 1970; Delpierre *et al.*, 1981) but in dogs this CO_2 excitatory effect is less significant (Coleridge *et al.*, 1978b). The problem, however, appears important as in the last years there has been an increasing interest and revival of an early concept that ventilation may be reflexly increased via vagal endings in the lungs sensitive to CO_2. The hypothesis is based on the correlation between PCO_2 in the mixed venous blood entering pulmonary circulation and magnitude of ventilation (among others Wasserman *et al.*, 1975; Tallman & Grodins, 1982; Sheldon & Green, 1982). The pulmonary-CO_2 chemoreflex has been shown in birds (for review see Bouverot, 1978); while in dogs with cardiopulmonary bypass an increase in PCO_2 in the vascularly isolated lungs augments both breathing frequency and minute ventilation, an effect abolished by bilateral vagotomy (Bartoli *et al.*, 1974). Similar results have been obtained by Banzett *et al.* (1978) who found a significant reflex increase in the magnitude and rate of phrenic nerve discharge when a vascularly isolated lung was separately ventilated with increasing PCO_2/oxygen mixtures. The effect was significant in the hypocapnic range of PCO_2 values. There is no clear understanding of which receptors and afferent fibres might mediate the pulmonary CO_2-ventilatory reflex. Bradley *et al.* (1976) have shown an inhibitory effect of increasing PCO_2 upon SAR discharge and suggested that a reduced responsiveness of the inhibitory SAR reflex could fully account for the pulmonary CO_2-ventilatory reflex. Mitchell *et al.* (1980) have shown in dogs that an end-tidal CO_2 of up to 2 per cent CO_2, but no more (i.e. by pocapnia) in an isolated and artificially ventilated lung, produced a decrease in SAR activity and a parallel increase in the phrenic nerve integrated activity. They concluded that the reflex effect of CO_2, only over the hypocapnic range of values, is brought about by inhibition of SAR activity and attenuation of the inhibitory SAR reflex; thus the pulmonary CO_2 reflex would be a consequence of disinhibition of the central respiratory drive. A similar interpretation had been proposed previously by Coleridge *et al.* (1978b). To support this conclusion they found that vagal cooling to $7°-8\ °C$, blocking traffic in the myelinated fibres, abolished the reflex response (Banzett *et al.*, 1978). As they pointed out, however, frequency of discharge in non-myelinated vagal afferents may be also impaired by that range of temperature, and thus some contribution by non-myelinated fibres could not be excluded. The latter possibility has been supported by the recent finding of Raybould & Russel (1982), that an increased respiratory frequency in hypercapnic rabbits was not abolished by differential d.c. anodal block of the myelinated vagal fibres, (known to leave intact the traffic in the non-myelinated fibres, Guz & Trenchard, 1971b). Their conclusion is consistent with that of Bartoli *et al.* (1974), who also suggested that CO_2 exerts an excitatory influence on some pulmonary receptors responsible for the reflex hyperventilation and is also consistent with the results of Delpierre *et al.* (1981) who found increased activity in the non-myelinated vagal afferents recorded in the nodose ganglion following increasing end-tidal CO_2 concentration. This extended over the hypercapnic range of CO_2 concentration in some fibres.

It appears, therefore, that both mechanisms, CO_2 induced inhibition of the SAR inhibitory reflex with consequent respiratory disinhibitions and stimulation of J-receptors with increasing PCO_2 might contribute to the pulmonary-CO_2 reflex.

Reflex bronchoconstriction mediated by non-myelinated vagal afferents. Stimulation of J type receptors in rabbits by phenyl diguanide produces a pronounced reflex bronchoconstriction (Karczewski & Widdicombe, 1969b); similar effects have been shown in dogs following stimulation of J type receptors by capsaicine injected into the right ventricle (Russel & Lai-Fook, 1979). The vagally mediated bronchoconstrictor effect is maintained in a tonic fashion as a differential cold block of vagal myelinated fibres provokes a persistent bronchoconstriction in rabbits (Karczewski & Widdicombe, 1969c), and tracheal constriction in dogs (Roberts *et al.*, 1981), apparently due to the withdrawal of the antagonistic, tonic bronchodilator influence of the SAR reflex (see above). Only a further drop in temperature to $2\ °C$, sufficient to block conduction in non-myelinated vagal fibres, decreases reflexly the tension of tracheal smooth muscle (Roberts *et al.*, 1981).

Jammes & Mei (1979) provided convincing evidence that total, resting, bronchoconstrictor tone is maintained reflexly by the activity of non-myelinated vagal afferents. Vagal blockade with procaine, which preferentially blocks non-myelinated afferent fibres, produced a bronchodilatation and a decrease in the airway resistance identical with that following total, bilateral, sensory vagotomy. Myelinated vagal fibres mediating the RAR bronchoconstrictor reflex were not blocked by procaine and the RAR reflex could be provoked. These results clearly indicate that RARs do not play any significant role in the maintenance of resting bronchomotor tone.

The bronchoconstrictor effect of hypercapnia has been shown to be of vagal reflex origin as sensory vagotomy at the level of the nodose ganglion, or block of the non-myelinated vagal afferents by procaine, abolished it whereas anodal block of myelinated vagal fibres was ineffective (Delpierre *et al.*, 1980). The stimulatory effect of CO_2 upon the pulmonary and/or bronchial receptors innervated by the non-myelinated vagal afferents has been suggested as the mechanism of CO_2 induced bronchoconstriction (Delpierre *et al.*, 1981).

Reflexes from Bronchial Receptors Mediated by Non-myelinated Vagal Afferents

Coleridge & Coleridge (1977a) identified a separate group of sensory endings innervated by bronchial non-myelinated afferent vagal fibres and situated in the walls of the intrapulmonary airways. They differ from the pulmonary, type J receptors as they are supplied by the bronchial (systemic) circulation and not by the pulmonary circulation. Although the reflex effects following stimulation of either type of receptors are similar – tachypnoea and bronchoconstriction – there are some differences due to their different location and to a greater chemosensitivity of the bronchial receptors. Bronchial receptors are less sensitive to mechanical stimuli, lung deflation, hyperinflation or congestion, and they exhibit a pronounced and specific sensitivity to bradykinin which is ineffective in the stimulation of type J or rapidly adapting receptors (Kaufmann *et al.*, 1980). Bradykinin injected into the bronchial artery provokes a reflex tracheal contraction mediated by the bronchial non-myelinated fibres (Roberts *et al.*, 1981). Inhalation of bradykinin aerosol in humans, especially in asthmatic patients, causes a reflex bronchoconstriction which may be abolished by atropine (Simonsson *et al.*, 1973). Bradykinin has been classified as one of the 'lung autocoids' (Brockelhurst, 1976) which, together with histamine, the prostaglandins and other chemicals, are formed and released locally during pulmonary anaphylaxis and asthma (see Kellemeger & Schwartz, 1976). Bronchial endings of non-myelinated vagal afferents have been shown to be very sensitive to 'lung autocoids' (Coleridge *et al.*, 1978a). Thus non-myelinated bronchial vagal afferents appear as a link in the reflex mechanism of lung anaphylaxis, asthma and inflammation, involved in the reflex bronchoconstriction. The important role of non-myelinated rather than myelinated afferents in the mechanism of the respiratory reflex responses to lung anaphylaxis or to inflammation in rabbits has been already suggested by Karczewski & Widdicombe (1969a), who applied differential vagal cold block and by Guz & Trenchard (1971b) and Trenchard, Gardner & Guz, 1972) who used d.c. anodal block to disrupt the traffic in the myelinated vagal afferents. The degree of cold block of the vagus nerves which is needed to prevent the respiratory response in conscious dogs — rapid shallow breathing — to the inhalation of histamine aerosol is greater than that which blocks merely thick myelinated fibres (Bleecker *et al.*, 1976) and hence may also impair conduction in non-myelinated afferents (for discussion see Kaufman *et al.*, 1980).

Functional Significance of the Reflexes Originating from the Sensory Endings of Non-myelinated Vagal Fibres
Afferent vagal input gives rise to respiratory sensations as dyspnoea (breathlessness) occurring in patients with lung diseases and unpleasant sensations (tickling and burning), referred to the airways and may be provoked by inhaled irritants or by chemicals of endogenous origin. Local anaesthesia of the vagal trunk (Guz *et al.*, 1970) or vagotomy (Bradley *et al.*, 1982) have been shown to bring a relief to human patients. The excitation of both RARs and the endings of non-myelinated vagal afferents have been considered as the source of these unpleasant sensations. Their relative contribution and importance are still very little known. On the ground of indirect evidence Paintal (1977a) refuted the role of RARs in the mechanism of respiratory sensations and assigned it to the sensory endings of non-myelinated vagal fibres, in particular to J type receptors. Widdicombe (1974) suggested that the distress of breathing in extreme or pathological conditions is due to a multifactorial system. There may be an analogy with nocioceptive cutaneous sensation, where pain is mediated by sensory endings of non-myelinated and myelinated nerve fibres. In pathological conditions usually both kinds of receptors are simultaneously excited and, therefore, the interaction of different heterogeneous inputs appears as the most common situation associated with respiratory

sensations. There are, however, some conditions facilitating stimulation of only one type of receptors. Physical exercise and increased cardiac output resulting in the increase in pulmonary capillary pressure has been proposed as a specific situation for the stimulation of J type receptors. Thus the J receptor reflex has been suggested as a reflex of exercise (Paintal, 1970), responsible for dyspnoeic sensations both in exercise and in pathological conditions involving the pulmonary circulation (Paintal, 1973, 1977a). The J receptor reflex, by its inhibitory influence upon the somatic monosynaptic reflexes of both flexor and extensor muscles, (Deshpande & Devanandan, 1970), has been regarded as a kind of brake, causing cessation of the exercise itself (Paintal, 1969, 1973, 1977a). More recently, however, Paintal and his associates (Anand *et al.*, 1982) pointed out a facilitatory influence of the J receptor reflex upon ventilation and suggested that it may contribute to exercise hyperpnoea and possibly mediate the pulmonary-CO_2 ventilatory reflex (see above). According to this recent suggestion, pulmonary receptors of non-myelinated vagal afferents would play a role of positive feedback augmenting ventilation in exercise and in other conditions in which pulmonary capillary pressure increases. The full physiological significance of such a mechanism remains, however, still to be elucidated.

In contrast, the SAR reflex seems to produce a relief of dyspnoeic sensations, presumably by its inhibitory influence upon ventilation. It has been shown that differential vagal block of the SAR reflex increases the unpleasant sensations of dyspnoea on breathing CO_2-enriched mixtures (Cross *et al.*, 1976).

Reflexes mediated by non-myelinated vagal afferents constitute an important mechanism for the respiratory defence responses in various pathophysiological conditions (see Coleridge & Coleridge, 1981). The typical protective reflex pattern — rapid shallow breathing and bronchoconstriction — accompanies most of the pathophysiological changes (e.g. oedema, pulmonary congestion, embolism) within the lungs and pulmonary circulation. Asthmatic attacks provoked by exercise in asthmatic patients, as well as chronic bronchoconstriction in patients with congestive heart failure and increased pulmonary interstitial volume, have been related to J type receptor stimulation (Paintal, 1977a).

Specific chemosensitivity of bronchial receptors of non-myelinated vagal afferents to endogenous chemicals, 'lung autocoids', released in the lungs in anaphylaxis and by local irritation supports the suggestion on the role of the reflexes mediated by non-myelinated bronchial vagal afferents in pulmonary anaphylaxis and in asthma (Roberts *et al.*, 1981) as well as in other pathological conditions including unpleasant sensations (Coleridge & Coleridge, 1981). The cough reflex in humans has also been attributed to the pulmonary receptors innervated by non-myelinated vagal afferents (Paintal, 1977b) as lobeline, a stimulant for them, provokes cough on its passage through the pulmonary circulation (Jain *et al.*, 1972). Recent years have seen considerable discussion and controversy (see above and Paintal, 1977a, b; Coleridge & Coleridge, 1981) over the contribution of the reflexes mediated by thin myelinated vagal fibres to the mechanism of the respiratory defence responses. To the present author it seems reasonable to suggest that the experimental data are not mutually exclusive and that an interaction of both kinds of reflexes rather than a separate reflex pathway, determines the pattern of the defence responses from the lungs and tracheobronchial tree. The RAR reflex could play a role in initiating the response, whereas the reflexes from the pulmonary and bronchial receptors, mediated by non-myelinated vagal afferents, appear to be involved in the maintenance of long-lasting effects.

Circulatory Reflexes from the Lungs

The first report on a circulatory reflex response to lung inflation appeared almost as early as the discovery of the Hering–Breuer reflex and dates back to 1871 when Hering published his observation on the cardioacceleration produced by moderate lung inflation and on decrease in heart rate following hyperinflation of the lungs (Hering, 1871). This observation was confirmed and extended by Anrep, Pascual & Rossler (1936a) who showed that reflex increase in the heart rate was due to a withdrawal of cardio-vagal tone. Lung deflation produced reflex bradycardia (Anrep *et al.*, 1936a; Daly & Scott, 1958). Lung inflation in the eupnoeic range is also accompanied by the peripheral vasodilatation (Salisbury *et al.*, 1959; Daly *et al.*, 1967; Mancia, Shepherd & Donald, 1975) caused by the reflex withdrawal of sympathetic vasoconstrictor tone (Daly & Robinson, 1968; Glick, Wechster & Epstein, 1969; Mancia *et al.*, 1973). Reflex inhibition of the sympathetic activity is integrated at the spinal level and is not secondary to the suppression of inspiration, as a short-lasting lung inflation delivered during expiration brings about a hyperpolarisation of single sympathetic preganglionic spinal neurones (Lipski, Coote & Trzcbski, 1977) and reduces spontaneous activity in the efferent splanchnic and cervical sympathetic nerves (Gootman, Feldman & Cohen, 1980).

There is a serious difficulty in studying circulatory reflexes from the lungs in isolation from the whole cardiopulmonary area for both areas receive a widespread innervation from the same group on non-myelinated vagal fibres and reflexes from the ventricles, atria and lungs have a similar pattern (for review see Thoren, Donald & Shepherd, 1976; Thoren, 1979). Most of the experimental procedures used in the study of pulmonary reflexes produce secondary mechanical effects on the heart and pulmonary circulation, changing the pulmonary blood flow, the pressures in the heart chambers and thus affecting reflexogenic areas outside the lungs. On the other hand, any change in the ventricular or atrial pressure due to changes in the central blood volume, in addition to eliciting reflex responses from carciac receptors, may influence pulmonary blood flow, change pulmonary capillary pressure and influence J receptors. The difficulty in separating the reflexogenic areas of the lung, heart and pulmonary vessels, and the similarity of the respective reflexes mediated by non-myelinated vagal afferents, led to a concept that cardiopulmonary receptors may be thought of as being functionally united (see for review Thoren, 1979).

The association of the information from the cardiovascular and respiratory apparatus transmitted to the central nervous system is even more direct as the same single afferent fibre may be used. Thus it has been shown that discharges in large, myelinated fibres mediating the SAR reflex are rhythmically modulated over the cardiac cycle by mechanical deformation of mechanosensitive endings by each cardiac contraction or vascular pulsation within the thorax (Adrian, 1933; Whitteridge, 1948; Pearce & Whitteridge, 1951; Mei, 1970). More detailed computer analysis of such signals revealed that almost all SAR afferents transmit information synchronous with cardiac function (Lipski, Trzebski & Merrill, 1981). The significance of this rhythmic circulatory input for the central processing of the SAR respiratory reflex is unknown.

Mancia & Donald (1975) compared the contribution of lung, cardiac ventricle and atrial receptors separately to the tonic reflex inhibition of the vasomotor tone in dogs maintained on cardiopulmonary bypass with the carotid sinus and aortic receptors denervated. Bilateral vagal cold block was applied following removal of the heart, leaving the ventilated lungs, or following removal of the ventricles together with lungs, leaving only the atria, or following the removal of lungs, leaving the ventricles and denervated atria. Cold vagal block produced comparable arterial blood pressure rises following each of these procedures indicating that heart and lung receptors were equally responsible for the tonic inhibition of the sympathetic vasoconstrictor tone. The inhibitory influence from the cardiopulmonary receptors is exerted on the resistance vessels in skeletal muscle, kidney, intestine and on the capacitance splanchnic vessels but not on the cutaneous veins (Thoren *et al.*, 1976; see for review Shepherd, 1981). The reflex tonic inhibition exerted separately by the heart and lung receptors on the hindlimb and renal vascular resistance is about the same (Mancia *et al.*, 1975). Pulmonary artery baroreceptors and respective reflexes (Coleridge & Kidd, 1963) may also have some contribution to the depressor reflex from the whole lung (Lloyd, 1977).

There has been much less information on the types of receptors involved in the circulatory than in the respiratory reflexes from the lung. The question of which receptors and afferent fibres mediate cardiovascular reflex responses is largely unsolved. Reflex withdrawal of the cardiac vagal tone and cardioacceleration follows moderate lung inflation in the range of resting tidal volume and apparently is mediated by SARs (Angell-James & Daly, 1978; Angell-James, Elsner & de Durgh Daly, 1981). A characteristic feature of Hering–Breuer respiratory inflation reflex is its slow adaptation. In contrast, the cardiac response is inflation rate sensitive and the reflex 'escapes' with static lung inflation (Gandevia *et al.*, 1980a; Trzebski, 1980a). This difference need not imply separate receptors for each reflex but may depend on different central organisation of both reflexes (for discussion see Gandevia *et al.*, 1980a). The pulmonary-cardiac reflex could possibly involve preferentially rapidly adapting I-beta inspiratory neurones (Marino *et al.*, 1981) excited by lung inflation. I-beta inspiratory neurones have already been suggested as the central neurones mediating reflex inhibition of the cardiac vagal tone (Lopes & Palmer, 1976).

Decrease in the heart rate following hyperinflation of the lung is presumably due to the stimulation of high-threshold lung mechano-receptors: RARs, pulmonary J type or bronchial receptors. Phenyl diguanide provokes bradycardia and hypotension (Paintal, 1955) but this effect appears to be related to cardiac receptors (Anand & Paintal, 1980). Capsaicine, another stimulant of J type and bronchial receptors, produces also reflex bradycardia and hypotension (Coleridge *et al.*, 1965). Pathological situations, known to stimulate pulmonary J type receptors, like microembolism or lung congestion bring about a reflex bradycardia and hypotension (for review see Paintal, 1973; Widdi-combe, 1974).

The threshold of the pulmonary-vasodepressor tonic reflex is very low as the reflex inhibition of the sympathetic vasomotor tone is still present during apnoea in dogs with their lungs collapsed at atmospheric pressure (Mancia *et al.*, 1975), a finding suggesting that low-threshold SARs are involved. Daly & Robinson (1968) found that the reflex vasodilator response to lung inflation in dogs was abolished together with the Hering–Breuer reflex by bilateral vagal cooling to 8 °C and concluded that thick myelinated fibres from SARs mediated both reflex responses. Yet, in rabbits with the carotid and sinus nerves cut cooling of the vagi below 6 °C down to 0 °C and blocking of the traffic in non-myelinated afferents produced a further increase of the systemic blood pressure reaching about 40 per cent of the total pressor response. D.C. anodal block of myelinated vagal fibres brings about a pressor response in cats of only 20 per cent of the magnitude of the response obtained by interruption of all afferent vagal traffic by deep cooling (Thoren *et al.*, 1977). Thus a major part of the tonic inhibition of the sympathetic vasomotor tone depends on the non-myelinated vagal afferents, a finding consistent with the tonic, irregular activity of the pulmonary and bronchial endings of the non-myelinated fibres (Coleridge & Coleridge, 1977a, b). However, in these experiments all non-myelinated vagal afferents from the whole cardiopulmonary area were blocked and no distinction was made between the heart and the lung receptors.

The problem of which type of tracheobronchial or pulmonary receptors have to be assigned to a particular cardiovascular reflex response is still far from being solved. It has been shown by applying lung inflations of different magnitude that reflex circulatory responses may be dissociated from the respiratory responses (Hainsworth, 1974). There might be a group of receptors involved more specifically in the circulatory reflexes from the lungs but, more probably, the dissociation may be accounted for by a different central organisation of the input from the same receptors, processed separately for the respiratory and cardiovascular responses. The last possibility deserves more research as the processing of information from the lungs by the central neurones involved in the circulatory control is so far an almost unexplored area.

There is little known on the circulatory reflexes from the lungs in humans. Stimulation of vagal afferents in conscious human subjects brings about a fall of the arterial blood pressure which seems to be, however, secondary to mechanical changes within the thorax and to impairment of venous return (Guz *et al.*, 1964). Lung inflation in anaesthetised patients with cardiopulmonary bypass did not affect heart rate or total peripheral resistance (Ott, Tarhan & McGoon, 1975). In human patients during sleep apnoea a pronounced bradycardia has been shown to be abruptly reversed to reflex tachycardia with lung inflation (Meacock, 1982). Drugs known to stimulate pulmonary receptors, phenyl diguanide or lobeline, have been shown to produce cardiac slowing and fall of the arterial blood pressure in human subjects, a response attributed to the stimulation of J type pulmonary receptors. This mechanism appears, however, to be limited only to lobeline which produces a reflex bradycardia and apnoea on its passage through pulmonary circulation in man (Jain *et al.*, 1972).

Functional Significance of the Circulatory Reflexes from the Lungs

Early studies (Anrep *et al.*, 1936b) indicated that the pulmonary cardiac reflex constitutes a reflex mechanism of respiratory sinus arrhythmia, which, in addition to central inhibition of vagal tone by inspiratory activity, withdraws rhythmically the cardiac vagal tone and increases the heart rate during inspiration.

The functional significance of the respiratory sinus arrhythmia, related to resting vagal tone (Katona & Jih, 1975), consists apparently in the rhythmical adjustment of the cardiac output to the increase in the venous return following each inspiration.

Sinus arrhythmia, marked in children, is attentuated although present in adults, perhaps in relation to the reduced responsiveness of the SAR reflex with maturation (see above), but such a possibility has not been systematically checked. Whatever the central mechanism, the reflex threshold of the respiratory sinus arrhythmia in humans is apparently below that of the Hering–Breuer respiratory reflex, which is not active over the range of resting tidal volume.

Melcher (1976, 1980) questioned the view that lung inflation in humans contributes to inspiratory cardioacceleration and presented indirect evidence that reflex increase in the heart rate originates in the low-pressure cardiac mechanoreceptors stimulated by increased filling of the right atrium during thoracic expansion in inspiration.

The mechanism of respiratory sinus arrhythmia in humans is evidently complex and centrally mediated as it is not entirely abolished even in the absence of any lung or thorax expansion or any expressed respiratory activity during voluntary apnoea (Trzebski *et al.*, 1980b). A rhythmical decrease of the responsiveness of the baroreceptor cardiac vagal reflex, accompanying each inspiration has been shown. both in animals (Koepchen, Wagner & Luz, 1961; Haymet & McCloskey, 1975; Neil & Palmer, 1975; Davidson, Goldner & McCloskey, 1976; Potter,

1981) and in humans (Eckberg & Orsham, 1977; Eckberg *et al.*, 1980; Trzebski *et al.*, 1980b, c). This mechanism contributes also to respiratory sinus arrythmia in humans (Trzebski *et al.*, 1980b; Melcher, 1980).

Tonic influences from lung receptors contribute to the cardiopulmonary reflex inhibition of vasoconstrictor sympathetic activity (Thoren *et al.*, 1976; Thoren, 1979) complementing the arterial baroreceptor reflexes in the control of the heart rate and blood pressure (Vatner *et al.*, 1975; Mancia *et al.*, 1976; Guo, Thames & Abboud, 1982). A possible contribution of pulmonary receptors to the control of blood volume exerted via reflex inhibition of ADH and renin secretion (Kiowski & Julius, 1978; Donald, 1979; Thames & Schmid, 1979, 1981) has not been checked. Lung interstitial volume and J receptor activity appear to be sensitive detectors of the volume load imposed upon the pulmonary circulation (Anand & Paintal, 1980) and their involvement in the neurohormonal reflex control of central blood volume seems an attractive hypothesis. So far, however, reflex control of ADH and renin secretion has been analysed in terms of total vagal input from the cardiopulmonary area without separating the input from lung receptors.

Pulmonary Reflexes as a Link in the Cardiovascular-Respiratory Reflex Interactions

A major role of the lung inflation circulatory reflex consists in the modulation of the other circulatory reflexes in such a way as to adjust the final pattern of integrated response to the increase in pulmonary ventilation and in oxygen demand. Lung inflation elicits a vagally mediated inhibition of the baroreceptor cardio-inhibitory reflex. The effect is related to the rate of inflation and exhibits a pronounced adaptation (Guandevia *et al.*, 1978a). Reflexly evoked discharge in the efferent vagal cardiac fibres is inhibited, but tonic vagal activity is less sensitive to lung inflation. In contrast, the central inspiratory activity inhibits both reflex and tonic cardiac vagal activity (Potter, 1981).

Inhibition of cardiodepressor reflexes by lung inflation seems to be a general phenomenon as a similar inhibitory effect on other cardiac vagal responses provoked by ocular pressure, trigeminal nasopharyngeal receptor stimulation or immersion of the face ('diving') in human subjects have been shown (Angell-James & Daly, 1978; Gandevia *et al.*, 1978b; Angell-James *et al.*, 1981).

The most significant effect for physiological adjustment seems to be the pulmonary inflation dependent inhibition of the cardio-inhibitory and vasoconstrictor components of the arterial chemoreceptor reflex (Daly & Scott, 1958, 1962; Daly & Hazzledine, 1963; Scott, 1966; Angell-James & Daly, 1969; Daly, 1972; Haymet & McCloskey, 1975; Rutheford & Vatner, 1978; Gandevia *et al.*, 1978a; Angell-James *et al.*, 1981; Potter, 1981). In conscious dogs the pulmonary inflation reflex attenuates or reverses chemoreceptor-induced sympathetic vasoconstrictor response in the iliac, renal, mesenteric and cerebral vascular beds and significantly enhances reflex vasodilator response in the coronary circulation (Vatner & Rutheford, 1981). The lung inflation reflex also counteracts the centrally mediated vasoconstrictor effect of CO_2, especially in the kidney (Ott, Lorenz & Shepherd, 1972; Ott & Shepherd, 1975).

During systemic hypoxia, accompanied by augmented ventilation, the pulmonary inflation reflex and central inspiratory activity constitute two mechanisms which work together in overriding the primary chemoreceptor reflex. Both mechanisms, reflex and central, are responsible for the reversal of primary bradycardia to tachycardia in dogs and in conscious rabbits (Daly & Scott, 1958; Crocker *et al.*, 1968; Korner *et al.*, 1973) and for the attenuation of the reversal of the reflex vasoconstriction to vasodilatation (Kontos *et al.*, 1967; Daly, 1972; Korner, 1980). The pattern of the integrated cardiovascular response to hypoxia is, therefore, reflexly adjusted to match augmented ventilation with the appropriate increase in cardiac output and with a vasodilatation to facilitate oxygen supply to the tissues.

In an entirely different physiological situation when hypoxia or asphyxia are accompanied by reflex cessation of breathing (e.g. during diving and during stimulation of trigeminal and upper airway receptors which override the excitatory effect of chemoreceptors upon breathing, inhibiting central inspiratory activity, facilitating expiration, and bringing the lungs to deflation) an uninhibited primary chemoreceptor reflex is fully expressed (Angell-James & Daly, 1969, 1978; Daly, 1972; Daly & Angell-James, 1975; Daly *et al.*, 1978; Angell-James *et al.*, 1981). Lung deflation, by itself, reflexly facilitates cardiac vagal responses (Angell-James & Daly, 1969, 1978; Haymet & McCloskey, 1975; Gandevia *et al.*, 1978a). The pattern of the uninhibited primary chemoreceptor induced bradycardiac response and predominant vasoconstriction, except of coronary, cerebral and some cutaneous vessels, has a potential physiological advantage by conserving blood oxygen and redistributing blood flow and oxygen supply preferentially to the vital organs. This response has been regarded as the pattern of circulatory adjustment to diving (Daly & Angell-James, 1975).

Thus reflexes from the lung mechanoreceptors constitute an

important mechanism of cardiovascular-respiratory interactions determining, together with $PaCO_2$ level (Daly & Scott, 1962; Scott, 1966) a switch from the oxygen supplying to oxygen conserving pattern of the integrated reflex response.

The lung inflation reflex seems to play a similar role in humans. It has been shown that during sleep apnoea progressive asphyxia provoked slowing of the heart rate which was abruptly reversed to tachycardia with resumption of lung inflation. This effect is due to the pulmonary inflation reflex as it was abolished in a patient who had undergone bilateral pulmonary vagotomy (Meancock, 1982). Similar observations have been made in quadriplegic patients during cessation and subsequent return of mechanical ventilation. Pathophysiological situations involving irritation of trigeminal and upper airway receptors and apnoeic asphyxia may cause pronounced bradycardia and cardiac arrest in the absence of the inhibitory modulation of the arterial chemoreceptor reflex by lung inflation and by central inspiratory activity. This mechanism seems to be responsible for the causes of severe bradycardia and cardiac arrest during underwater swimming and in a number of clinical situations (e.g. sleep apnoea syndrome, pickwickian syndrome, apnoea-stridor-bradycardia syndrome, adenoidal-hypertrophy-pulmonary-hypertension syndrome in children, accidents including intubations, bronchoscopy or laryngoscopy — for reference see Daly *et al.*, 1979).

There appears to be some species difference in the magnitude of the two main mechanisms inhibiting arterial chemoreceptor reflex: the pulmonary inflation reflex and the central inspiratory activity (Korner, 1980). The former is predominant in subprimate species like the rabbit and in diving animals like seals (Angell-James *et al.*, 1981), yet less important in dogs, the latter seems to play a major role in the primate (Daly *et al.*, 1978).

In summary, circulatory reflexes from the lungs serve two main functions: a tonic one, similar to other cardiopulmonary baroreflexes regulating to arterial blood pressure and blood volume, and a phasic one, modulating other reflexes involved in cardiovascular-respiratory reflex interactions.

References

Adrian, E.D. (1933) 'Afferent Impulses in the Vagus and their Effect on Respiration', *J. Physiol. (Lond.)*, *79*, 332-58

Agostoni, E., Chinnock, J.E., Daly, M. De B. & Marray, J.G. (1957) 'Functional and Histological Studies of the Vagus Nerve and its Branches to the Heart, Lungs and Abdominal Viscera in the Cat', *J. Physiol. (Lond.)*, *135*, 182-205

Anand, A., Loeschcke, H.H., Marek, W. & Paintal, A.S. (1982) 'Significance of the Respiratory Drive by Impulses from J Receptors', *J. Physiol. Lond.*, *325*, 14P

Anand, A. & Paintal, A.S. (1980) 'Reflex Effects Following Selective Stimulation of J Receptors in the Cat', *J. Physiol. (Lond.)*, *299*, 553-72

Angell-James, J.E. & Daly, M. de B. (1969) 'Cardiovascular Responses in Apnoeic Asphyxia: Role of Arterial Chemoreceptors and the Modification of their Effects by Pulmonary Vagal Inflation Reflex', *J. Physiol. (Lond.)*, *201*, 87-104

Angell-James, J.E. & Daly, M. de B. (1978) 'The Effects of Artificial Lung Inflation on Reflexly Induced Bradycardia Associated with Apnoea in the Dog', *J. Physiol. (Lond.)*, *274*, 349-66

Angell–James, J.E., Elsner, R. & de Durgh Daly, M. (1981) 'Lung Inflation: Effects on Heart Rate, Respiration and Vagal Afferent Activity in Seals', *Am. J. Physiol.*, *240*, H190-8

Anrep, G.V., Pascual, W. & Roessler, R. (1936a) 'Respiratory Variations of the Heart Rate. I The Reflex Mechanism of the Respiratory Arrhythmia', *Proc. R. Soc. Biol.*, *119*, 191-217

Anrep, G.V., Pascual, W. & Roessler, R. (1936b) 'Respiratory Variations of the Heart Rate. II The Central Mechanism of the Respiratory Arrhythmia and the Interrelations between the Central and Reflex Mechanism', *Proc. R. Soc. Biol.*, *119*, 219-30

Armstrong, D.J. & Luck, J.C. (1974) 'A Comparative Study of Irritant and Type J Receptors in the Cat', *Resp. Physiol.*, *21*, 47-60

Arita, H. & Bishop, B. (1982) 'Responses of Internal Intercostal Muscles of Cat to Hypercapnia and Lung Inflation', in J.L. Feldman & A.J. Berger (eds), *International Symposium Central Neural Production of Periodic Respiratory Movements*, Harrison Conference Center, Lake Bluff, Ill. pp. 120-1

Banzett, R.B., Coleridge, H.M. & Coleridge, J.C.G. (1978) 'I Pulmonary – CO_2 Ventilatory Reflex in Dogs: Effective Range of CO_2 and Results of Vagal Cooling', *Resp. Physiol.*, *34*, 121-34

Barillot, J.C. & Bianchi, A.L. (1971) 'Activité de Motoneurones Laryngées Pendant les Reflexes de Hering-Breuer', *J. Physiol. (Paris)*, *63*, 783-92

Bartlett, D. Jr. (1980) 'Effects of Vagal Afferents on Laryngeal Responses to Hypercapnia and Hypoxia', *Resp. Physiol.*, *42*, 189-98

Bartlett, D. Jr., Jeffery, P., Sant'Ambrogio, G. & Wise, J.C.M. (1976) 'Location of Stretch Receptors in Trachea and Bronchi of the Dog', *J. Physiol. Lond.*, *258*, 409-20

Bartlett, D. Jr., Knuth, S.L. & Knuth, K.K. (1981) 'Effects of Pulmonary Stretch Receptor Blockade on Laryngeal Responses to Hypercapnia and Hypoxia', *Resp. Physiol.*, *45*, 67-77

Bartlett, D. Jr., Remmers, J.E. & Gauthier, H. (1973) 'Laryngeal Regulation of Respiratory Airflow', *Resp. Physiol.*, *18*, 194-204

Bartlett, D. & Sant'Ambrogio, G. (1976) 'Effects of Local and Systemic Hypercapnia on the Discharge of Stretch Receptors in the Airway of the Dog', *Resp. Physiol.*, *26*, 91-9

Bartlett, D. Jr., Sant'Ambrogio, G. & Wise, J.C.M. (1976) 'Transduction Properties of Tracheal Stretch Receptors', *J. Physiol. (Lond.)*, *258*, 421-32

142   Respiratory Reflexes

Bartoli, A., Bystrzycka, E., Guz, A., Jain, S.K., Noble, M.I.M. & Trenchard, D. (1973) 'Studies of the Pulmonary Vagal Control of Central Respiratory Rhythm in the Absence of Breathing Movements', *J. Physiol. (Lond.), 230*, 449-65

Bartoli, A., Cross, B.A., Guz, A., Jain, S.K., Noble, M.I.M. & Trenchard, D. (1974) 'The Effect of Carbon Dioxide in the Airways and Alveoli on Ventilation: a Vagal Reflex Studied in the Dog', *J. Physiol. (Lond.), 240*, 91-109

Bartoli, A., Cross, B.A., Guz, A., Huszczuk, A. & Jefferies, R. (1975) 'The Effect of Varying Tidal Volume on the Associated Phrenic Motoneurone Output: Studies of Vagal and Chemical Feedback', *Resp. Physiol., 25*, 135-55

Berger, A.J. (1977) 'Dorsal Respiratory Group Neurones in the Medulla of Cat: Spinal Projections, Responses to Lung Inflation and Superior Laryngeal Nerve Stimulation', *Brain Res., 135*, 231-45

Bergren, D.R. & Sampson, S.R. (1982) 'Characterization of Intrapulmonary, Rapidly Adapting Receptors of Guinea Pigs', *Resp. Physiol., 47*, 83-95

Bleecker, E.R., Cotton, D.J., Fischer, S.P., Graf, P.D., Gold, W.M. & Nadel, J.A. (1976) 'The Mechanism of Rapid, Shallow Breathing after Inhaling Histamine Aerosol in Exercising Dogs', *Am. Rev. Resp. Dis., 114*, 909-16

Bodegard, G., Schwieller, G.H., Skogland, S. & Zetterstrom, R. (1969) 'Control of Respiration in Newborn Babies. I The Development of the Hering–Breuer Inflation Reflex', *Acta Paediatr. Scand., 58*, 567-71

Bohn, D.J., Migusaka, K., Marchak, B.E., Thompson, W.K., Froese, A.B. & Bryan, A.C. (1981) 'Ventilation by High-Frequency Oscillation', *J. Appl. Physiol. Resp. Environ Exercise Physiol., 48*, 710-6

Bouverot, P. (1978) 'Control of Breathing in Birds Compared with Mammals', *Physiol. Rev., 58*, 604-55

Bouverot, P., Crance, P.J. & Dejours, P. (1970) 'Factors Influencing the Intensity of the Breuer–Hering Inspiration-Inhibiting Reflex', *Resp. Physiol., 8*, 376-84

Bradley, G.W. (1977) 'Control of Breathing Pattern', in J.G. Widdicombe (ed.) *Inter. Rev. Physiol. Respiratory Physiol., II*, University Park Press, Baltimore, pp. 185-217

Bradley, G.W., Euler, C. von, Martilla, I. & Roos, B. (1975) 'A Model of the Central and Reflex Inhibition of Inspiration in the Cat', *Biol. Cybernetics, 19*, 105-16

Bradley, G.W., Hale, T., Pimble, J., Rowlandson, R. & Noble, M.I.M. (1982) 'Effect of Vagotomy on the Breathing Pattern and Exercise Ability in Emphysematous Patients', *Clin. Sci., 62*, 311-9

Bradley, G.W., Noble, M.I.M. & Trenchard, D. (1976) 'The Direct Effect on Pulmonary Stretch Receptor Discharge Produced by Changing Lung Carbon Dioxide Concentration in Dogs on Cardiopulmonary Bypass and its Action on Breathing', *J. Physiol. (Lond.), 261*, 359-73

Breuer, J. (1868) 'Die Selbststeuerung der Atmung durch den Nervous vagus', *Sitzungsberichte d. Wien. Akad. Wissensch. math-naturwiss. Kl., 58*, 909-37

Brockelhurst, W.E. (1976) 'Pharmacodynamics and Mechanisms of Asthma' in E.B. Weiss & M.S. Segal (eds), *Bronchial Asthma*, Little, Brown, Boston, MA, pp. 117-36

Bruce, E.N., Euler von C. & Yamashira, S.M. (1979) 'Reflex and Central Chemoceptive Control of the Time Course of Inspiratory Activity', in C. von Euler & H. Langecrantz (eds), *Central Nervous Control Mechanisms in Breathing*, Pergamon Press, Oxford, pp. 177-84

Budzinska, K., Karczewski, W.A., Naslonska, E. & Romaniuk, J.R. (1979) 'Poststimulus Effects and Their Possible Role for Stabilizing Respiratory Output', in *Wener-gren Center Internat. Symposium Series, vol. 32, Central Nervous Control Mechanisms in Breathing*, Pergamon Press, pp. 115-27

Butler, W.J., Bohn, D.J., Bryan, A.C. & Froese, A.B. (1980) 'Ventilation by High Frequency Oscillation in Humans', *Anesth. Analg. Cleveland*, *59*, 577-84

Bystrzycka, E., Gromysz, H., Huszczuk, A. & Grotek, A. (1972) 'Studies on the Hering-Breuer Inflation and Deflation Reflexes in the Rabbit', *Acta Physiol. Pol.*, *23*, 539-55

Bystrzycka, E.K. & Nail, B.S. (1980) 'CO_2 Sensitivity of Stretch Receptors in the Marsupid Lung', *Resp. Physiol.*, *39*, 111-19

Callannan, D., Dixon, M. & Widdicombe, J.G. (1975) 'The Acute Effects of SO_2 on Pulmonary Mechanics, Breathing Patterns and Pulmonary Vagal Afferent Receptors in the Rabbit', *J. Physiol. (Lond.)*, *247*, 23-4P

Chinn, R.A., Johnson, H.G., Davis, B. & Nadel, J.A. (1981) 'Reflex Effects of Lung Inflation on Submucosal Gland Secretion in Canine Trachea', *Cystic Fibrosis Club. Abstr.*, *22*, 102

Clark, F.J. & Euler, C. von (1970) 'Regulation of Depth and Rate of Breathing in Cat and Man', *Acta Physiol. Scand.*, *80*, 20-1A

Clark, F.J. & Euler, C. von (1972) 'On the Regulation of Depth and Rate of Breathing', *J. Physiol. (Lond.)*, *222*, 267-95

Cohen, M.I. (1964) 'Respiratory Periodicity in the Paralysed, Vagotomised Cat: Hypocapnic Polypnea', *Am. J. Physiol.*, *206*, 845-54

Cohen, M.I. (1969) 'Discharge Patterns of Brain-Stem Respiratory Neurons during Hering–Breuer Reflex Evoked by Lung Inflation', *J. Neurophysiol.*, *32*, 356-74

Cohen, M.I. (1979) 'Neurogenesis of Respiratory Rhythm in the Mammal', *Physiol. Rev.*, *59*, 1105-73

Cohen, M.I. (1975) 'Phrenic and Recurrent Laryngeal Discharge Patterns and the Hering-Breuer Reflex', *Am. J. Physiol.*, *228*, 1489-96

Cohen, M.I., Feldman, J.L. & Donally, D.F. (1982a) 'Comparison of Responses to Lung Inflation Sown by Dorsal Medullary Inspiratory Neurons and by Phrenic Motoneurones', in J.L. Feldman & A.J. Berger (eds), *International Symposium Central Neural Production of Periodic Respiratory Movements*, Harrison Conference Center, Lake Bluff, Ill, pp. 139-40

Cohen, M.I., Sommer, D. & Feldman, J.L. (1982b) 'Relation of Expiratory Neuron Discharge to Lung inflation and Expiratory Duration', in J.L. Feldman & A.J. Berger (eds), *International Symposium Central Neuronal Production of Periodic Respiratory Movements*, Harrison Conference Center, Lake Bluff, Ill., pp. 141-2

Colebath, J.G., Gandevia, S.C. & McCloskey, D.I. (1977) 'Reduction in Inspiratory Activity in Response to Sternal Vibration', *Resp. Physiol.*, *29*, 327-38

Coleridge, H.M. & Coleridge, J.C.G. (1977a) 'Impulse Activity in Afferent Vagal C-fibers with Endings in the Intrapulmonary Airways of Dogs', *Resp. Physiol.*, *29*, 125-42

Coleridge, H.M. & Coleridge, J.C.G. (1977b) 'Afferent C-fibers in the Dog Lung: their Discharge During Spontaneous Breathing and their Stimulation by Alloxan and Pulmonary Congestion', in A.S. Paintal & P. Gill-Kumar (eds), *Krogh Centenary Symposium on Respiratory Adaptations, Capillary Exchange and Reflex Mechanisms*, Delhi, Vallabhbhai Patel Chest Institute, pp. 393-406

Coleridge, H.M. & Coleridge, J.C.G. (1981) 'Afferent Fibers Involved in the Defense Reflexes from the Respiratory Tract', in J. Hutas & L.A. Debreczeni (eds), *Adv. Physiol. Sci.*, vol. 10, *Respiration*, Pergamon Press, Akademiai Klado, pp. 467-77

Coleridge, H.M., Coleridge, J.C.G., Baker, D.G., Ginzel, K.H. & Morrison, M.A. (1978a) 'Comparison of the Effect of Histamine and Prostaglandins on Afferent C-fiber Endings and Irritant Receptors in the Intrapulmonary Airways', in R.S. Fitzgerald, H. Gauthier & S. Lahiri (eds), *Regulation of Respiration during Sleep and Anaesthesia*, Plenum Press, *Adv. Expt. Med. Biol.*, *99*, 291-306

Coleridge, H.M., Coleridge, J.C.G. & Banzett, R.B. (1978b) 'Effect of CO_2 on Afferent Vagal Endings in the Canine Lung', *Resp. Physiol.*, *34*, 135-41

Coleridge, H.M., Coleridge, J.C.G. & Luck, J.C. (1965) 'Pulmonary Afferent Fibers of Small Diameter Stimulated by Capsicin and by Hyperinflation of the Lungs', *J. Physiol. (Lond.)*, *179*, 248-62

Coleridge, J.C.G., & Kidd, C. (1963) 'Reflex Effects of Stimulating Baroreceptors in the Pulmonary Artery', *J. Physiol. (Lond.)*, *166*, 197-210

Crocker, E.F., Johanson, R.O., Korner, P.I., Uther, J.B. & White, S.W. (1968) 'Effects of Hyperventilation on the Circulatory Response of the Rabbit to Arterial Hypoxia', *J. Physiol. (Lond.)*, *199*, 267-82

Cross, K.W. (1961) 'Head's Paradoxical Reflex', *Brain*, *84*, Part 4, 529-34

Cross, K.W., Klaus, M., Tooley, W.H. & Weisser, K. (1960) 'The Response of the Newborn to Inflation of the Lungs', *J. Physiol. (Lond.)*, *151*, 55-565

Cross, B.A., Guz, A., Jain, S.K., Archer, S., Stevens, J. & Reynold, F. (1976) 'The Effect of Anesthesia on the Airway in Dog and Man: a Study of Respiratory Reflexes, Sensations and Lung Mechanics', *Clin. Sci. Mol. Med.*, *50*, 439-54

Cross, B.A., Jones, P.W. & Guz, A. (1980) 'The Role of Vagal Afferent Information during Inspiration in Determining Phrenic Motoneurone Output', *Resp. Physiol.*, *39*, 149-67

Cross, B.A., Guz, A. & Jones, P.W. (1981) 'The Summation of Left and Right Lung Volume Information in the Control of Breathing in Dogs', *J. Physiol. (Lond.)*, *321*, 449-67

Daly, M. de B. (1972) 'Interaction of Cardiovascular Reflexes', *Sci. Basis of Med. Ann. Dev.*, Chapt. *27*, 307-32

Daly, M. de B. & Angell-James, J.E. (1975) 'Role of the Arterial Chemoreceptors in the Control of the Cardiovascular Responses to Breath-hold Diving', in M.J. Purves (ed.), *Peripheral Arterial Chemoreceptors*, Cambridge University Press, pp. 387-407

Daly, M. de B., Angell-James, J.E. & Elsner, R. (1979) 'Role of Carotid-body Chemoreceptors and their Reflexes Interactions in Bradycardia and Cardiac Arrest', *Lancet*, 764-7

Daly, M. de B. & Hazzledine, J.L. (1963) 'The Effects of Artificially Induced Hyperventilation on the Primary Cardiac Reflex Response to Stimulation of the Carotid Bodies in the Dog', *J. Physiol. (Lond.)*, *168*, 782-889

Daly, M. de B., Hazzledine, J.L. & Ungar, A. (1967) 'The Reflex Effects of Alterations in Lung Volume on Systemic Vascular Resistance in the Dog', *J. Physiol. (Lond.)*, *188*, 331-51

Daly, M. de B., Korner, P.I., Angell-James, J.E. & Oliver, J.R. (1978) 'Cardiovascular-Respiratory Reflex Interactions between Carotid Bodies and Upper-airways Receptors in the Monkey', *Am. J. Physiol.*, *264*, H293-9

Daly, M. de B. & Robinson, B.H. (1968) 'An Analysis of the Reflex Systemic Vasodilator Response Elicited by Lung Inflation in the Dog', *J. Physiol. (Lond.)*, *195*, 387-406

Daly, M. de B. & Scott, M.J. (1958) 'The Effect of Stimulation of the Carotid Body Chemoreceptors on Heart Rate in the Dog', *J. Physiol. (Lond.)*, *144*, 148-66

Daly, M. de B. & Scott, M.J. (1962) 'An Analysis of the Primary Cardiovascular Reflex Effects of Stimulation of the Carotid Body Chemoreceptors in the Dog', *J. Physiol. (Lond.)*, *162*, 55-73

D'Angelo, E. (1978) 'Central and Direct Vagal Dependent of Expiratory Duration in Anaesthetized Rabbits', *Resp. Physiol.*, *34*, 103-19

Davies, A., Dixon, M., Callannan, D., Huszczuk, A., Widdicombe, J.G. & Wise, J.C.M. (1978) 'Lung Reflexes in Rabbits during Pulmonary Stretch Receptor Block by Sulphur Dioxide', *Resp. Physiol.*, *34*, 83-101

Davies, A. & Roumy, M. (1982) 'The Effect of Transient Stimulation of Lung Irritant Receptors on the Pattern of Breathing in Rabbits', *J. Physiol. (Lond.)*, *324*, 389-401

Davies, A., Sant'Ambrogio, F.B. & Sant'Ambrogio, G. (1980) 'Control of Postural Changes of End-expiratory Volume (FRC) by Airway Slowly Adapting Mechanoreceptors', *Resp. Physiol*, *41*, 211-26

Davies, A., Sant'Ambrogio, F. & Sant'Ambrogio, G. (1981) 'Onset of Inspiration in Rabbits during Artificial Ventilation', *J. Physiol. (Lond.)*, 17-23

Davidson, N.S., Goldner, S. & McCloskey, D.I. (1976) 'Respiratory Modulation of Baroreceptor and Chemoreceptor Reflexes Affecting Heart Rate and Cardiac Vagal Efferent Nerve Activity', *J. Physiol. (Lond.)*, *259*, 523-30

Davis, H.L., Fowler, W.S. & Lambert, E.H. (1956) 'Effect of Volume and Rate of Inflation and Deflation on Transpulmonary Pressure and Response of Pulmonary Stretch Receptors', *Am. J. Physiol.*, *187*, 558-66

Davis, B., Marin, M.G., Fisher, S., Graf, P., Widdicombe, J.G. & Nadel, J.A. (1976) 'New Method for Study of Canine Mucus Gland Secretion *in Vivo* Cholinergic Regulation', *Am. Rev. Resp. Dis.*, *113*, 257-8

Delpierre, S., Grimaud, Ch., Jammes, Y. & Mei, E. (1981) 'Changes in Activity of Vagal Bronchopulmonary C-fibers by Chemical and Physical Stimuli in the Cat', *J. Physiol. (Lond.)*, *316*, 61-74

Delpierre, S., Jammes, Y., Mei, N., Mathiot, M.J. & Grimaud, Ch. (1980) 'Mise en evidence de l'origine vagale reflexe des effects bronchoconstricteurs de CO_2 chez le chat', *J. Physiol. (Paris)*, *76*, 889-91

Deshpande, S.S. & Devanandan, M.S. (1970) 'Reflex Inhibition of Monosynaptic Reflexes by Stimulation of Type J Pulmonary Endings', *J. Physiol. (Lond.)*, *206*, 345-57

Dickinson, C.J. & Paintal, A.S. (1970) 'Stimulation of Type-J Pulmonary Receptors in the Cat by Carbon Dioxide', *Clin. Sci.*, *38*, 33P

Di Marco, A.F., Euler, C. von, Romaniuk, J.R. & Yamamoto, Y. (1981) 'Positive Feedback Facilitation of External and Phrenic Inspiratory Activity by Pulmonary Stretch Receptors', *Acta Physiol. Scand.*, *113*, 375-86

Dixon, M., Jackson, D.M. & Richards, J.M. (1979) 'The Effects of Histamine, Acetylocholine and 5-Hydroxytryptamine on Lung Mechanics and Irritant Receptors in the Dog', *J. Physiol. (Lond.)*, *287*, 393-403

Dixon, M., Jackson, D.M. & Richards, J.M. (1980) 'The Effects of 5-Hydroxytryptamine, Histamine and Acetylocholine on the Reactivity of the Lung of the Anaesthetized Dog', *J. Physiol. (Lond.)*, *307*, 85-96

Dixon, M., Szereda-Przetaszewska, M., Widdicombe, J.G. & Wise, J.C.M. (1974) 'Studies on Laryngeal Calibre during Stimulation of Peripheral and Central Chemoreceptors, Pneumothorax and Increased Respiratory Loads', *J. Physiol. (Lond.)*, *239*, 347-63

Donald, D.E. (1979) 'Studies on the Release of Renin by Direct and Reflex Activation of Renal Sympathetic Nerves', *Physiologist*, *22*, no. 3, 39-42

Donald, D.E. & Shepherd, J.T. (1978) 'Reflexes from the Heart and Lungs: Physiological Curiosities or Important Regulatory Mechanisms', *Cardiovasc. Res.*, *12*, 449-69

Eckberg, D.L., Kifle, Y.T. & Roberts, V.I. (1980) 'Phase Relationship between Normal Human Respiration and Baroreflex Responsiveness', *J. Physiol. (Lond.)*, *304*, 489-502

Eckberg, D.L. & Orshan, C.R. (1977) 'Respiratory and Baroreceptor Reflex Interactions in Man', *J. Clin. Invest.*, *59*, 716-80

Eldridge, F.L. & Gill-Kumar, P. (1980) 'Central Respiratory Effects of Carbon Dioxide, Carotid Sinus Nerve and Muscle Afferents Stimulation', *J. Physiol. (Lond.)*, *300*, 75-87

England, S.J., Onayemi, B. & Bryan, A.C. (1982) 'Stimulation of Phrenic Nerve Activity by Neuromuscular Blockade during High Frequency Ventilation', in J.L. Feldman & A.J. Berger (eds), *Central Neural Production of Periodic Respiratory Movements*, *Internat. Symposium*, Lake Bluff, Illinois, pp. 126-7

Euler, C. von (1977) 'The Functional Organization of the Respiratory Phase-switching Mechanism', *Fed. Proc.*, *36*, 2375-80

Euler, C. von (1980) ' 'Off-switching' of Inspiration — Stockholm View', in *The Role of the Nuclei Tractus Solitarii in the Central Regulation of the Respiratory and Cardiovascular Systems*, *Proc. Int.* Workshop on the Central Control of Circulation and Respiration. Heidelberg/Hirschorn

Euler, C. von, Hervero, F. & Wexler, I. (1970) 'Control Mechanisms Determining Rate and Depth of Respiratory Movements', *Resp. Physiol.*, *10*, 93-108

Euler, C. von & Trippenbach, T. (1976) 'Excitability Changes of the Inspiratory 'Off-switch' Mechanisms Tested by Electrical Stimulation in Nucleus Para-brachialis in the Cat', *Acta Physiol. Scand.*, *97*, 175-88

Evans, D.H.L. & Murray, J.G. (1954) 'Histological and Functional Studies on the Fibre Composition of the Vagus Nerve of the Rabbit', *J. Anat.*, *88*, 320-37

Farber, J.P. (1982) 'Pulmonary Receptor Discharge and Expiratory Muscle Activity', *Resp. Physiol.*, *47*, 219-29

Feldman, J.L. & Cohen, M.J. (1978) 'Relation between Expiratory Duration and Rostral Medullary Expiratory Neuronal Discharge', *Brain Res.*, *141*, 172-8

Feldman, J.L., Cohen, M.I. & Wolotsky, P. (1976) 'Powerful Inhibition of Pontine Respiratory Neurons by Pulmonary Afferent Activity', *Brain Res.*, *104*, 341-6

Feldman, J.L. & Gauthier, H. (1976) 'Interaction of Pulmonary Afferents and Pneumotaxic Center in Control of Respiratory Pattern in Cats', *J. Neurophysiol.*, *39*, 31-44

Fishman, H.H., Phillipson, E.A. & Nadel, J.A. (1973) 'Effect of Differential Vagal Cold Blockade on Breathing Patterns in Conscious Dogs', *J. Appl. Physiol.*, *34*, 754-63

Fleming, P.J., Bryan, A.C. & Bryan, M.H. (1982) 'Functional Immaturity of Pulmonary Irritant Receptors and Apnea in Newborn Preterm Infants', *Pediatrics*, *61*, 515-18

Gandevia, S.C. & McCloskey, D.I. (1976) 'Changes in the Pattern of Breathing Caused by Chest Vibration', *Resp. Physiol.*, *26*, 163-72

Gandevia, S.C., McCloskey, D.I. & Potter, E.K. (1978a) 'Inhibition of Baroreceptor and Chemoreceptor Reflexes on Heart Rate by Afferents from the Lungs', *J. Physiol. (Lond.)*, *276*, 369-81

Gandevia, S.C., McCloskey, D.I. & Potter, E.K. (1978b) 'Reflex Bradycardia Occurring in Response to Diving, Nasopharyngeal Stimulation and Ocular Pressure, and its Modification by Respiration and Swallowing', *J. Physiol. (Lond.)*, *276*, 313-94

Gauthier, H., Remmers, J.E. & Bartlett, D. Jr. (1973) 'Control of the Duration of Expiration', *Resp. Physiol.*, *18*, 205-21

German, V.F., Ueki, J.F. & Nadel, J.A. (1980) 'Micropipette Measurements of Airway Submucosal Gland Secretion: Laryngeal Reflex', *Am. Rev. Resp. Dis.*, *122*, 413-6

Glick, G., Wechsler, A.S. & Epstein, S.E. (1969) 'Reflex Cardiovascular Depression Produced by Stimulation of Pulmonary Stretch Receptors in the Dog', *J. Clin. Invest.*, *48*, 467-73

Glogowska, M., Richardson, P.M., Widdicombe, J. & Winning, J.G. (1972) 'The Role of Vagus Nerves, Peripheral Chemoreceptors and Other Afferent Pathways in the Genesis of Augmented Breaths in Cats and Rabbits', *Resp. Physiol.*, *16*, 179-96

Glogowska, M., Stransky, A. & Widdicombe, J.G. (1974) 'Reflex Control of Discharge in Motor Fibres to the Larynx', *J. Physiol. (Lond.)*, *239*, 365-79

Gold, W.M. (1975) 'The Role of the Parasympathetic Nervous System in Airways Disease', *Postgrad. Med. J.*, *51*, Suppl. 7, 53-62

Gold, W.M., Kessler, G.F. & Yu, D.Y.C. (1972) 'Role of Vagus Nerve in Experimental Asthma in Allergic Dogs', *J. Appl. Physiol.*, *33*, 719-25

Goldman, M., Grimby, G. & Mead, J. (1976) 'Mechanical Work of Breathing Derived from Rib Cage and Abdominal V-P Partioning', *J. Appl. Physiol.*, *41*, 764-75

Gootman, Ph. M., Feldman, J.L. & Cohen, M.I. (1980) 'Pulmonary Afferent Influences on Respiratory Modulation of Sympathetic Discharge, in H.P. Koepchen, S.M. Hilton & A. Trzebski (eds), *Central Interaction between Respiratory and Cardiovascular Control Systems*, Springer, Berlin, Heidelberg, New York, pp. 172-9

Graham, K. & Duffin, J. (1982) 'Cross-correlation of Medullary Dorso-medial Inspiratory Neurons in the Cat', *Exp. Neurol.*, *75*, 627-43

Grimby, G. (1977) 'Pulmonary Mechanics: The Load', in J.A. Dempsey & C.E. Reed (eds), *Muscular Exercise and the Lung*, The University of Wisconsin Press, Madison, pp. 17-24

Gromysz, H., Karczewski, W.A., Naslonska, E. & Romaniuk, J.R. (1982) 'The Role of Timing and Magnitude of the Vagal Input in Controlling the Phrenic Output in Rabbits and Baboons', *Acta Neurobiol. Exp.*, *40*, 563-74

Guo, G.B., Thames, M.D. & Abboud, F.M. (1982) 'Differential Baroreflex Control of Heart Rate and Vascular Resistance in Rabbits. Relative Role of Carotid, Aortic and Cardiopulmonary Baroreceptors', *Circ. Res.*, *50*, 554-65

Guz, A., Noble, M.I.M., Eisele, J.H. & Trenchard, D. (1970) 'Experimental Results of Vagal Block in Cardiopulmonary Disease', in *Breathing: Hering-Breuer Centenary Symposium*, Churchill, London, pp. 315-29

Guz, A. & Trenchard, D.W. (1971a) 'Pulmonary Stretch Receptor Activity in Man: Comparison with Dog and Cat', *J. Physiol. (Lond.)*, *213*, 329-43

Guz, A. & Trenchard, D. (1971b) 'The Role of Non-myelinated Vagal Afferents Fibers from the Lungs in the Genesis of Tachypnea in the Rabbit', *J. Physiol. (Lond.)*, *213*, 345-71

Guz, A., Noble, M.I.M., Trenchard, D., Cochrane, H.L. & Makey, A.R. (1964) 'Studies on the Vagus Nerve in Man: Their Role in Respiratory Control', *Clin. Sci.*, *27*, 293-304

Guz, A., Noble, M.I.M., Trenchard, D., Smith, A.J. & Makey, A.J. (1966a) 'The Hering–Breuer Inflation Reflex in Man. I Studies of Unilateral Lung Inflation and Vagus Nerve Block', *Resp. Physiol.*, *1*, 382-9

Guz, A., Noble, M.I.M., Widdicombe, J.G., Trenchard, D. & Mushin, W.W. (1966b) The Effect of Bilateral Block of Vagus and Glossopharyngeal Nerves on the Ventilatory Responses to CO_2 of Conscious Man', *Resp. Physiol.*, *1*, 206-10

Hainsworth, R. (1974) 'Circulatory Responses from Lung Inflation in Anesthetized Dog', *Am. J. Physiol.*, *226*, 247-55

Hammouda, M., & Wilson, W.H. (1935) 'The Presence in the Vagus of Fibers Transmitting Impulses Augmenting the Frequency of Respiration', *J. Physiol. (Lond.)*, *83*, 292-331

Hammouda, M. & Wilson, W.H. (1939) 'Reflex Acceleration of Respiration Arising from Excitation of the Vagus or its Termination in the Lungs', *J. Physiol. (Lond.)*, *94*, 497-524

Haymet, B.T. & McCloskey, D.I. (1975) 'Baroreceptor and Chemoreceptor Influences on the Heart Rate during the Respiratory Cycle in the Dog', *J. Physiol. (Lond.)*, *245*, 699-712

Hering, E. (1871) 'Uber den Einfluss der Atmung auf den Kreislauf Zwiete Mitteilung über eine reflektorische Beziehung zwischen Lung und Herz', *Sitzungsberichte d. Wien. Adad. Wissensch., math.-naturwiss. Kl.*, *64*, 333-53

Hering, E. & Breuer, J. (1868) 'Die Selbsteuerung der Atmung durch den Nervous Vagus', *Stizungsberichte d. Wien. Akad. Wissensch., math.-naturwiss. Kl.*, 672-7

Holmes, R. & Torrance, R.W. (1959) 'Afferent Fibers of the Stellate Ganglion', *Q.J. Exp. Physiol.*, *44*, 271-81

Homma, I. (1980) 'Inspiratory Inhibitory Reflex Caused by Chest Wall Vibrations in Man', *Resp. Physiol.*, *39*, 345-53

Howard, P., Clayton, R.M., Brennan, S.R. & Andersson, P.B. (1977) 'Lignocaine Aerosol and Persistent Cough', *Br. J. Dis. Chest.*, *71*, 19-24

Hugelin, A. & Cohen, M.I. (1963) 'The Reticular Activating System and Respiratory Regulation in the Cat', *Ann. N.Y. Acad. Sci.*, *109*, 586-603

Huszczuk, A., Jankowska, L., Kulesza, J. & Ryba, M. (1977) 'Studies on Reflex Control of Breathing in Pigs and Baboons', *Acta Neurobiol. Exp.*, *37*, 275-98

Irvin, C.G., Boileau, R., Tremblay, J., Marten, R.R. & Macklem, P.T. (1980) 'Bronchodilatation: Non-cholinergic, Non-adrenergic Mediation Demonstrated *in Vivo* in the Cat', *Science*, *207*, 791-2

Iscoe, A., Young, R.B. & Jennings, D.B. (1982) 'Control of Respiratory Pattern in Conscious Mammals – a Modified Clark von Euler Model', in J.L. Feldman & A.J. Berger (eds), *International Symposium Central Neural Production of Periodic Respiratory Movements*, Harrison Conference Center, Lake Bluff, Ill., pp. 147-9

Jaffe, H.J. & Martin, W.R. (1980) 'Opioid Analgesic and Antagonist', in A.G. Gilman, L.S. Goodman & A. Gilman (eds), *The Pharmacological Basis of Experimental Therapeutics*, 6th edn. Macmillan, New York, pp. 494-511

Jain, S.K., Subramamanian, S., Julka, D.B. & Guz, A. (1972) 'Search for Evidence by Lung Chemoreflex in Man: Study of Respiratory and Circulatory Effects of Phenyldiguanide and Lobeline', *Clin. Sci.*, *42*, 163-77

Jammes, Y. & Mei, N. (1979) 'Assessment of the Pulmonary Origin of Bronchoconstrictor Vagal Tone', *J. Physiol. (Lond.)*, *291*, 305-16

Jammes, Y., Fornaris, E., Mei, N. & Barrat, E. (1982) 'Afferent and Efferent Component of the Bronchial Vagal Branches in Cats', *J. Aut. Ner. Syst.*, *5*, 165-76

Karczewski, W., Budzinska, K., Gromysz, H., Herczynski, R. & Romaniuk, J.R. (1976) 'Some Responses of the Respiratory Complex to Stimulation of its Vagal and Mesencephalic Inputs', in B. Duron (ed.), *Respiratory Centres and Afferent System*, vol. 59, INSERM, pp. 107-15

Karczewski, W.A., Budzinska, K., Naslonska, E., Jazowiecka, E., Romaniuk, J.R. & Ryba, M. (1978) 'Rate of Rise of Inspiration at Various Levels of CNS Excitability', in R.S. Fitzgerald, H. Gauthier & S. Lahiri (eds), *Adv. Exp. Med. Biol., The Regulation of Respiration During Sleep and Anaesthesia*, Plenum Press, New York 99, pp. 23-4

Karczewski, W., Karczewska, E. & Ryczymbel, Z. (1972) 'Respiratory Responses to CO_2 in Normo- and Hyperthermic Rabbits', *Arch. Fisiol.*, *69*, supl. 479-87

Karczewski, W.A., Naslonska, E. & Romaniuk, J.R. (1980a) 'Respiratory Responses to Stimulation of Afferent Vagal Fibers in Rabbits', *Acta Neurobiol. Exp.*, *40*, 543-62

Karczewski, W.A., Naslonska, E. & Romaniuk, J.R. (1980b) 'Inspiratory Facilitatory and Inhibitory Vagal Influences during Apnoea in Rabbits', *Acta

Neurobiol. Exp., *40*, 575-92

Karczewski, W. & Widdicombe, J.G. (1969a) 'The Role of the Vagus Nerves in the Respiratory and Circulatory Reactions to Anaphylaxia in Rabbits', *J. Physiol. (Lond.)*, *201*, 293-304

Karczewski, W. & Widdicombe, J.G. (1969b) 'The Role of the Vagus Nerve in the Respiratory and Circulatory Responses to Intravenous Histamine and Phenyl Diguanide in Rabbits', *J. Physiol. (Lond.)*, *201*, 271-91

Karczewski, W. & Widdicombe, J.G. (1969c) 'The Effect of Vagotomy Vagal Cooling and Efferent Vagal Stimulation on Breathing and Lung Mechanics of Rabbit', *J. Physiol. (Lond.)*, *201*, 259-270

Kashani, M. & Haig, A.L. (1975) 'The Effect of Vagotomy on Ventilation and Blood Gas Composition in Dog, Sheep and Rabbit', *Q.J. Exp. Physiol.*, *60*, 285-98

Katona, P.G. & Jih, F. (1975) 'Respiratory Sinus Arrhythmia; Non-invasive Measure of Parasympathetic Cardiac Control', *J. Appl. Physiol.*, *39*, 801-5

Kaufmann, M.P., Coleridge, H.M., Coleridge, J.C.G. & Baker, G. (1980) 'Brady-kinin Stimulates Afferent Vagal C-fibers in Intrapulmonary Airways in Dogs', *J. Appl. Physiol.*, *Respirat. Environ. Exercise*, *48*, 511-7

Kellermeger, R.W. & Schwartz, H.J. (1976) 'The Kinins: Basic Chemistry, Biological Action and Implications in Human Asthma', in E.B. Weiss & M.S. Segal (eds), *Bronchial Asthma*, Little, Brown, Boston, pp. 217-30

Kiowski, W. & Julius, S. (1978) 'Renin Response to Stimulation of Cardiopulmonary Mechanoreceptors in Man', *J. Clin. Invest.*, *62*, 656-63

Kiwull, P. & Wiemer, W. (1971) 'Der Einfluss der Vagusauschaltung auf die Wirkung der elektrischen Sinusnervenreizung', *Pflüg. Arch.*, *330*, 15-27

Kiwull-Schöne, H. & Kiwull, P. (1979) 'The Role of the Vagus Nerve in the Ventilatory Response to Lowered $PaCO_2$ with Intact and Eliminated Chemoreceptor Reflexes', *Pflüg. Arch.*, *381*, 1-9

Kiwull-Schöne, H., Ward, S.A. & Kiwull, P. (1981) 'The Involvement of Expiratory Termination in the Vagally Mediated Facilitation of Ventilatory CO_2 Responsiveness during Hyperoxia', *Pflüg. Arch.*, *390*, 63-9

Knowlton, G.C. & Larabbee, M.G. (1946) 'A Unitary Analysis of Pulmonary Volume Receptors', *Am. J. Physiol.*, *147*, 100-14

Knox, C.K. (1973) 'Characteristics of Inflation and Deflation Reflexes during Expiration in the Cat', *J. Neurophysiol.*, *36*, 284-95

Knox, C.K. (1979) 'Reflex and Central Mechanisms Controlling Expiratory Duration', in C. von Euler & H. Lagercrantz (eds), *Wenner-Gren Center Internat. Symp.* vol. 32, *Central Nervous Control Mechanisms in Breathing*, Pergamon Press, pp. 203-10

Koepchen, H.P., Wagner, P.H. & Lux, H.D. (1961) 'Uber die Zusammenhange zwischen zentraler Erregbarkaeit, reflektorischen Tonus und Atemrhythmus bei der nervösen Steuerung der Herzfrequenz', *Pflüg. Arch. ges. Physiol.*, *273*, 443-65

Koepchen, H.P., Sommer, D., Frank, Ch., Klüssendorf, D., Kramer, A., Rosin, P. & Forstreuter, K. (1979) 'Characteristics and Functional Significance of the Expiratory Bulbar Neurone Pools', in C. von Euler & H. Langecrantz (eds), *Wenner-gren Center International Symposium Series*, vol. 32, *Central Nervous Control Mechanisms in Breathing*, pp. 217-32

Koller, E.A. & Ferrer, P. (1973) 'Discharge Patterns of Lung Stretch Receptors and Activation of Deflation Fibers in Anaphylactic Bronchial Asthma', *Resp. Physiol.*, *17*, 113-26

Kontos, H.A., Goldin, D., Richardson, D.W. & Patterson, J.L. (1967) 'Contribution of Pulmonary Vagal Reflexes to Circulatory Response to Hypoxia', *Am. J. Physiol.*, *212*, 1441-8

Korner, P.I. (1980) 'Operation of the Central Nervous System in Reflex Circulatory Control', *Fed. Proc.*, *39*, 2504-12

Korner, P.I., Shaw, J., West, M.J., Oliver, J.R. & Hilder, R.G. (1973) 'Integrative Reflex Control of Heart Rate in the Rabbit during Hypoxia and Hyperventilation', *Circ. Res.*, *33*, 63-73

Korpas, J. & Tomori, Z. (1979) *Cough and Other Respiratory Reflexes*, Karger, p. 358

Kostreva, D.R., Zuperka, E.J., Hess, G.L., Coon, R.L. & Kampine, J.P. (1975) 'Pulmonary Afferent Activity Recorded from Sympathetic Nerves', *J. Appl. Physiol.*, *39*, 37-90

Kostreva, D.R., Hopp, F.A., Zuperka, E.J., Igler, F.O., Coon, R.L. & Kampine, J.P. (1978) 'Respiratory Inhibition with Sympathetic Afferent Stimulation in the Canine and Primate', *J. Appl. Physiol.: Resp. Environ. Exercise Physiol.*, *44*, 718-24

Kubin, L. & Lipski, J. (1979) 'Properties of Reversible Graded Inhibition of Phrenic Nerve Activity by Pulmonary Afferents', *Acta Physiol. Polon.*, *30*, 571-9

Langrehr, D. (1964) 'Rezeptor-Afferenzen im Halsvagus des Menschen', *Klin. Wochenschr.*, *42*, 239-44

Larabbee, M.G. & Knowlton, G.C. (1946) 'Excitation and Inhibition of Phrenic Motoneurones by Inflation of the Lungs', *Am. J. Physiol.*, *147*, 90-9

Lim, T.P.K., Luft, U.C. & Grodins, F.S. (1958) 'Effects of Cervical Vagotomy on Pulmonary Ventilation and Mechanics', *J. Appl. Physiol.*, *13*, 317-24

Lipski, J., Coote, J. & Trzebski, A. (1977) 'Temporal Patterns of Antidromic Invasion Latencies of Preganglionic Sympathetic Neurons Related to Central Inspiratory Activity and Pulmonary Stretch Receptor Reflex', *Brain Res.*, *135*, 162-6

Lipski, J., Kubin, L. & Jodkowski, J. (1983) 'Synaptic Action of R_β Neurons on Phrenic Motoneurons Studied with the Spike-triggered Averaging', *Brain Res.* (in press)

Lipski, J., Trzebski, A. & Kubin, L. (1979) 'Excitability Changes of Dorsal Inspiratory Neurons during Lung Inflation as Studied by Measurement of Antidromic Invasion Latencies', *Brain Res.*, *161*, 25-38

Lipski, J., Trzebski, A. & Merrill, E.G. (1981) 'Do Vagal Pulmonary Stretch Receptor Afferents Transmit Information on Cardiac Function?' *Eur. J. Clin. Invest.*, *11*, Abstr. 111, p. 19

Lloyd, T.C., Jr. (1977) 'Cardiopulmonary Baroreflex: Effects of Pulmonary Congestion and Edema', *J. Appl. Physiol.: Resp. Environ. Exercise Physiol.*, *43*, 107-13

Lloyd, T.C. Jr. (1979) 'Effects of Extrapulmonary Airway Distension on Breathing in Anesthetized Dogs', *J. Appl. Physiol.: Resp. Environ. Exercise Physiol.*, *46*, 890-6

Looffbourrow, G.N., Wood, W.B. & Baird, I.L. (1957) 'Tracheal Constriction in the Dog', *Am. J. Physiol.*, *191*, 411-5

Lopes, D.U. & Palmer, J.F. (1976) 'Proposed Respiratory 'Gating' Mechanism for Cardiac Slowing', *Nature*, *264*, 454-6

Luck, J.C. (1970) 'Afferent Vagal Fibres with an Expiratory Discharge in the Rabbit', *J. Physiol. (Lond.)*, *211*, 63-71

Mancia, G. & Donald, D.E. (1975) 'Demonstration that Atria, Ventricles and Lungs each are Responsible for a Tonic Inhibition of the Vasomotor Center in the Dog', *Circ. Res.*, *36*, 316-8

Mancia, G., Donald, D.E. & Shepherd, J.T. (1973) 'Inhibition of Adrenergic Outflow to Peripheral Blood Vessels by Vagal Afferents from the Cardiopulmonary Region in the Dog', *Circ. Res.*, *33*, 713-21

Mancia, G., Shepherd, J.T. & Donald, D.E. (1975) 'Role of Cardiac, Pulmonary and Carotid Mechanoreceptors in the Control of Hind Limb and Renal Circulation in Dogs', *Circ. Res.*, *37*, 200-8

Mancia, G., Shepherd, J.T. & Donald, D.E. (1976) 'Interplay among Carotid Sinus, Cardiopulmonary and Carotid Body Reflexes in Dogs', *Am. J. Physiol.*, *230*, 19-24

Marino, R., Davies, R.O. & Pack, A.I. (1981) 'The Responses of Iβ Cells to Increases in the Rate of Lung Inflation', *Brain Res.*, *219*, 289-306

Marshall, R. & Widdicombe, J.G. (1958) 'The Activity of Pulmonary Stretch Receptors during Congestion of the Lungs', *Q. J. Exp. Physiol.*, *43*, 320-30

McCaffrey, T.V. & Kern, E.B. (1980) 'Laryngeal Regulation of Airway Resistance, II Pulmonary Receptor Reflexes', *Ann. Otol.*, *89*, 462-6

Mead, J. (1960) 'Control of Respiratory Frequency', *J. Appl. Physiol.*, *15*, 325-36

Meancock, C.I. (1982) 'Influence of the Vagus Nerve on Changes in Heart Rate during Sleep Apnoea in Man', *Clin. Sci.*, *62*, 163-7

Mei, N. (1970) 'Mecanorecepteurs vagaux cardio-vasculaires et respiratoires chez le chat', *Exp. Brain Res.*, *11*, 480-501

Melcher, A. (1976) 'Respiratory Sinus Arrhythmia in Man: A Study in Heart Rate Regulating Mechanism', *Acta Physiol. Scand.*, (suppl.), *435*, 1-31

Melcher, A. (1980) 'Carotid Baroreflex Heart Rate Control during the Active and Assisted Breathing Cycle in Man', *Acta Physiol. Scand.*, *108*, 165-71

Mills, J.E., Sellick, H. & Widdicombe, J.G. (1969) 'Activity of Lung Irritant Receptors in Pulmonary Microembolism, Anaphylaxis and Drug-induced Bronchoconstriction', *J. Physiol. (Lond.)*, *203*, 337-57

Mills, J.E. & Widdicombe, J.G. (1970) 'Role of the Vagus Nerves in Anaphylaxis and Histamine-induced Bronchoconstrictions in Guinea Pigs', *Br. J. Pharm.*, *39*, 724-31

Mills, J.E., Sellick, H. & Widdicombe, J.G. (1970) 'Epithelial Irritant Receptors in the Lungs', in R. Porter (ed.), *Ciba Found. Symp., Breathing: Hering-Breuer Centenary Symposium*, Churchill, London, pp. 77-92

Miserocchi, G. (1976) 'Role of Peripheral and Central Chemosensitive Afferents in the Control of Depth and Frequency of Breathing', *Resp. Physiol.*, *26*, 101-11

Miserocchi, G., Mortola, J. & Sant'Ambrogio, G. (1973) 'Localization of Pulmonary Stretch Receptors in the Airways of the Dog', *J. Physiol. (Lond.)*, *235*, 775-82

Miserocchi, G. & Sant'Ambrogio, G. (1974) 'Distribution of Pulmonary Stretch Receptors in the Intrapulmonary Airways of the Dog', *Resp. Physiol.*, *21*, 71-5

Miserocchi, G. & Trippenbach, T. (1981) 'The Role of the Pneumotaxic Mechanism in the Tachypnea of Pulmonary Vagal Origin', *Resp. Physiol.*, *43*, 275-85

Miserocchi, G., Trippenbach, T., Mazarelli, M., Jasper, N. & Hazucha, M. (1978) 'The Mechanism of Rapid Shallow Breathing due to Histamine and Phenyldiguanide in Cats and Rabbits', *Resp. Physiol.*, *32*, 141-53

Mitchell, G.S., Cross, B.A., Hiramoto, T. & Scheid, P. (1980) 'Effects of Intrapulmonary CO_2 and Airway Pressure on Phrenic Activity and Pulmonary Stretch Receptor Discharge in Dogs', *Resp. Physiol.*, *40*, 29-48

Mortola, J.P. & Sant'Ambrogio, G. (1979) 'Mechanics of the Trachea and Behaviour of its Slowly Adapting Stretch Receptors', *J. Physiol. (Lond.)*, *286*, 577-90

Nadel, J.A. (1973) 'Neurophysiological Aspects of Asthma', in K.F. Austen & L.N. Lichtenstein (eds), *Asthma, Physiology, Immunopharmacology and Treatment*, Academic Press, New York, pp. 29-37

Nadel, J.A. (1977) 'Autonomic Control of Airway Smooth Muscle and Airway

Secretion', *Ann. Rev. Resp. Dis.*, *115*, (Suppl.) 117-26

Neil, E. & Palmer, J.F. (1975) 'Effects of Spontaneous Respiration on the Latency of Reflex Cardiac Chronotropic Responses to Baroreceptor Stimulation', *J. Physiol. (Lond.)*, *175*, 193-202

Newsom Davis, J. & Stagg, D. (1975) 'Interrelationship of the Volume and Time Components of Individual Breaths in Resting Man', *J. Physiol. (Lond.)*, *245*, 481-98

Otis, A.B., Fenn, W.O. & Rahn, H. (1950) 'Mechanics of Breathing in Man', *J. Appl. Physiol.*, *2*, 592-607

Ott, N.T., Lorenz, R.R. & Shepherd, J.T. (1972) 'Modification of Lung Inflation Reflex in Rabbits by Hypercapnia', *Am. J. Physiol.*, *223*, 812-9

Ott, N.T. & Shepherd, J.T. (1975) 'Modification of Vagal Depressor Reflex by CO_2 in Spontaneously Breathing Rabbits', *Am. J. Physiol.*, *228*, 530-5

Ott, N.T., Tarhan, S. & McGoon, D.C. (1975) 'Circulatory Effects of Vagal Inflation Reflex in Man', *Z. Kardiol.*, *64*, 1066-70

Pack, A.I., DeLaney, R.G. & Fishman, A.P. (1971) 'Augmentation of Phrenic Neural Activity by Increased Rates of Lung Inflation', *J. Appl. Physiol.: Resp. Environ. Exercise Physiol.*, *50*, 149-61

Pack, A.I. (1982) 'Pulmonary Vagal Receptors', in J.L. Feldman & A.J. Berger (eds), *Internat. Symposium Central Neural Production of Periodic Respiratory Movements*, Harrison Conference Center, Lake Bluff, Ill., pp. 30-4

Paintal, A.S. (1955) 'Impulses in Vagal Afferent Fibers from Specific Pulmonary Deflation Receptors. The Responses of these Receptors to Phenyldiguanide, Potato Starch, 5-Hydroxytryptamine and Nicotine, and their Role in Respiratory and Cardiovascular Reflexes', *Q. J. Exp. Physiol.*, *40*, 89-111

Paintal, A.S. (1969) 'Mechanism of Stimulation of Type J Pulmonary Receptors', *J. Physiol. (Lond.)*, *203*, 511-32

Paintal, A.S. (1970) 'The Mechanism of Excitation of Type J Receptors and the J Reflex', in R. Porter (ed.), *Ciba Found. Symp. Breathing: Hering–Breuer Centenary Symposium*, Churchill, London, pp. 59-71

Paintal, A.S. (1973) 'Vagal Sensory Receptors and their Reflex Effects', *Physiol. Rev.*, *53*, 159-227

Paintal, A.S. (1977a) 'Thoracic Receptors Connected with Sensation', *Br. Med. Bull.*, *33*, 169-74

Paintal, A.S. (1977b) 'The Nature, and Effect of Sensory Inputs into the Respiratory Centers', *Fed. Proc.*, *36*, 2428-32

Pearce, J.W. & Whitteridge, D. (1951) 'The Relation of Pulmonary Arterial Pressure Variations to the Activity of Afferent Pulmonary Vascular Fibers', *Q.J. Exp. Physiol.*, *36*, 117-88

Phillipson, E.A., Hickey, R.F., Bainton, C.R. & Nadel, J.A. (1970) 'Effects of Vagal Blockade on Regulation of Breathing in Conscious Dogs', *J. Appl. Physiol.*, *29*, 475-9

Phillipson, E.A., Fishman, N.H., Hickey, R.F. & Nadel, J.A. (1973) 'Effect of Differential Vagal Blockade on Ventilatory Response to CO_2 in Awake Dogs', *J. Appl. Physiol.*, *34*, 759-63

Phipps, R.J. & Richardson, P.S. (1976) 'The Effects of Irritation at Various Levels of the Airway upon Tracheal Mucus Secretion in the Cat', *J. Physiol. (Lond.)*, *261*, 563-81

Potter, E.K. (1981) 'Inspiratory Inhibition of Vagal Responses to Baroreceptor and Chemoreceptor Stimuli in the Dog', *J. Physiol. (Lond.)*, *316*, 177-90

Raybould, H.E. & Russel, J.W. (1982) 'Afferent Activity in Pulmonary Vagal C-fibers Reflexly Increases Respiratory Rate during Hypercapnia in the Anaesthetized Rabbits', *J. Physiol. (Lond.)*, *326*, 60P-61P

Rees, P.J., Chowienczyk, P.J. & Clark, T.J.H. (1980) 'Immediate Response to

Inhaled Lung Irritant in Man', *Clin. Sci.*, *58*, 5p

Remmers, J.E. (1970) 'Inhibition of Inspiratory Activity by Intercostal Muscle Afferents', *Resp. Physiol.*, *10*, 358-83

Remmers, J. & Bartlett, D. (1977) 'Neural Control Mechanisms', in J.A. Dempsey & C.F. Reed (eds), *Muscular Exercise and the Lung*, University Wisconsin Press, Madison, 41-56

Remmers, J.E., Baker, J.P. Jr., & Younes, M.K. (1979) 'Graded Inspiratory Inhibition: the First Stage of Inspiratory 'Off-switching'', in *Wener-gren Center Internat. Symposium Series*, vol. 32, *Central Nervous Control Mechanisms in Breathing*, Pergamon Press, pp. 195-201

Reynolds, L.B. (1962) 'Characteristics of an Inspiration-augmenting Reflex in Anaesthetized Cats', *J. Appl. Physiol.*, *17*, 683-8

Richardson, P.S. & Widdicombe, J.G. (1969) 'The Role of the Vagus Nerve in the Ventilatory Responses to Hypercapnia and Hypoxia in Anesthetized and Unanesthetized Rabbits', *Resp. Physiol.*, *7*, 122-35

Roberts, A.M., Kaufman, M.P., Baker, D.G., Brown, J.K., Coleridge, H.M. & Coleridge, J.C.G. (1981) 'Reflex Tracheal Contraction induced by Stimulation of Bronchial C-fibers in Dogs', *J. Appl. Physiol.: Resp. Environ. Exercise Physiol.*, *51*, 485-93

Robertson, D.G., Warrell, D.A., Newton-Howes, J.S. & Fletcher, C.M. (1969) 'Bronchial Reactivity to Cigarette and Cigar Smoke', *Br. J. Med.*, *3*, 269-71

Roumy, M. & Leitner, L.M. (1980) 'Localization of Stretch and Deflation Receptors in the Airways of the Rabbit', *J. Physiol. (Paris)*, *76*, 67-70

Russel, J.A. & Bishop, B. (1976) 'Vagal Afferents Essential for Abdominal Muscle Activity during Lung Inflation in Cats', *J. Appl. Physiol.*, *41*, 310-5

Russel, J.A. & Lai-Fook, S.J. (1979) 'Reflex Bronchoconstriction induced by Capsaicin in the Dog', *J. Appl. Physiol.: Resp. Environ. Exercise Physiol.*, *47*, 961-7

Rutheford, J.D. & Vatner, S.F. (1978) 'Integrated Carotid Chemoreceptor and Pulmonary Inflation Reflex Control of Peripheral Vasoactivity in Conscious Dogs', *Circ. Res.*, *43*, 200-8

Salisbury, P.F., Galletti, P.M., Levin, R.J. & Rieben, P.A. (1959) 'Stretch Reflexes from the Dog's Lung to the Systemic Circulation', *Circ. Res.*, *7*, 62-7

Sampson, S.R. & Vidruk, E.H. (1975) 'Properties of 'Irritant' Receptors in Canine Lung', *Resp. Physiol.*, *25*, 9-22

Sant'Ambrogio, G. (1982) 'Information Arising from the Tracheobronchial Tree of Mammals', *Physiol. Rev.*, *62*, 531-69

Sant'Ambrogio, G. & Miserocchi, G. (1973) 'Functional Localization of Pulmonary Stretch Receptors in the Airways of the Cat', *Arch. Fisol.*, *70*, 3-9

Sant'Ambrogio, G., Miserocchi, G. & Mortola, J. (1974) 'Transient Responses of Pulmonary Stretch Receptors in the Dog to Inhalation of Carbon Dioxide', *Resp. Physiol.*, *22*, 191-7

Sant'Ambrogio, G. & Mortola, J.P. (1977) 'Behaviour of Slowly Adapting Stretch Receptors in the Extrathoracic Trachea of the Dog', *Resp. Physiol.*, *31*, 377-85

Sant'Ambrogio, G. & Widdicombe, J.G. (1965) 'Respiratory Reflexes Acting on the Diaphragm and Inspiratory Intercostal Muscles of the Rabbit', *J. Physiol. (Lond.)*, *180*, 766-79

Schwieler, G.H. (1968) 'Respiratory Regulation during Postnatal Development in Cats and Rabbits and some of its Morphological Substrate', *Acta Physiol. Scand.*, *304*, (Suppl.), 72

Scott, M.J. (1966) 'The Effects of Hyperventilation on the Reflex Cardiac Response from the Carotid Bodies in the Cat', *J. Physiol. (Lond.)*, *186*, 307-20

Sellick, H. & Widdicombe, J.G. (1969) 'The Activity of Lung Irritant Receptors during Pneumothorax, Hyperpnoea and Pulmonary Vascular Congestion', *J. Physiol. (Lond.)*, *203*, 359-81

Sellick, H. & Widdicombe, J.G. (1970) 'Vagal Deflation and Inflation Reflexes Mediated by Lung Irritant Receptors', *Q. J. Exp. Physiol.*, *55*, 153-63

Sheldon, M.I. & Green, F.J. (1982) 'Evidence for Pulmonary CO_2 Chemosensitivity: Effects on Ventilation', *J. Appl. Physiol.*, *52*, 1192-7

Shepherd, J.T. (1981) 'The Lungs as Receptor Sites for Cardiovascular Regulation', *Circulation*, *63*, 96-101

Simonsson, B.G., Skoogh, B.E., Bergh, N.P., Andersson, R. & Svedmyr, N. (1973) '*In Vivo* and *in Vitro* Effect of Bradykinin on Bronchial Motor Tone in Normal Subjects and Patients with Airways Obstruction', *Respiration*, *30*, 378-88

Spann, R.W. & Hyatt, R.E. (1971) 'Factors Affecting Upper Airways Resistance in Conscious Man', *J. Appl. Physiol.*, *32*, 460-6

St. John, W.M. (1979) 'An Analysis of Respiratory Frequency Alterations in Vagotomized Decerebrate Cats', *Resp. Physiol.*, *36*, 167-86

Stransky, A., Szereda-Przetaszewska, M. & Widdicombe, J.G. (1973) 'The Effects of Lung Reflexes on Laryngeal Resistance and Motoneurone Discharge', *J. Physiol. (Lond.)*, *231*, 417-38

Szereda-Przetaszewska, M. & Widdicombe, J.G. (1973) 'The Effect of Intravascular Injections of Veratrine on Laryngeal Resistance to Airflow in Cats', *Q. J. Exp. Physiol.*, *58*, 379-85

Tallman, R.D. Jr. & Grodins, F.S. (1982) 'Intrapulmonary CO_2 Receptors and Ventilatory Response to Lung CO_2 Loading', *J. Appl. Physiol.*, *52*, 1271-6

Thames, M.D. & Schmid, P.G. (1979) 'Cardiopulmonary Receptors with Vagal Afferents Tonically Inhibit ADH Release in Dogs', *Am. J. Physiol.*, *237*, H299-H304

Thames, M.D. & Schmid, P.G. (1981) 'Interaction between Carotid and Cardiopulmonary Baroreflexes in Control of Plasma ADH', *Am. J. Physiol.*, *241*, H431-4

Thompson, W.K., Marchak, B.E., Bryan, A.C. & Froese, A.B. (1981) 'Vagotomy Reverses Apnea Induced by High Frequency Oscillatory Ventilation', *J. Appl. Physiol.: Resp. Environ. Exercise Physiol.*, *51*, 1484-7

Thoren, P.N. (1979) 'Role of Cardiac Vagal C-fibers in Cardiovascular Control', *Rev. Physiol. Biochem. Pharmac.*, *86*, 1-94

Thoren, P.N., Donald, D.E., Shepherd, J.T. (1976) 'Role of Heart and Lung Receptors with Non-medullated Vagal Afferents in Circulatory Control', *Circ. Res.*, *38* (Suppl. II), II-2-II-9

Thoren, P.N., Shepherd, J.T. & Donald, D.E. (1977) 'Anodal Block of Medullated Cardiopulmonary Vagal Afferents in Cats', *J. Appl. Physiol.*, *42*, 461-5

Trenchard, D., Gardner, D. & Guz, A. (1972) 'Role of Pulmonary Vagal Afferent Nerve Fibres in the Development of Rapid Shallow Breathing in Lung Inflammation', *Clin. Sci.*, *42*, 251-63

Trippenbach, T., Zinman, R., Mozes, R. & Murphy, L.M. (1979) 'Differences and Similarities in the Control of Breathing Pattern in the Adult and Neonate', in von C. Euler & H. Langecrantz (eds), *Wener-gren Center International Symposium Series*, vol. 32, *Central Nervous Control Mechanisms in Breathing*, Pergamon Press, pp. 313-26

Trzebski, A. (1980a) 'Discussion', in H.P. Koepchen, S.M. Hilton & A. Trzebski (eds) *Central Interaction between Respiratory and Cardiovascular Control Systems*, Springer, Berlin, Heidelberg, New York, p. 195

Trzebski, A., Raczkowska, M. & Kubin, L. (1980b) 'Influence of Respiratory Activity and Hypocapnia on the Carotid Baroreceptor Reflex in Man', in

P. Sleight (ed.), *Arterial Baroreceptors and Hypertension*, Oxford University Press, 282-90

Trzebski, A., Raczkowska, M. & Kubin, L. (1980c) 'Carotid Baroreceptor Reflex in Man, its Modulation over the Respiratory Cycle', *Acta Neurobiol. Exp.*, *40*, 807-20

Vatner, S.F., Bottcher, D.H., Heyndrick, G.R. & McRitchie, R.J. (1975) 'Reduced Baroreflex Sensitivity with Volume Loading in Conscious Dogs', *Circ. Res.*, *37*, 231-42

Vatner, S.F. & Rutheford, J.D. (1981) 'Interaction of Carotid Chemoreceptor and Pulmonary Inflation Reflexes in Circulatory Regulation in Conscious Dogs', *Fed. Proc.*, *40*, 2188-93

Wasserman, K., Whipp, B.J., Casaturi, R., Huntoman, D.J., Castagna, J. & Lugliani, R. (1975) 'Regulation of Arterial PCO_2 during Intravenous CO_2 Loading', *J. Appl. Physiol.*, *38*, 651-6

Whitteridge, D. (1948) 'Afferent Nerve Fibers from the Heart and Lungs in the Cervical Vagus', *J. Physiol. (Lond.)*, *107*, 496-512

Widdicombe, J.G. (1954a) 'Respiratory Reflexes from the Trachea and Bronchi of the Cat', *J. Physiol. (Lond.)*, *123*, 55-70

Widdicombe, J.G. (1954b) 'Receptors in the Trachea and Bronchi of the Cat', *J. Physiol. (Lond.)*, *128*, 71-104

Widdicombe, J.G. (1961) 'Respiratory Reflexes in Man and Other Mammalian Species', *Clin. Sci.*, *21*, 163-70

Widdicombe, J.G. (1963) 'Regulation of Tracheobronchial Smooth Muscle', *Physiol. Rev.*, *43*, 1-37

Widdicombe, J.G. (1966) 'Action Potentials in Parasympathetic and Sympathetic Nerve Efferent Fibers to the Trachea and Lungs of Dogs and Cats', *J. Physiol. (Lond.)*, *186*, 56-88

Widdicombe, J.G. (1974) 'Reflexes from the Lungs in the Control of Breathing', in R.J. Linden (ed.), *Recent Advances in Physiology*, Churchill, vol. 9, 239-78

Widdicombe, J.G. & Glogowska, M. (1973) 'Relative Roles of Irritant, Type J and Pulmonary Stretch Receptors in Lung Reflexes', *Acta Neurobiol. Exp.*, *33*, 21-31

Widdicombe, J.G., Kent, D.C. & Nadel, J.A. (1962) 'Mechanism of Bronchoconstriction during Inhalation of Dust', *J. Appl. Physiol.*, *17*, 613-21

Widdicombe, J.G. & Nadel, J.A. (1963) 'Reflex Effects of Lung Inflation on Tracheal Volume', *J. Appl. Physiol.*, *18*, 681-6

Widdicombe, J.G. & Sterling, G.M. (1970) 'The Autonomic Nervous System and Breathing', *Arch. Int. Med.*, *126*, 311-29

Widdicombe, J.G. & Winning, A.J. (1976) 'The Effect of Lung Reflexes on the Pattern of Breathing in Cats', *Resp. Physiol.*, *27*, 253-66

Wiemer, W. & Kiwull, P. (1972) 'The Role of the Vagus Nerves in the Respiratory Response to CO_2 under Hyperoxic Conditions', *Pflüg. Arch.*, *336*, 147-70

Willette, R.N. & Sapru, H.N. (1982) 'Pulmonary Opiate Receptor Activation Evokes a Cardiorespiratory Reflex', *Eur. J. Pharm.*, *78*, 61-70

Winning, A.J. & Widdicombe, J.G. (1976) 'The Effect of Lung Reflexes on the Pattern of Breathing', *Resp. Physiol.*, *27*, 253-66

Yu, D.Y., Galant, S.P. & Gold, W.M. (1972) 'Inhibition of Antigen Induced Bronchoconstriction by Atropine in Asthmatic Patients', *J. Appl. Physiol.*, *32*, 823-8

Younes, M.K., Remmers, J.E. & Baker, J. (1978) 'Characteristics of Inspiratory Inhibition by Phasic Volume Feedback in Cats', *J. Appl. Physiol.: Resp. Environ. Exercise Physiol.*, *45*, 80-6

Younes, M. & Riddle, W. (1981) 'A Model for Relation between Respiratory Neural and Mechanical Outputs', *J. Appl. Physiol.: Resp. Environ. Exercise*

Physiol., *51*, 963-78
Zechman, F.W. Jr., Salzano, J. & Hall, F.G. (1958) 'Effect of Cooling the Cervical
   Vagi on the Work of Breathing', *J. Appl. Physiol.*, *12*, 301-4

5 TISSUE OXYGEN TRANSPORT IN HEALTH AND DISEASE

H. Acker

Introduction

The human organism needs a continuous supply of oxygen to maintain its specific cell functions under different working conditions (for example suitable heart rate, a well-modulated excretory capacity of the kidneys and appropriate activity of the central nervous system). Oxygen is a vital substrate for the human body, for no metabolic pathway can generate this substance for the organism. The human body has developed different mechanisms to transport oxygen to cells and to transport the CO_2 which is produced as a waste product of O_2 consumption away from the cells. Figure 5.1 shows the separate steps of (a) oxygen transport from the outer environment to the cell and (b) CO_2 transport out of the body. The composition of respiratory gases in the lung alveoli is determined (a) by the respiratory movements which transport oxygen from the air to the lung alveoli and CO_2 in the opposite direction, and (b) by the blood circulation in the lung which transports CO_2 and O_2. By external respiration, we mean the gas exchange between lung alveoli and lung capillaries. Cardiac output and its relative distribution to the organs and tissue regulates the transport of respiratory gases from lung capillaries into body capillaries. The binding properties of blood play a decisive role in respiratory gas transport. Internal respiration, oxidative cell metabolism with its O_2 consumption and formation of CO_2 and energy, concludes this circle (Piiper, 1972).

Oxygen Transport to Tissue

The oxygen transport to the tissue and thus, a sufficient supply of oxygen for the tissue, mainly depends on the following:
1. An adequate capillary network and microcirculation in the tissue and the O_2 transport capacity of the blood.
2. Transport of oxygen by diffusion.
3. Oxygen pressure in tissue and its distribution.

157

Figure 5.1: Route of O_2 and CO_2 in the Organism with the Aid of External Respiration, Circulation and Internal Respiration.

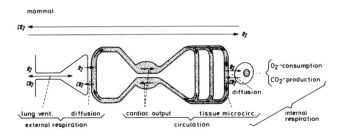

Source: Piiper, 1972.

4. Critical mitochondrial oxygen pressure.
5. Local oxygen consumption.

Since the oxygen is utilised in the mitochondria, it has to be transported from the capillaries into the tissue to the mitochondria. This mainly happens by diffusion. A concentration equilibrium in general appears if freely moveable particles in different concentrations (activities) are available at different parts of a medium; i.e. particles move in the direction of the concentration gradient until the concentration difference has been adjusted (Lübbers, 1972).

According to diffusion laws the quantity of oxygen moved by diffusion amounts to:

$$V = -D\text{grad } u = -D \cdot \alpha \text{ grad } P = -K \text{ grad } p \qquad (1)$$

where V = flow (ml cm^{-2} s^{-1}); D = diffusion coefficient (cm^2 s^{-1}); u = concentration (g cm^{-3}); α = solubility coefficient (ml cm^{-3} torr^{-1}); P = gas pressure (torr); K = diffusion conductivity (ml cm^{-2} s^{-1} torr^{-1}). Therefore, the quantity depends on the diffusion coefficient or diffution conductivity and the concentration or pressure gradient.

It is essential to understand that when gases diffuse in a liquid phase, it is the gas pressure, not gas concentration, which determines diffusion. The following example helps clarify the relation (Lübbers, 1970). Two fluids, F_1 and F_2, with a gas solubility, a_1 and a_2, are contained in two tanks. In fluid F_1 the gas concentration (u_1) amounts to $a_1 \cdot P$ (Henry's law), in fluid F_2, the gas concentration (u_2) amounts to $a_2 \cdot P$. If a gas-permeable membrane is put up between the two fluids, then the concentrations, u_1 and u_2, do not change as long as the gas pressure, P, is the same in both fluids.

If gas solubility, a_2, for example, is twice as high as a_1 the gas concentration in fluid F_2 is then twice as high as in fluid F_1; despite the high concentration differences between F_1 and F_2, no concentration change takes place as the gas pressure in both fluids is the same. In systems with different gas solubilities, therefore, it is not sufficient to determine the oxygen concentration, but one must also know the gas pressure. In the organism such differences in solubility occur, for example in the water and lipid phases. They are even more marked in blood-containing tissue. Here, oxygen combines with haemoglobin and myoglobin and thereby produces an apparently increased solubility (apparent-solubility $\alpha^\star$ (Opitz & Thews, 1952)), which can be a hundred times higher than in serum or tissue. Therefore, at a given oxygen pressure difference the effectiveness of oxygen transport is determined by the product $\alpha \times D$, where D is the oxygen diffusion coefficient. One and the same oxygen gradient transports an increased quantity of oxygen if α or D have increased. At increasing solubility- or diffusion-coefficients the gradient can become smaller without changing diffusion flow. Oxygen is transported to the cells by diffusion through the capillary wall and the interstitium. Since oxygen is consumed in the tissue, gradients develop between tissue and capillary. For reasons of high solubility, especially in the lipid phase, oxygen can diffuse into the tissue through the whole surface of the capillary wall. There is no reason to believe that diffusion conditions are different in capillaries and in the surrounding tissue (Lübbers, 1978a). Zander & Land (1975) found that despite the lipid-fraction oxygen solubility is proportional to the water content of the tissue. Solubility becomes very small, if the water content falls below 60 per cent. Acker, Lübbers & Purves (1971) have shown the presence of oxygen-impermeable zones in the organism. The diffusion coefficient of oxygen also depends on the protein concentration (Kreuzer, 1950); thus an interaction with protein changes the diffusion coefficient for oxygen (Lübbers, 1978a), and changing the albumin-concentration in plasma produces a greater effect on oxygen diffusion than does changing solutes (Navari, Gainer & Hall, 1970). In biological structures like muscle, brain, kidney and liver the diffusion coefficient only changes to a small extent. In Table 5.1 some values of diffusion coefficient, D, are given for different tissues. The temperature coefficient of D (Q_{10}), amounts to 1.2 for water and 1.3 for lung tissue (Lübbers, 1978). Other transport mechanisms, for example convective transport and the diffusion of myoglobin molecules loaded with oxygen, are not accounted for as it is not yet known

Table 5.1: Oxygen (O_2) Diffusion Coefficient, Conductivity and Solubility in Different Biological Materials (Lübbers, 1978)

	O_2 diffusion coefficient $D \times 10^5$ ($cm^2\ s^{-1}$)	O_2 solubility coefficient $\alpha \times 10^2$ ($mlO_2\ (ml \cdot atm)^{-1}$)	O_2 conductivity coefficient $K \times 10^5$ ($mlO_2\ (cm \cdot min \cdot atm)^{-1}$)	References
			T = 20 °C	
Water	2.3	3.1	4.25	Grote, 1967;
Red cell	0.80	3.0	1.1	Gertz & Loeschcke, 1954
Heart muscle (rat)	1.0	2.0	1.6	Grote & Thews, 1962
Grey matter of brain (rat)	1.2	2.7	2.0	Thews, 1960
Lung (rat)	1.5	2.2	2.0	Grote, 1967
			T = 37 °C	
Water	3.3	2.4	4.7	Grote, 1967
Red cell	0.8	2.6	1.3	Grote & Thews, 1962
Heart muscle	1.55	2.1	1.9	
Grey matter of brain (rat)	1.7	2.2	2.3	Thews, 1960
Lung (rat)	2.3	1.8	2.5	Grote, 1967

how far they inflence the critical boundary conditions. Oxygen pressure is the main influence when transport occurs by diffusion. By applying diffusion laws the PO_2 decrease in tissue, ΔPO_2 [tis] can be calculated (Lübbers, 1972; Lübbers & Leniger-Follert, 1978; Lübbers, 1981; Thews, 1960).

The dimension of the PO_2 gradient depends on the following:

1. The diffusion properties of tissue.
2. The spatial distribution of the tissue and oxygen uptake.
3. The capillary distance and capillary structure.

This can be expressed by the following equation

$$\Delta PO_2 \text{ [tis]} = \frac{A[O_2]}{2\alpha . D} g(r_t, r_c, r_z) \tag{2}$$

with

$$A[O_2] = [a[O_2] - v[O_2]] . F \tag{3}$$

and

$$g(r_t, r_c, r_z) = r_z^2 \ln \left(\frac{r_t}{r_z}\right) - (r_c^2 - r_t^2) \tag{4}$$

where $a[O_2]$ = arterial O_2 content; $v[O_2]$ = venous O_2 content; F = blood flow; $A[O_2]$ = oxygen uptake; $2\alpha . D$ = O_2 solubility coefficient; r_c = radius of capillary; r_t = distance of PO_2 measuring point from capillary centre; r_z = radius of tissue cylinder.

Equation (3) (Krogh-Erlang) shows that the O_2 consumption is linearly related to the oxygen pressure difference which is necessary to transport the oxygen into the tissue and that the geometry of the tissue enters approximately as a squared function (Lübbers & Leniger-Follert, 1978; Lübbers, 1981). The actual tissue PO_2 (P_tO_2) depends on the arterial PO_2, (P_aO_2), on the PO_2 decrease along the capillary, (ΔPO_2 [cap]) and on the PO_2 decrease within the tissue (ΔPO_2 [tis])

$$P_tO_2 = P_aO_2 - \Delta PO_2 \text{ [tis]} - \Delta PO_2 \text{ [cap]} \tag{5}$$

One can recognise from equations (2, 3, and 5) that the tissue PO_2 is a function of the blood flow and the haemoglobin dissociation curve when P_aO_2, $A[O_2]$, $\alpha . D$ and $g(r_t, r_c, r_z)$ are constant.

Measurements of oxygen pressure in tissue can be carried out with the Pt-electrodes using a polarographic measuring method (Barankay

Figure 5.2: PO_2 Multiwire Electrode for Measuring Local PO_2 in Tissue. In the left upper side the measuring field of the electrode is shown in relation to seven liver lobules.

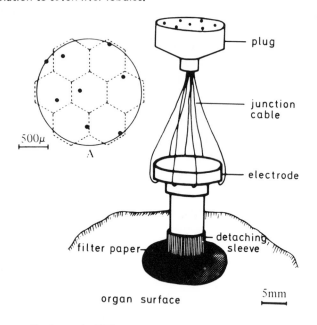

Source: Kessler *et al.*, 1974.

et al., 1976; Baumgärtl *et al.*, 1972; Baumgärtl & Lübbers, 1973; Bicher & Knisely, 1970; Lübbers, 1968b, 1969, 1977; Lübbers *et al.*, 1974; Rodenhäuser *et al.*, 1971; Silver, 1965, 1969; Whalen, Riley & Nair, 1967). If, by fitting a membrane, care is taken that oxygen can reach the platinum surface only by diffusion, then the O_2 flux through the membrane, and, consequently, the O_2 reduction current are dependent on the PO_2 of the outer membrane surface. Such an arrangement allows for PO_2 measurements on the outer membrane surface (Gleichmann & Lübbers, 1960; Grunewald, 1971; Heyrovsky & Kuta, 1965; Lübbers, 1970). For *in situ* measurements platinum multi-wire electrodes are available (Figure 5.2), which may be used on organ surfaces and platinum microelectrodes with a tip diameter of 1-3 μm can (Figure 5.3) be used for measuring the PO_2 within tissues. Figure 5.4 shows as an example the local PO_2 in brain measured by vertically inserting a platinum microelectrode into the grey matter. The local PO_2 varies; it is low between the capillaries and can reach arterial

Figure 5.3: Platinum Micro Electrode for Measuring PO_2 in Tissue. The electrode consists of a platinum wire, the corroded tip of which is melted into glass. The overall diameter amounts to 1–3 μm.

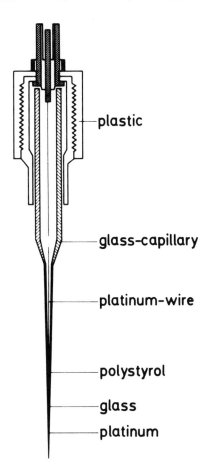

plastic

glass-capillary

platinum-wire

polystyrol

glass

platinum

Source: Schuchhardt & Lübbers, 1969.

values when the electrode tip approaches the arterial part of a capillary. This demonstrates, that in tissue oxygen is mainly transported by diffusion. To characterise the O_2 supply of an organ, measured PO_2 values are divided into classes, (e.g. classes of 5 mm Hg). The PO_2 values can then be plotted as a frequency distribution histogram. Under normal supply conditions and undisturbed microcirculation most

Figure 5.4: PO_2 Distribution in the Cerebral Cortex of Guinea Pig. Tissue PO_2 was measured with a platinum microelectrode by penetrating the cortex (solid line). Ordinate: PO_2, abscissa: penetration depth. For control, PO_2 on the surface of cortex was measured throughout the experiment (dotted line, abscissa: time in min).

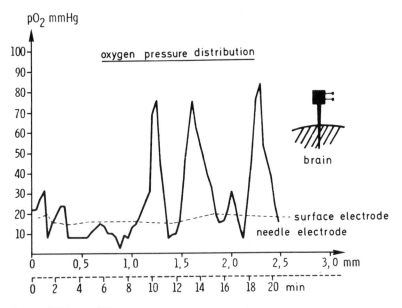

Source: Lübbers, 1972.

organs show a similar PO_2 histogram (Lübbers, 1978). High and low PO_2 values appear side by side. Figure 5.5 shows the PO_2 histogram of rat brain. The high PO_2 values correspond to the arterial PO_2, the low values lie below the venous PO_2, and in a normally perfused brain can be as low as 1–2 mm Hg. The PO_2 measurements clearly show that it is impossible to quote a PO_2 for an organ which represents the O_2 supply to the tissue; this can only be appreciated by reference to a PO_2 histogram such as that of Figure 5.5. Thus even mean PO_2 can simply give evidence of an extremely deficient O_2 supply. Boundary problems cannot be registered because in this case high and low oxygen pressures average out (Lübbers, 1970).

The oxygen supply of an organ can only be quantitatively described, therefore, by quoting the oxygen pressure field, because local oxygen pressure is determined by both, local oxygen supply and local oxygen consumption.

Figure 5.5: PO_2 Frequency Distribution of Rat Cortex (PO_2 Histogram). Oxygen pressures are combined in classes of 5 mm Hg.

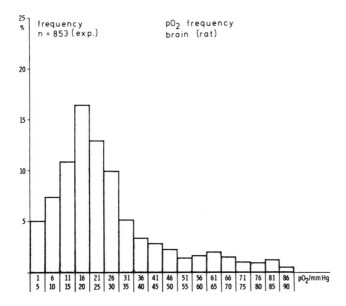

Source: Lübbers, 1972.

Aerobic metabolism produces CO_2, the amount corresponding to the respiratory quotient which lies between 0.7 and 1.0 ($RQ = CO_2/O_2$). According to Henry's law, at increased CO_2 concentration increased PCO_2 results.

$$PCO_2 = \frac{[CO_2]}{\alpha CO_2} \tag{6}$$

$[CO_2]$ = CO_2 concentration; αCO_2 = solubility coefficient for CO_2.

For the same increase in concentration of oxygen and CO_2, the increase of the PCO_2 is much smaller than of PO_2, because the solubility coefficient of CO_2 is 24 times higher than of O_2. ($\alpha CO_2 = 0.545$, $\alpha O_2 = 0.232$ ml O_2 or CO_2 per ml water and 760 ml Hg at 38 °C.) For the same concentration increase in CO_2 and O_2 ($\Delta CCO_2 = \Delta CO_2$) we find the following relation (Lübbers, 1978):

$$\frac{\Delta PCO_2}{\Delta PO_2} = \frac{\alpha O_2}{\alpha CO_2} \frac{1}{24} \tag{7}$$

If a respiratory quotient of 1 is assumed, then one can only measure a very small PCO_2 gradient in tissue. Siesjö (1961) and Gleichmann *et al.* (1962) measured the PCO_2 on the brain surface. The mean tissue PCO_2 was about 1 mm Hg higher than the mean capillary PCO_2. The mean capillary PCO_2 was calculated from the arithmetic mean of arterial and venous PCO_2.

The PCO_2 increase mainly depends on the CO_2 dissociation curve of haemoglobin and the ion concentration in blood. PCO_2 measurements suggest, that CO_2 in tissue is mainly transported by diffusion. For CO_2 the same diffusion laws apply as for O_2 (eqn (1)).

Influence of Capillary Architecture on Oxygen Supply to the Tissues

Apart from the oxygen consumption, oxygen pressure distributions are determined by the size of the capillary supply area. This is a complex problem. Determining the mean capillary distance of an organ by counting the capillaries and measuring the surrounding cells is inadequate since it is not only the arrangement of venous and arterial capillary portions which decide PO_2 distribution in tissue but also the perfusion direction in the capillaries (Lübbers, 1967; 1968b). Different models have been devised to enable better analysis. The Krogh model with parallel running equidirectionally perfused capillaries (cylinder model), the Diemer model with parallel running, counter current perfused capillaries (conic model) and the Grunewald–Lübbers model with asymmetrical and opposingly perfused capillaries. Krogh (1918/1919, 1924) suggested that tissues could be described by parallel running and parallel perfused capillaries which supply a cylindrical space (Figure 5.6). It is characteristic of this model that the PO_2 is the same in all capillaries at the same distance from their arteriolar end. Tissue at the venous end of the capillary may well be anoxic in spite of enough oxygen at the arteriolar end of the cylinder. Diemers' (1963, 1965a, b) studies on the capillaries in newborn brain cast doubt on the general validity of this model and suggested that capillaries are not perfused homogeneously.

A large supplying circuit exists at the arterial capillary and a small one at the venous capillary, which results in a supplying cone instead of a supplying cylinder (Figure 5.6). The advantage of this model is, that the existing arterial PO_2 can be completely used for the oxygen transport in the tissue and under certain conditions the PO_2 at the venous end can even be as low as the lowest PO_2 in the tissue. Therefore,

Figure 5.6: O_2 Supply Areas in Tissue with Different Capillary Loops (PO_2 Field).

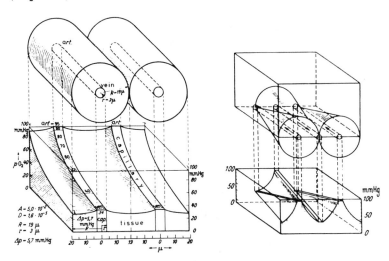

Sources: cylinder model according to Krogh (left) (see Opitz & Schneider, 1950); cone model according to Diemer (Diemer, 1965; Lübbers, 1972) (right).

in the Diemer cone more of the available oxygen pressure is utilised for oxygen supply than in the Krogh cylinder, and there is a high arterio-venous PO_2 difference. It then becomes apparent that, due to this high arterio-venous PO_2 difference, and with arterial and venous capillary ends lying directly opposite each other (capillary distance 50-60 μm), oxygen may diffuse past the tissue into the venous capillary. In such a case, an increase of capillary PO_2 produces but little improvement in tissue oxygen supply, because the oxygen is immediately shunted from the tissue via the vein. Lübbers (1970) and Grunewald (1968) arranged the efflux of capillaries asymmetrically. This results in capillary structures which allow an especially favourable utilisation of PO_2 for O_2 transport (Figure 5.7). In the asymmetric capillary model, O_2 shunt diffusion is less important because oxygen having reached a capillary area does not diffuse into the venous capillary and is therefore not lost for the O_2 supply to tissue.

Figure 5.7 shows PO_2 fields for the Diemer cone and for the asymmetric capillary structure, calculated according to Grunewald (1969). The capillaries pass at the edges of the tissue square. Above each section through the tissue square, the respective PO_2 field has been registered.

Figure 5.7: PO_2 Distribution in Tissue with Symmetric and Asymmetric Capillary Loops. The tissue square is drawn in five series of sections. The PO_2 profile is inserted above each section. The upper half of the figure shows a symmetric capillary structure with antiparallel perfusion, whilst the lower half shows the effect of moving the arterial and venous ends of the capillary towards each other under identical boundary conditions (asymmetric capillary loop). An anoxic zone only occurs with the symmetric antiparallel capillary structure.

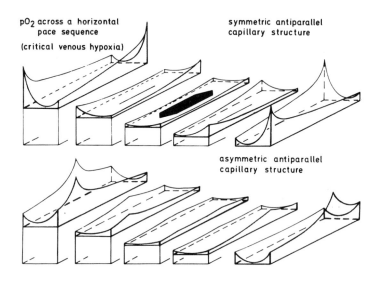

Source: Lübbers, 1972.

Under identical conditions it may be seen that in the Diemer model there is a hypoxic region in the clear countercurrent structure (black hatching), whereas no hypoxic zone is found in the asymmetric capillary structure model. It is clearly shown in these models, that not only capillary distance but also spatial arrangement and perfusion of capillaries are decisive in regulating oxygen supply. Schuchhardt *et al.* (1978) have attempted to relate capillary structure to the PO_2 histogram of heart which had been measured by them (Figure 5.8). The lower part of the figure schematically shows a microscopic cutout of myocardial capillaries according to Fabel (1968). Oxygen pressures existing in the tissue have been drawn in relief in the upper half of the diagram. This relief represents only a snap shot in the spatial oxygen flow which continuously interpasses all living tissue, and naturally, can be altered by change of microcirculation, i.e. capillary flow, O_2 consumption or arterial PO_2.

Figure 5.8: Relief of Oxygen Pressure Occurring in a Two-dimensional Block of Myocardium in a Working Heart. Lower part: schematic drawing of a block of myocardium (dimension ca. 300 μm by 250 μm) showing muscle cells and capillaries (perfusion direction marked with arrows). Upper part: PO_2 values from a myocardial PO_2 histogram have been distributed on the tissue block according to the capillary pattern (arrows = arterial end) and projected to the PO_2 relief as values of altitude. (Base of relief: 0 mm Hg, highest points: ca. 90 mm Hg.)

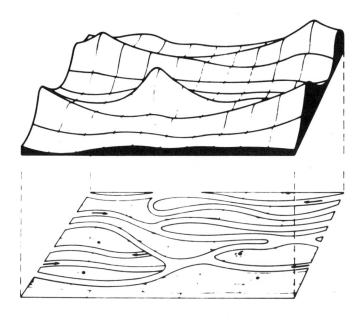

Sources: Fabel, 1968 (lower part), Schuchhardt *et al.*, 1978 (upper part).

Regulation of Oxygen Supply to Tissue

Figure 5.9 shows a range of PO_2 histograms from different tissues. Comparing the mean value of the PO_2 histograms to the venous PO_2, marked by an arrow, one can see that a great number of PO_2 values are smaller than the venous PO_2. Grunewald (1968, 1969) showed, that this cannot be explained by the Krogh cylinder model; the explanation lies in the inhomogeneity of the microcirculation and the fact that capillary length is variable. Kessler (1974a, b) and Kessler *et al.* (1974) have ascertained the relative flow distribution in liver sinusoids. In short capillaries flow was high and the arterio-venous PO_2 decrease little, while in long capillaries flow was low and the corresponding PO_2 decrease high; this is to be expected according to the Hagen-Poiseuill law.

Figure 5.9: PO_2 Histograms of Different Organs Measured with the Needle Electrode and Surface Electrode. Abscissa: oxygen pressures are combined in classes of 10 mm Hg. Ordinate: frequency of measurements. Carotid body: needle electrode, animal; kidney: surface electrode, animal; cortex: needle electrode, human; duodenum: surface electrode, animal; pancreas: surface electrode, animal; liver: surface electrode, animal; heart: needle electrode, animal. Black arrows mark the venous oxygen pressures.

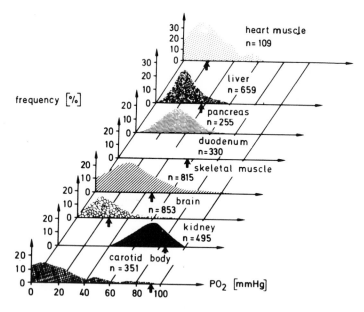

Source: Kessler *et al.*, 1974.

$$I = \frac{\pi r^4}{8 \eta L} \, dp \tag{8}$$

dp = pressure decrease; I = flow velocity, or blood volume (per unit of time); L = capillary length; η = viscosity coefficient; r = radius of vessel. Incidentally from this it may be seen that changing the vessel radius is the most effective method of influencing the flow velocity; doubling the vessel radius from r to 2r increases this parameter by 16 times. The structure of the capillary bed with different capillary lengths may explain why low PO_2 values are found in histograms together with relatively high venous PO_2 values. The carotid body is an outstanding example of this. Acker *et al.* (1971) could show that the maximum tissue PO_2 is 15–20 Torr despite a very high venous PO_2 of nearly 90 Torr. Lübbers (1978) assumed that the terminal flow pattern is caused by a compound of regionally arranged high-flow and normally adapted flow-capillaries. This combination allows a prompt adaptation to local metabolic changes, without influencing the main circulation. These high flow capillaries may possibly be compared with the thorough-fare channels described by Zweifach & Kossmann (1937). Schmidt-Schönbein (1974) in conformity with the findings of Zweifach & Kossmann (1937) suggested a common principle of order for micro-circulation. The nutritive capillary which is responsible for supplying metabolic needs is characteristically a muscle-free endothelial tube with an inner diameter of 3–7 μm. The thoroughfare channels are lying in parallel to the nutritive capillary. An adequate exchange rate of meta-bolic products under normal flow conditions is maintained by the small diameter and the great number and density of the nutritive capillaries. However, in conditions of low flow such as in shock, blood may easily be misplaced and flow through the thoroughfare channels rather than nutritive channels. Whilst it is mainly the precapillary arterioles which control microcirculation, this danger is reduced as the thoroughfare channels possess a permanent muscular wall, while the nutritive capillaries consist of a mainly non-contractile endothelial tube; activity of the muscular component of the thoroughfare channel wall can hence regulate, and to some extent control, flow in this compartment.

Figure 5.10 shows the suggested arrangement thoroughfare channels according to Schmidt-Schönbein (1974). The departure of nutritive capillaries takes place in a relative blunt branching angle and shows functional sphincters (precapillary sphincter). If microcirculation is disturbed, blood flow stops in the nutritive capillaries and most venules, while the thoroughfare channels show a rather well-preserved circulation. Therefore, the thoroughfare channel becomes a preferred channel or, a functional anastomosis. Under rheological favourable conditions and through a relative overperfusion of the thoroughfare channels the tissue possesses a reserve amount of blood which is always available. If required, it can immediately be directed into neighbouring capillary districts without any readjustment in the main or regional circulation. Tissue perfusion can therefore be regulated in such a way that the O_2 supply is adjusted to the O_2 requirement.

Figure 5.10: Thoroughfare Channels of the Microcirculation: Suggested Arrangement (after Schmidt-Schönbein, 1974).

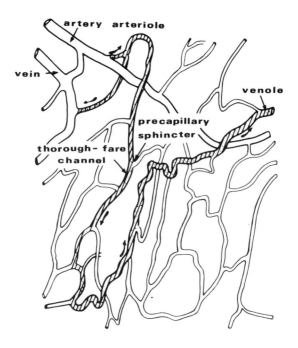

The entrance of red cells into the capillary is a function of the relationship between shear stresses acting either in the direction of the parent vessel flow or towards the capillary orifice. Alterations of the capillary flow are accompanied by changes in the fraction of capillary flow occupied by red cells due to plasma skimming. The distribution of red cells and plasma within a microvascular network is inhomogenous, remarkably low haematocrit values occur in the true capillaries and high haematocrit values in the shunt vessels. Model studies have demonstrated that the most important parameters governing red cell distribution at an arterial bifurcation are the diameter ratio of the cells and vessels, the shape and deformability of the cells, the haematocrit in the main feeder vessels and the ratio of the flow velocities in the two downstream vessels. In the case where the two downstream vessels are of equal diameter, the ratio of haematocrits in the two downstream vessels is a linear function of the ratio of the two blood velocities up to a critical velocity ratio, above which all the cells are essentially swept into the vessel with higher velocity (Cotzelet, 1980). A significant redistribution of haematocrit results from an alteration of volume flow distribution in the microcirculation. This redistribution leads to changes in local oxygen availibility which cannot be predicted on the basis of the flow changes alone (Pries, Gaethgens & Kanzow, 1981; Schmidt-Schönbein, 1981). Figure 5.11 illustrates some of the mechanisms which determine the width of arterioles and opening and closing of the precapillary sphincters in muscle. Apart from adrenergic vasoconstrictor sympathetic fibres, other regulatory mechanisms exist which are of local chemical nature. Substances like adenosine triphosphate, acetycholine and potassium ions are candidates for this local role.

Figure 5.11: The Neural and Local Chemical Regulation of Muscle Perfusion (+, increase; —, decrease in perfusion).

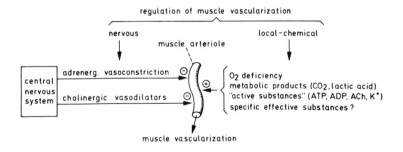

Source: Piiper, 1972.

An O_2 deficiency produces an accumulation of acid metabolic products or specific effective substances which also lead to a vasodilation. The resulting additional flow eliminates or decreases the O_2 deficiency and therewith the concentration of effective substances (Zweifach &

Kossman, 1937). In principle, the above-mentioned regulatory mechanisms are also being discussed in regard to the adaptation of tissue circulation to O_2 requirements in other organs like heart and brain. By electrically stimulating the capillary wall in the rabbit mesentery, Lübbers, Hauck & Weigelt (1976) and Weigelt *et al.* (1981) could reproducibly bring about direct flow changes in the respective capillaries. This suggests that flow regulation also exists at the capillary level and that apart from the smooth muscle cell and the pericyte, a third regulatory element, the specialised endothelial cell (Weigelt *et al.*, 1981) exists in the capillary network, which permits an individual regulation of blood flow in response to neuronal, humoral or metabolic activation. If these regulatory mechanisms are exhausted, further increase of blood circulation is only possible by increasing blood pressure; here the function of heart and circulation indicate the limit attainable. Under such conditions certain flow redistributions are necessary as the total oxygen transport capacity is insufficient for a maximum working efficiency of all organs (Lübbers, 1974).

Transport and Storage Possibilities for Oxygen and their Influence on Oxygen Supply

Oxygen is delivered to the cell via the microcirculation and during this transport is chemically bound to the haemoglobin of erythrocytes and also dissolved in plasma. The presence of haemoglobin influences the oxygen supply in different ways. The solubility of oxygen in plasma at $37\,^{\circ}C$ amounts to 0.003 vol. per cent/mm Hg. In normal blood chemically bound oxygen reaches a maximal value of 20 vol. per cent, at a PO_2 of 150 mm Hg and is 0 at a PO_2 of 0 Torr; in other words haemoglobin increases the oxygen transport capacity of the capillary blood. Above a PO_2 of 150 mm Hg chemically bound oxygen remains approximately constant. The concentration of chemically bound oxygen at complete oxygen saturation of haemoglobin is called the oxygen capacity of blood; it is proportional to the haemoglobin concentration in blood, because haemoglobin binds O_2 in a stoichiometric relation (1 mol O_2/1 g-atom-haemoglobin-iron = 22.4 litres O_2/16200 g Hb = 1.38 ml O_2/1 g haemoglobin) (Piiper, 1972). The haemoglobin concentration of blood is dependent on the number of erythrocytes and the haemoglobin concentration in erythrocytes. Normal values for human blood are: number of erythrocytes: $5,000,000/mm^3$; haemoglobin concentration: 15 g/100 ml blood; O_2 capacity: 20 vol. per cent.

The binding curve of haemoglobin is specially adjusted to oxygen transport. The haemoglobin molecules consist of four units with four iron atoms, which react with oxygen. The S-shaped oxygen-dissociation curve is achieved because the reaction of haemoglobin with oxygen has a series of constants. It is the 'effective O_2 binding curve' between arterial blood and venous blood which is decisive for gas exchange at rest (Thews, 1976).

Figure 5.12: Dependence of the O_2 Binding Curve of Blood on Different Parameters. (A) Temperature dependence, (B) pH dependence (Bohr effect), (C) dependence on CO_2-partial pressure, (D) dependence on intra-erythrocyte concentration of 2, 3-diphosphoglycerate (2, 3-DPG).

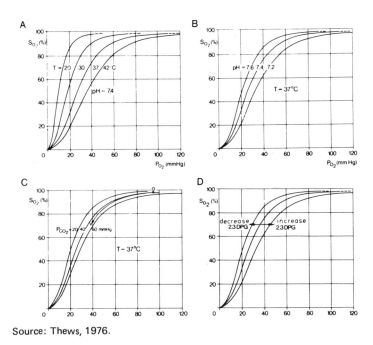

Source: Thews, 1976.

Figure 5.12 shows the oxygen dissociation curve of haemoglobin. In a PO_2 range of 100-40 mm Hg O_2, there is little change in saturation whereas lowering the PO_2 below 40 mm Hg produces a marked desaturation. This arrangement guarantees that PO_2 in capillary and venous blood stays high thus aiding O_2 supply to the tissue by diffusion. In a PO_2 range of 0-40 mm Hg the O_2 binding curve is steep, thus a small decrease in the oxygen pressure in the tissue leads to a relatively large amount of oxygen being released from haemoglobin. A decrease in temperature increases the affinity of haemoglobin for oxygen (left shift), whilst increasing PCO_2 or H^+ concentration leads to a right shift. A decrease of PCO_2 or H^+ concentration leads to a left shift of O_2 binding curve (Bohr effect), mainly caused by hydrogen ions (Piiper, 1972). The Bohr effect aids oxygen uptake in the lung, and also the oxygen release in tissue. CO_2 release and pH increase in the lung leads to a left shift of the binding curve thus allowing additional oxygen uptake at a constant PO_2; in the tissue, where CO_2 is taken up, there is a decreased affinity of haemoglobin and hence at a given PO_2, more oxygen is set free from the haemoglobin. Figure 5.12C shows the oxygen saturation of haemoglobin which is most important for the gas exchange at rest. This figure is based on blood gas values shown in Table 5.2. 2, 3-Diphosphoglycerol, produced on a side path of glycolysis, stabilises the deoxygenated form of haemoglobin and therefore reduces the oxygen affinity of haemoglobin (Figure 5.12D). Acute and chronic hypoxia induce an increase of 2, 3-DPG which may, by this mechanism improve the oxygen release to tissue. The immediate reaction of oxygen binding and oxygen release are decisive for the oxygen supply. The time needed for a 50 per cent desaturation of HbO_2 in erythrocytes amounts to 0.05 s (Lübbers, 1978). At normal temperature, the erythrocyte spends enough time in the capillary to release oxygen in the way described by the O_2 binding curve.

Table 5.2: Blood Gas Data and pH Values in Arterial and Venous Blood for a Healthy Juvenile at Rest (Thews, 1976)

	P_{O_2} (mm Hg)	S_{O_2} (%)	$[O_2]$ (Vol –%)	P_{CO_2} (mm Hg)	$[CO_2]$ (Vol –%)	pH
Arterial Blood	95	97	20	40	50	7.40
Venous Blood	40	73	15	46	54	7.37
Arteriovenous Difference			5		4	

Erythrocytes flowing through the capillaries are being deformed. This results in an intercellular convection increasing the oxygen exchange; it has been suggested that this convective transport of oxygen is quantitatively more important than the diffusion (Zander & Schmidt-Schönbein, 1973).

Oxygen diffusion can be facilitated by simultaneous diffusion of oxyhaemoglobin (for review, see Kreuzer, 1976; Lübbers, 1978). The diffusion of a substance with a higher speed than can be expected from the nature of the substance and diffusion medium is generally known as facilitated diffusion. Possibly, this happens through the participation of a carrier molecule reacting reversibly with the diffusing substance. The facilitation of oxygen diffusion comes about through simultaneous diffusion of oxygen in physical solution and in its haemoglobin bound form. Whilst the physiological significance in the intact organism is uncertain, facilitated diffusion could accelerate uptake and release of O_2 from haemoglobin at the end of the exchange procedure, especially at low oxygen pressures (Kreuzer, 1976).

Figure 5.13: Comparison of CO_2 and O_2 Binding Curve in the Same Proportion. Abscissa: $CO_2 - O_2$ pressure respectively, Ordinate: $CO_2 - O_2$ content, respectively. In general the CO_2 binding curve runs much steeper than the O_2 binding curve. The points entered for arterial and venous blood (a and v) show that at venosity of arterial blood PO_2 decreases much more than CO_2 pressure increases.

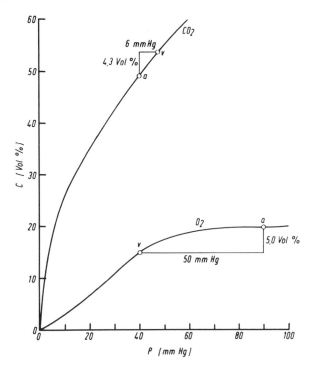

Source: Piiper, 1972.

Since CO_2 mainly influences the position of the haemoglobin dissociation curve through pH-effect, the CO_2 binding curve of blood will be shown in comparison to the O_2 binding curve on the same scale (Figure 5.13). CO_2 is also found in blood in physical solution and chemically bound. The physical solubility of CO_2 in blood at 37 °C is about 0.065 vol. per cent/mm Hg. The main part of CO_2 is reversibly bound as bicarbonate in the erythrocytes, the smaller part as carbamino-binding to free amino groups of haemoglobin. Using the same scale, the course of the CO_2 binding curve is steeper than the O_2 binding curve and does not reach a marked saturation value. An increase in temperature causes a right shift of CO_2 binding curve as does an increase in H^+ concentration. A diminished haemoglobin content flattens the CO_2 binding curve. Under the same PCO_2 conditions deoxygenated blood binds much more CO_2 than the same blood after saturating the haemoglobin with O_2 (Haldane-effect). This causes an additional release of CO_2 in the lung at increasing O_2 saturation. In tissue, the decrease of O_2 saturation allows an additional uptake of CO_2 in blood.

Figure 5.14: O_2 binding curve of haemoglobin Hb (pH = 7.4, T = 37 °C) and myoglobin (Mb).

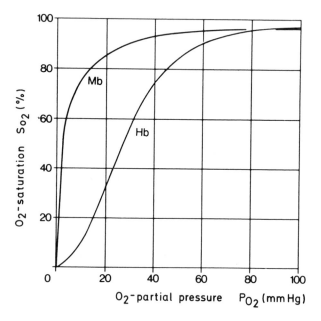

Source: Thews, 1976.

Species of haemoglobin which deviate from the normal structure have a different O_2 transport capacity. The O_2 binding curve of fetal blood, for example, is shifted to the left in comparison to that of the adult. Its affinity is also higher. Thus O_2 exchange in the placenta is favoured.

During the last intrauterine and first extrauterine months fetal haemoglobin is being exchanged by adult haemoglobin. Abnormal haemoglobin in humans (e.g. thalassaemia) is only compatible with life, if it constitutes a small fraction of the total haemoglobin as its O_2 transport capacity is much reduced.

Myoglobin is the second chemical storage mechanism available for O_2 in the organism. In its capacity of oxygen buffer it can increase oxygen supply in heart and skeletal muscle especially under working conditions; the red colour of muscle is mainly due to myoglobin. It has a molecular weight of ca. 17000, contains only one haem molecule and corresponds to a subunit of haemoglobin. Myoglobin has a hyperbolic binding curve, and takes part in the simple reaction $Mb + O_2 \rightleftharpoons Mb-O_2$. In the lower saturation region the O_2 affinity of myoglobin is higher than that of haemoglobin (Figure 5.14). Myoglobin is regarded as O_2 reservoir which releases oxygen at decreasing PO_2. 1 g myoglobin can maximally bind 1.34 ml of oxygen. Hence by means of their differing saturation kinetics, haemoglobin and myoglobin ensure that the PO_2 gradient in tissue is only little changed by O_2 transport, thus allowing a maximal transport rate (Leniger-Follert & Lübbers, 1973). It is known that myoglobin also has a facilitated diffusion (Kreuzer, 1976), and hence it too can theoretically improve oxygen supply. To what extent this facilitated diffusion actually participates in oxygen supply, is still unknown.

Oxygen Consumption of Tissue

The O_2 efflux to tissue per unit of blood flow is defined as O_2 supply, i.e. O_2 supply = blood flow $\times$ O_2 capacity $\times$ O_2 saturation. Depending on the rate of work, a certain quantity of this O_2 supply is used up by the tissue. The O_2 consumption is defined as blood flow $\times$ arterio-venous oxygen concentration difference.

Oxygen consumption is measured in ml $O_2/100$ g wet weight and minute. Table 5.3 shows some oxygen consumption values for different human organs at $36°C$. The table illustrates the variability between organs and the relationship between metabolic activity and O_2 con-

Table 5.3: Mean Values for Blood Flow (Q), the Arteriovenous Difference of O₂ Concentration in Blood (avDO_2) and O₂ Consumption ($\dot{V}O_2$) of Different Human Organs at 37 °C (Grote, 1976).

Organ	Blood flow Q (ml/100 g min)	art.-ven. difference avDO_2 (Vol.-%)	O₂ consumption $\dot{V}O_2$ (ml/100 g mm)	Reference
Blood			0.008 0.01	Lenfant & Aucutt (1965)
Skeletal muscle				
at rest	2 4	10 15	0.25 0.5	Kunze (1969); Golenhofen (1971)
at hard work	up to ca. 50		up to ca. 10	
Spleen	100 130	0.5 0.8	0.6 0.8	Lutz & Bauereisen (1971)
Brain	50 60	6 7	3.5	Vaupel et al. (1973)
cerebral cortex	80 110	10	8 10	Grote (1967) / Hirsch (1971)
white matter	15 25	4 6	1	Lassen (1959)
Liver	100 (25% hepatica)	3 3.5 (V. portae V. hepatica) / 7 10 (A. hepatica V. hepatica)	4.5	Lutz & Bauereisen (1971) / Greenway & Stark (1971)
Kidney				
renal cortex	400 500	1.5 2.0	5.5 6.5	Deetjen (1968)
outer medulla	400	2 2.5	9 10	Kramer et al. (1960)
inner medulla	120	5	6 6.5	Thurau (1971)
	25	1 2	0.3 0.5	
Heart				
at rest	80 90	10 15	8 10	Bretschneider (1961, 1972) / Grote & Thews (1962) / Lochner (1971)
at hard physical stress	up to ca. 400	up to ca. 17	up to ca. 40	

sumption. During work, the oxygen consumption of the heart muscle increases 3-4 times, while working skeletal muscle can increase its O_2 consumption by 20 to 50 times. The relation of O_2 consumption to O_2 supply is called O_2 utilisation.

$$O_2 \text{ utilisation } = \frac{O_2 \text{ consumption}}{O_2 \text{ supply}}$$

In the kidney O_2 utilisation amounts to ca. 10 per cent, in coronary circulation ca. 60 per cent, in resting muscle 40 per cent and in working muscle 90 per cent. For the total blood circulation, under resting conditions, it amounts to 25 per cent, but at maximal physical work utilisation can reach 75 per cent (Grote, 1976).

In the cell, energy is produced by the reaction of O_2 and H_2 which produces water and energy.

$$O_2 + 4e^- + 4H^+ = 2H_2O + \text{free energy}$$

This reaction takes place in the mitochondrial respiratory chain, and is catalysed by the cytochrome aa_3 complex. Hydrogen is supplied as substrate-hydrogen $(S-H_2)$ from carbohydrates, fats and proteins. The energy which is set free is stored in the form of a chemical binding. Thus, adenosine triphosphate (ATP) is formed from adenosine diphosphate (ADP) and phosphate (PO_4^{2-}), and it is this ATP which is available for energy utilisation. When the energy is consumed, ADP and PO_4^{2-} are produced. The magnitude of ATP production is determined by the quantity of ADP and PO_4^{2-} if sufficient oxygen and substrate hydrogen are available, thus, the derivative products regulate the synthesis of ATP. In Figure 5.15 the abscissa schematically shows the single components of the respiratory chain. Oxygen oxidises the cytochrome oxidase (cytochrome $a+a_3$), the cytochrome oxidase oxidises the cytochrome c, cytochrome c the cytochrome b, etc. Experiments with isolated mitochondria have shown that oxidation of cytochrome oxidase is independent of O_2 pressure and that consumption is not reduced before PO_2 has reached 0.05 Torr (critical mitochondrial pressure) (Starlinger & Lübbers, 1973).

Oxidised and reduced conditions show characteristically different extinction spectra and these enable photometric determination of the proportion of total cytochrome in the oxidised state (Chance, 1976; Harbig *et al.*, 1976; Ji *et al.*, 1977; Jöbsis *et al.*, 1971; Oshino *et al.*, 1974).

The respective reduction ratio develops a flow balance, because the oxidation of cytochrome oxidase by O_2 and its reduction by cytochrome c and c_1 proceeds in parallel. The flow balance is dependent on the relative speed of all the participating reactions. In isolated mito-chondria 90-95 per cent of the cytochrome oxidases are oxidised. The other members of the respiratory chain are more reduced, the more they are associated to NAD. 80-90 per cent of NAD^+ is available as reduced NADH. A supply of ADP and PO_4^{2-} activates the respiratory chain. The redox relation is shifted so that the oxidised (= ad.) form clearly predominates (Lübbers, 1974). Wilson *et al.* (1967) showed, that mitochondrial oxidative phosphorylation in intact cells is dependent on oxygen tensions throughout the psysiological range.

Figure 5.15: Redox Pattern of Respiration Chain in Heart Muscle Mitochondria. The redox pattern shows the connection between the stationary redox condition in the respiration chain and the functional condition of mitochondria. Changing from rest (contr.) to active condition (act.) the respiration chain is oxidised (Cyt = cytochrome; UQ = ubiquinone; NAD = nicotinamide-adeninenucleotide).

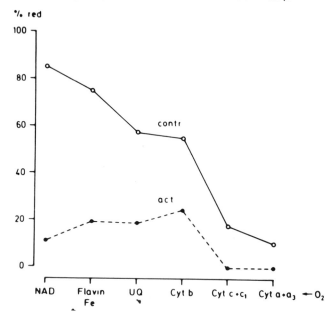

Source: Lübbers, 1974.

This dependence is expressed as a decreased rate of ATP synthesis at a given $NAD^+/NADH$ and $[ATP]/[ADP][P_i]$ value as oxygen tension decreases. The cellular ATP-utilising reactions are in contrast independent of oxygen tension and this causes the rate of ATP utilisation to remain essentially constant despite declining oxygen tension. As a result, the $[ATP]/[ADP][P_i]$ falls until a new steady state is attained in which the rate of ATP synthesis is increased to where it equals the rate of ATP utilisation. The decrease in $[ATP]/[ADP][P_i]$ with decreasing oxygen tension has direct effects on most aspects of cellular metabolism and thus provides an effective cellular measurement of the oxygen tension. It results in increased intracellular AMP, intra- and extracellular adenosine, increased cytoplasmic Ca^{2+}, and modified plasma membrane ion fluxes.

Lübbers & Schuchhardt (1971) have studied a situation where the energy demand changes rhythmically. Figure 5.16 shows the energy demand, time, respiration and tissue oxygen transport in the heart.

Figure 5.16: Oxygen Supply, Tissue Respiration and Energy Need of Heart.

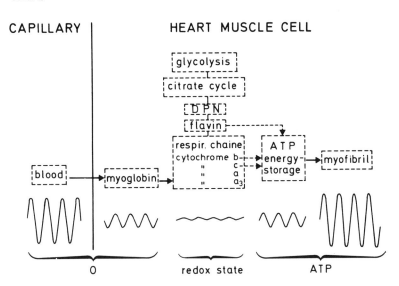

Source: Lübbers, 1974.

In the course of one single cycle, the blood flow and the energy demand of the myocardium changes. Due to the increased pressure, blood flow decreases during systole and therefore the oxygen supply reaches a minimum. Conversely, O_2 supply is maximal in diastole. The greatest energy demand exists during the contractile phase, the smallest at rest. During contraction there is a sufficient supply of oxygen for the respiratory chain derived from haemoglobin in the capillaries and myoglobin in muscle cells, and hence O_2 consumption remains constant despite phasic blood flow. Moreover, additional energy demand can be met by the energy store of the heart muscle (ATP, creatine phosphate).

If difficulties arise in supplying the organism with oxygen, then for a short time the energy need can be met by (1) utilisation of stored oxygen; (2) production of lactic acid by anaerobic glycolysis; (3) utilisation of energy stored in the form of energy-rich phosphates (ATP, creatine phosphate).

In the human body approximately 1.5 litres of oxygen are stored (ca. 400 ml in lung air, 50 ml physically dissolved in tissue, 800 ml chemically bound to haemoglobin and 250 ml chemically bound to myoglobin) (Piiper, 1972).

Figure 5.17: O_2 Uptake and Energy Conversion (Muscle or Total Body) at very hard Muscle Work. Ordinate: Energy conversion — equivalent of marked proceedings. The O_2 debt is entered at the beginning of muscle work. After finishing work this 'O_2 debt' is paid for. The different components of O_2 are marked by different dotted lines.

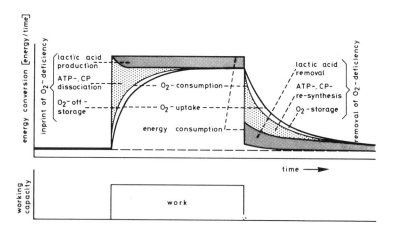

Source: Piiper, 1972.

The only quantitatively important process by which anaerobic energy can be produced is glycolysis, where lactic acid is produced from glycogen and glucose. A considerable part of resulting H^+ ions are buffered by bicarbonate with concomitant release of CO_2. Glycolysis can cause an increase of the RQ, to a level above 1.0 during extreme muscular exercise. Only 6 per cent of the energy which is yielded by the oxidative reduction of glycogen or glucose is released by anaerobic glycolysis, and the output of energy-rich phosphates is correspondingly low. Despite this low energy output anaerobic metabolism is utilised by skeletal muscle. During heavy work a physiological anoxia occurs in the musculature, the O_2 stores are depleted and the production of lactic acid begins. Consequently, less oxygen is taken up than appropriate to the level of metabolism. This deficiency of O_2 is called O_2 deficiency or O_2-debt (Figure 5.17). After finishing work the majority of the lactic acid is resynthesised – mainly in the liver – to glycogen.

A smaller part is oxidatively reduced to CO_2 and H_2O to supply the energy for resynthesis. Oxygen which is needed for this appears as removed O_2-debt. The component of O_2 debt which is combined with the production of lactic acid is called lactacid debt (Piiper, 1972). In addition to anaerobic metabolism the muscle also depletes its energy store, ATP is degraded to ADP and inorganic phosphate, and creatine phosphate degraded to creatine and inorganic phosphate.

Oxygen Deficiency (Hypoxia)

Hypoxia exists when the O_2 supply is insufficient to meet the requirements of the O_2-consuming metabolism. Such O_2 deficiency in tissue causes a characteristic shift to the left in the PO_2 histograms (i.e. a large increase in the number of low PO_2 values). The upper part of Figure 5.18 shows a PO_2 histogram of human tibialis anterior under normal conditions. This histogram has a similar form to those seen in animal experiments, suggesting that similar mechanisms of O_2 supply exist in humans. Kunze (1966, 1969) could show (Figure 5.18, lower part) that a leftshift of PO_2 histograms occurs in the tibialis anterior muscle when the muscle is pathologically changed and an O_2 deficiency occurs in the tissue. Note how even under normoxic conditions sporadic PO_2 values of 1 Torr occur. The critical mitochondrial PO_2 is some 0.05 Torr and hence, at 1 Torr a completely normal oxidative metabolism can be maintained.

Hypoxia in tissue can be produced by obstruction of cellular or tissue oxidation (Büchner, 1962) due to: (a) oxygen deficiency in tissue, (b) lack of substrate in tissue (e.g. due to hypoglycaemia), (c) obstruction of oxidation, (d) obstruction of enzymes regulating the oxidation, (e) uncoupling of oxidative phosphorylation. These different causes of hypoxia can be distinguished. An inadequate tissue PO_2 caused by hypoxia can be due to a variety of outer respiration problems. According to Herzog (1977) these include: (1) an anatomical right–left shunt (perfusion of non-ventilated lung units); (2) a functional right–left shunt (perfusion of hypoventilated lung units); (3) global hypoventilation of the alveolar region; (4) diffusion disturbances (thickening of the alveolar-capillary membrane or, an inadequate contact time between capillary blood and alveolar air to complete gas exchange because of capillary dysfunction.

Figure 5.18: PO_2 Frequency Distribution of Muscle Tibialis of Human. The upper part shows a normal frequency distribution at rest. The lower frequency distribution shows an increase of low values as they are measured at muscle disease: hypoxic histogram.

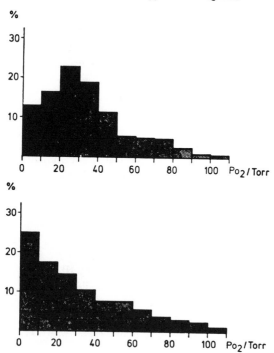

Source: Kunze, 1969.

Furthermore, oxygen deficiency can be triggered by climbing to altitude, where the atmospheric pressure (P_B) sinks nearly exponentially to zero with increasing altitude (Figure 5.19). The effective inspiratory O_2 pressure, P_IO_2, is given by 0.209 × ($P_B - 47$) mm Hg (Piiper, 1972). Oxygen deficiencies which are due to altitude can be imitated by breathing gas mixtures with reduced oxygen content at normal barometric pressure (Figure 5.19). Below a P_IO_2 of 100 mm Hg the tidal volume increases through stimulation of peripheral chemoreceptors. Functional disturbance of the central nervous system occurs with increased altitude which at a height of about 6000 m (corresponding to a P_IO_2 of 60 mm Hg), leads to unconsciousness.

Figure 5.19: O_2 Deficiency in the Atmospheric Air in Dependency of Height. From left to right: height above sea level, barometer pressure (P_B), O_2 pressure in inspired air saturated with steam at 37°C (P_IO_2), O_2 content of equivalent ventilating air mixture for sea level (F_IP_2 äq), physiological reaction etc. at acute exposition and at adjusted height.

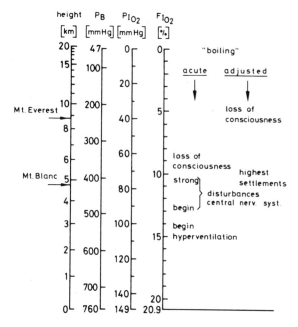

Source: Piiper, 1972.

The values given in Figure 5.19 may only be regarded as an approximation because apart from P_IO_2, many other factors play an important role in hypoxia tolerance. The main factors are: individual disposition, physical and mental condition, and the period and nature of hypoxia. The saturation pressure of water vapor at body temperature, 47 mm Hg, is the absolute limit of pressure decrease tolerance for the organism, because at a barometric pressure of 47 mm Hg which is reached at a height of 19 km, lung alveoli are exclusively filled with water vapour and boiling will start in body fluids. (Boiling point of water at P_B = 47 mm Hg is 37 $^\circ$C) (Piiper, 1972) (see also Chapter 9). Hypoxia is also caused by anaemia. Anaemia is especially frequent after acute blood loss. A deficiency of oxyhaemoglobin can also be caused by carbon monoxide poisoning, here carbon monoxide (CO) is bound to haemoglobin. Carboxihaemoglobin develops:

$$Hb + CO \rightleftharpoons HbCO$$

Figure 5.20: CO Binding in Comparison to O_2 Binding. Abscissa: O_2- or CO_2 pressure. Ordinate: saturation of haemoglobin with CO or with O_2. In the absence of O_2 (PO_2 = 0) the CO binding curve runs very steep, the saturation of 98 per cent is reached at PCO = 1 mm Hg. In the presence of O_2 it comes to a contest between CO and O_2 for Hb, and the CO binding curve is shifted to the right. At PO_2 = 100 mm Hg Hb not bound to CO is saturated with O_2 and a saturation of 98 per cent with CO is not reached before PCO amounting to ca. 25 mm Hg.

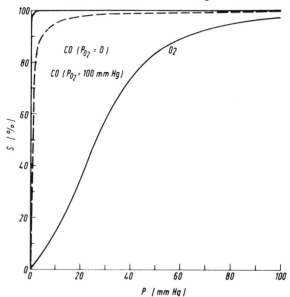

Source: Piiper, 1972.

Figure 5.20 compares the characteristics of CO binding and O_2 binding of haemoglobin, it may be seen that relatively the CO binding curve is extremely steep and shifted to the left. With an oxygen pressure of 100 mm Hg a 98 per cent saturation can be reached. Breathing O_2 shifts the equilibrium towards the formation HbO_2.

Met-haemoglobin which impairs the O_2 transport capacity of haemoglobin can develop after poisoning with a number of compounds including potassium chloride, nitrose gases, death cap poison, analgesics and the endotoxins of severe enteritis. In this condition the iron of haemoglobin is oxidised and unavailable for oxygen transport.

Circulatory abnormalities can also produce abnormal oxygen supply. This occurs mainly in congenital cardiac defects, especially Fallot's tetralogy (pulmonary stenosis, septal defect and overriding aorta). Here, venous blood of low PO_2 flows into the aorta so that O_2 tension in the arterial blood of the systemic circulation is considerably decreased. Thus, the capillary PO_2 is insufficient to supply oxygen to the tissues by diffusion.

The level of venous PO_2 in an organ is a good criterion of whether the intracapillary PO_2 is high enough for an adequate oxygen supply.

In brain the critical venous PO_2 amounts to 17-19 Torr, in heart muscle 7-8 Torr, in kidney about 13 Torr. Moving below these values causes oxygen deficiency in the appropriate organ.

In cases of a critical oxygen supply, when the oxygen supply limits the oxygen consumption, different criteria can be used to characterise the state (Lübbers, 1981). (1) In dog skeletal muscle a critical situation of oxygen supply was produced by reducing P_aO_2; it occurred during rest at a P_aO_2 of 60 mm Hg and a venous PO_2 (P_vO_2) of 25 mm Hg and during work at a P_aO_2 of 50 mm Hg and a P_vO_2 of 10 mm Hg. Although the O_2 consumption during work was 8 times higher than during rest, the blood PO_2 values during work were lower. This difference can be explained by the increased number of perfused capillaries in the working muscle which reduce the supply area of a single capillary, and by the increased flow (Granger & Shepherd, 1979). (2) As long as the O_2 supply of the muscle is sufficient, lactate is consumed, as can be measured by the arterio-venous lactate difference of the dog heart muscle (Bretschneider, 1958). Insufficient oxygen supply is accompanied by lactate production. The transition point from lactate consumption to lactate production is related to the magnitude of the venous PO_2. With an oxygen consumption of 150 μl O_2/g min the transition point is about $P_vO_2 = 6$ mm Hg, with an oxygen consumption of about 50-80 μl O_2/g min at about $P_vO_2 = 2$ mm Hg and at an oxygen consumption of 300 μl O_2/g min at about $P_VO_2 = 14$ mm Hg. (3) The normal venous PO_2 in the sinus sagittalis of the cortex can be decreased from a $P_VO_2 = 35$ mm Hg to a $P_VO_2 = 28$ mm Hg without any detectable reaction of the functional state of the brain. With a further decrease in P_VO_2 blood flow increases to maintain the P_VO_2 close to 28 mm Hg. Further reduction of P_VO_2 produces the first signs of changes in the ECG. With a P_VO_2 smaller than 17-19 Torr or mm Hg man loses consciousness reversibly; this becomes irreversible at a P_VO_2 of 12 Torr (Opitz & Schneider, 1950). Anaerobic glycolysis causes a drastic deterioration of the microcirculation because (as stated by Ardenne, 1979) a decreased pH leads to a stiffening of the erythrocytic membrane and therefore a loss of erythrocytic flexibility (Zander & Schmidt-Schönbein, 1973). Oxygen deficiency in tissue, therefore, strongly increases.

A local hypoxia can also be caused by inadequate perfusion, those resulting from disturbances of arterial influx and venous outflow can be distinguished. Disturbances of arterial influx can induce ischaemia, i.e. blood deficiency. An absolute ischaemia or a complete anoxia exists when blood flow is completely stopped. Under these conditions a number of functional changes occur in cells, e.g. necrosis, fatty degenerations and non-diffuse and diffuse formation of cytoplasmic vacuoles (Becker, 1959).

Some data as regards the tolerance or sensitivity to O_2 deficiency can be quoted for this interruption of O_2 supply (Figure 5.21) (Piiper, 1972).

(1) A short time after beginning of complete anoxia cell function is fully maintained = free interval.

(2) The period from beginning of anoxia up to extinction of normal function, is denoted as survival time (paralysing time, functional maintenance time).

(3) The reactivation time is the time from the beginning of anoxia to the renewal of normal function.

If the anoxia is shorter than the time of reactivation, the cell recovers without damage. A period of recovery is necessary for this. An anoxia lasting longer than the recovery period causes damage which may be partly repairable or induce cell death.

Survival time and recovery time differ markedly from cell to cell. The shortest periods are found for nerve cells, especially for cells of the cerebral cortex with survival times amounting to 10-20 s, and recovery times of 3-5 min. In the kidney the recovery time is 3 h, and 4 h for the liver, whereas the paralysed heart has a recovery period of several hours. In hypothermia the energy need is reduced to about two- to three-fold per 10 °C decrease in temperature and the recovery period is extended (RGT-rule van't Hoff) (Piiper, 1972). In cases where the blood supply is only restricted or insufficient to maintain full normal function, a relative ischaemia is produced. Ehrly & Schroeder (1978) could show, that in patients with intermittent claudication the average muscle PO_2 of 13.3 Torr in the tibialis anterior muscle was clearly lower than in the healthy control groups with an average PO_2 of 27.2 Torr. By administering Arwin, which decreases blood fibrinogen concentration and thereby blood viscosity, exercise tolerance was clearly improved and this was accompanied by an increase of mean tissue PO_2 to 22.9 Torr.

Figure 5.21: Time Course of Peracute Anoxia. The times entered are dimensionally valid for the cortex cerebri.

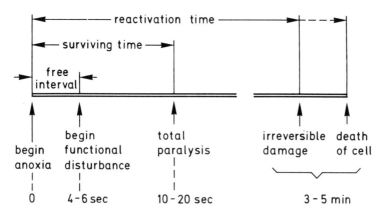

Source: Piiper, 1972.

Kessler *et al.* (1974) and Sunder-Plassmann & Messmer (1974) have shown that in perfusion insufficiency produced by haemorrhagic shock, (Figure 5.22), a combination of ischaemic and anoxic anoxia always exists. Failure to remove tissue metabolites during ischaemic anoxia induces extreme cellular acidosis and can cause irreversible damage to the respiratory chain. Anoxia with maintained perfusion can be tolerated for longer periods without distinct cell damage.

Apart from these perfusion insufficiencies, function perfusion disturbances occur due to contraction of arterial and arteriolar musculature. Functional narrowing or even occlusion can be produced; an example of such a condition is angina pectoris. Schuchhardt *et al.*

Figure 5.22: PO_2 Tissue Histogram of Dog Skeletal Muscle after Haemorrhage.

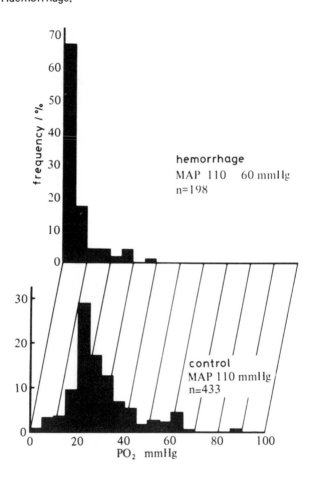

Source: Sunder-Plassmann & Messmer, 1974.

(1978) showed, that the usually used coronary dilatators increase the coronary flow, but that the tissue PO_2 in heart muscle does not increase; this phenomenon may be explained by a preferred perfusion of thoroughfare channels. This is in accordance with findings that the marginal microcirculation of striated muscle shows a reduction in flow velocity in the capillaries when the total flow to the muscle is enhanced by the infusion of vasodilators. A dissociation of total blood flow and capillary exchange capacity is induced by increasing the former and decreasing the latter (Duling & Klitzman, 1980).

During hypoxaemic hypoxia a number of regulatory changes are induced. An increase in respiratory frequency occurs reflexly due to stimulation of peripheral chemoreceptors (Chapter 1) and increased systolic ejection at increased heart frequency followed by an increase of cardiac output is induced. The plasma volume decreases through an increased sodium secretion of the kidney (Honig *et al.*, 1975). In addition the number of erythrocytes increases thus increasing the oxygen transport capacity of blood. It has been suggested that recruitment of capillaries occurs in the heart as a means of locally enhancing oxygen delivery capacity during hypoxia (Duling & Klitzman, 1980). This idea is the subject of much discussion.

Figure 5.23: Dependency of Cerebral Perfusion on Arterial Oxygen and CO_2 Pressures. The arterial CO_2 pressures are shown to the right side of the curve.

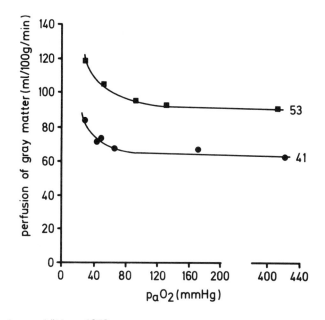

Source: Lübbers, 1972.

Under hypoxia cerebral circulation is increased in comparison to other organs. In Figure 5.23 the dependency of cerebral circulation on oxygen pressure is clearly shown. Whereas in many organs a vaso-dilation occurs during hypoxia (Kontos, 1981; Olsson, 1981), blood flow in the lung actually falls by up to 33 per cent due to vasoconstric-tion (see for review Downing & Lee, 1980); Pittmann (1981) has recently reviewed the evidence that prostaglandins are involved in the anoxic vasoconstriction in bovine coronary and pulmonary arteries. It can be shown in coronary strips that the prostaglandin efflux increased as solution PO_2 was lowered from 112 to 53 and 38 mm Hg and vascular tone decreases in this PO_2 range. Further decreases in PO_2 led to a decline in prostaglandin output and a reversal of the relaxation.

Hypoxia can induce a number of functional disturbances in the circulatory and central nervous systems. The psychovisual disturbances, associated with altitude sickness, also appear in acute carbon monoxide poisoning, and severe blood loss. In the heart of healthy humans subjected experimentally to altitude changes some of the ECG changes typical of coronary insufficiency have been reported. Thus decrease in pressure produces a monophasic deformed ventricular complex instead of the normal ST-slope. If the animals (and human) are brought back to normal conditions quickly enough, these changes recede. In the central nervous system O_2 deficiency causes a disruption of the circulation due to vasomotor changes. Decreased spectral sensitivity and visual acuity fade during hypoxia. The top threshold of audibility is decreased, and the lower one increased. Disturbances of the cerebral cortex become apparent illustrated by impaired judgement and the feeling of intermittent euphoria and depression. These effects are similar to alcoholic poisoning. With increasing cerebral hypoxia reflex activity ceases due to disturbance of the deeper nuclear areas. Severe damage of the cerebral cortex is apparent in a general disturbance of motor activity with tonic–clonic seizures which increase the energy need of the brain by about 75 per cent. Thus, the glucose reserves of brain are rapidly exhausted; a status epilepticus can develop, which can dissolve into a general paralysis. Shortly after the occurrence of seizures severe disturbances of respiratory rhythm occur; respiration becomes periodic and finally ceases. After paralysis, if normal oxygen conditions are restored, a phase of increased motor excitability occurs at the same level of O_2 deficiency as the preasphyxic ones (Büchner, 1962).

Oxygen Excess (Hyperoxia)

To improve the O_2 supply in hypoxic patients the inspiratory PO_2 is

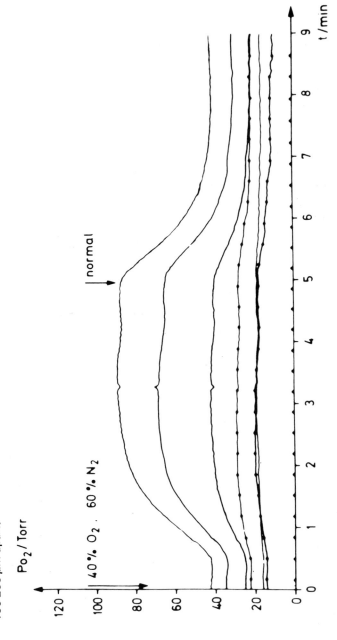

Figure 5.24: Reaction of Tissue PO_2 (Multiwire Surface Electrode) on Increasing Arterial O_2. Measuring points lie about 100-200 μm apart.

Source: Leniger-Follert *et al.*, 1975.

increased by O_2 therapy, thus enabling an increased tissue PO_2 by increasing P_aO_2. Respiration of pure oxygen increases the physically dissolved O_2 content of blood up to 2 vol. per cent. A further increase is only possible by hyperbaric oxygenation; at pressure in excess of 1 atm, the arterial blood contains 3–5 vol. per cent of dissolved oxygen. However, the use of such therapy is limited because of oxygen poisoning, the symptoms of which include dizziness and seizures. Increased vagal tone decreases cardiac output and brain- and kidney circulation are restricted.

In the lung, changes in the alveolar membrane which produce diffusion disturbances are apparent; in addition a pneumonoedema occurs. Oxygen therapy in premature births and newborns can induce pulmonary fibrosis and retrolental fibroplasia. To prevent oxygen poisoning P_aO_2 must not be increased above 450 mm Hg (ca. 60 vol. per cent) in adults and 300 mm Hg (ca. 40 vol. per cent) in the newborn (Grote, 1976). The development of transcutaneous PO_2 electrodes permits control of this difficult aspect of oxygen therapy (Huch *et al.*, 1972, 1975).

Figure 5.25: Course of PO_2 in the Aorta (Measured with a PO_2 Catheter Electrode, upper space) and in the Myocardium of the Left Ventricle (Measured with a PO_2 Needle Electrode, lower space) in Rabbit after Temporary Respiration with Oxygen.

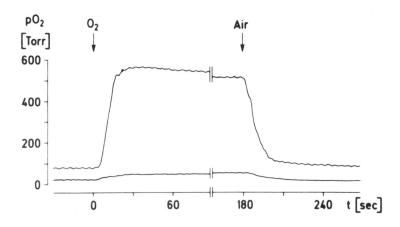

Source: Schuchhardt *et al.*, 1978.

With regard to oxygen therapy it is important to realise that tissue PO_2 does not change in some 60 per cent of all investigated sites (Figure 5.24) because hyperoxia can decrease local flow considerably (Lübbers, 1978). Schuchhardt *et al.* (1978) have demonstrated this effect in the heart (Figure 5.25); from their results the O_2 supply to heart is most efficiently improved by adding CO_2 to the inspired gas. Even during hyperbaric oxygen therapy an initial increase in tissue PO_2 is followed by a decrease (Lehmenkühler *et al.*, 1978) due entirely to a reduction in local flow. This is in agreement with the findings that increasing the blood oxygen tension or the oxygen tension of a fluid covering a microvessel preparation *in vitro* causes a constriction of arterioles and a decrease in the flow velocity in the capillaries (Duling & Klitzman, 1980).

Acknowledgement

I am obliged to Mrs Erika Menne for her assistance in translating this chapter.

References

Acker, H. (1976) 'Das Glomus caroticum. Ein Modell, um die Chemoreception zu verstehen', *Naturwissenschaften, 63,* 523-7

Acker, H., Holtermann, G., Carlsson, J. & Nederman, T. (1981) 'Methodological Aspects of Microelectrode Measurements in Cellular Spheroids', *ISOTT Meeting,* Detroit, (in press)

Acker, H., Lübbers, D.W. & Purves, M.J. (1971) 'Local Oxygen Tension Field in the Glomus Caroticum of the Cat and its Change at Changing Arterial PO_2', *Pflügers Arch., 329,* 136-55

Ardenne von, M. (ed.) (1979) *Physiologische und technische Grundlagen der Sauerstoff-Mehrschritt-Therapie,* Thieme, Stuttgart

Barankay, T., Baumgärtl, H., Lübbers, D.W. & Seidl, E. (1976) 'Oxygen Pressure in Small Lymphatics', *Pflügers Arch., 366,* 53-9

Baumgärtl, H., Leichtweiss, H.-P., Lübbers, D.W., Weiss, Ch. & Huland, H. (1972) 'The Oxygen Supply of the Dog Kidney: Measurements of Intrarenal PO_2', *Microvasc. Res., 4,* 247-57

Baumgärtl, H. & Lübbers, D.W. (1973) *Oxygen Supply. Theoretical and Practical Aspects of Oxygen Supply and Microcirculation of Tissue,* Urban & Schwarzenberg, München-Berlin-Wien

Becker, V. (1959) *Medizinische Grundlagenforschung,* Band II, Thieme, Stuttgart

Belmonte, C., Pallot, D., Acker, H. & Fidone, S. (1981) 'Arterial Chemoreceptors', *Proceedings of the VI International Meeting,* Leicester University Press

Bicher, H.J. & Knisely, M.H. (1970) 'Brain Tissue Reoxygenation Time, Demonstrated with a New Ultramicro Oxygen Electrode', *J. Appl. Physiol., 28,* 387-90

198   *Tissue Oxygen Transport in Health and Disease*

Bretschneider, H.J. (1958) *Probleme der Coronardurchblutung*, Springer, Berlin, Göttingen, Heidelberg

Bretschneider, H.J. (1961) 'Sauerstoffbedarf und -versorgung des Herzmuskels', *Verh. dtsch. Ges. Kreisl-Forsch.*, *27*, 32

Bretschneider, H.J. (1972) *Die therapeutische Anwendung β-sympathikolytischer Stoffe*, Schattauer, Stuttgart

Büchner, F. (1962) *Allgemeine Pathologie*, Urban & Schwarzenberg, München, Berlin, Wien

Carlsson, J., Stalnacke, C.G., Acker, H., Haji-Karim, M., Nilsson, S. & Larsson, B. (1979) 'The Influence of Oxygen on Viability and Proliferation in Cellular Spheroids', *Int. Rad-Onc. Biol. Phys.*, *5*, 2011-20

Chance, B. (1953) 'Dynamics of Respiratory Pigments of Ascites Tumor Cells', *Trans. N.Y. Acad. Sci.*, *16*, 74

Chance, B. (1976) 'Pyridine Nucleotide as an Indicator of the Oxygen Requirements for Energy-linked Functions of Mitochondria', *Circ. Res.*, *Suppl.*, *1*, I-31-38

Cotzelet, G.R. (1980) 'Rheology and hemodynamics', *Ann. Rev. Physiol.*, *42*, 311-24

Deetjen, P. (1968) *Oxygen Transport in Blood and Tissue*, Thieme, Stuttgart

Diemer, K. (1963) 'Eine verbesserte Modellvorstellung zur O_2-Versorgung des Gehirns', *Naturwissenschaften*, *50*, 617-8

Diemer, K. (1965a) 'Über die Sauerstoffdiffusion im Gehirn. I. Mitt. Räumliche Vorstellung und Berechnung der Sauerstoffdiffusion' *Pflügers Arch. ges. Physiol.*, *285*, 99

Diemer, K. (1965b) 'Über die Sauerstoffdiffusion im Gehirn II. Mitt. Die Sauerstoffdiffusion bei O_2-Mangelzuständen', *Pflügers Arch. ges. Physiol.*, *285*, 109-18

Downing, S.E. & Lee, J.C. (1980) 'Nervous Control of the Pulmonary Circulation', *Ann. Rev. Physiol.*, *42*, 199-210

Duling, B.R. & Klitzman, B. (1980) 'Local Control of Microvascular Function: Role in Tissue Oxygen Supply', *Ann. Rev. Physiol.*, *42*, 373-82

Ehrly, A.M. & Schroeder, W. (1978) *Oxygen Transport to Tissue III*, Plenum Press, New York, London

Fabel, H. (1968) *Oxygen Transport in Blood and Tissue*, Thieme, Stuttgart

Gertz, K.H. & Loeschcke, H.H. (1954) 'Bestimmung der Diffusionskoeffizienten von H_2, O_2, N_2 und He in Wasser und Blutserum bei konstant gehaltener Konvektion', *Z. Naturforsch.*, *9b*, 1-3

Gleichmann, U., Ingvar, D.H., Lübbers, D.W., Siesjö, B.K. & Thews, G. (1962) 'Tissue PO_2 and PCO_2 of the cerebral cortex, related to blood gas tensions', *Acta Physiol. Scand.*, *55*, 127-38

Gleichmann, U. & Lübbers, D.W. (1960) 'Die Messung des PO_2 in Gasen und Flüssigkeiten mit der Pt-Elektrode unter besonderer Berücksichtigung der Messung im Blut', *Pflügers Arch. ges. Physiol.*, *271*, 431-55

Golenhofen, K. (1971) *Lehrbuch der Physiologie in Einzeldarstellungen. Physiologie des Kreislaufs*, Band 1, Springer, Berlin, Heidelberg, New York

Granger, H.J. & Shepherd, A.P. (1979) *Advances in Biomedical Engineering*, vol. 7, Academic Press, New York, San Francisco, London

Greenway, C.V. & Stark, R.O. (1971) 'Hepatic vascular bed', *Physiol. Rev.*, *51*, 23

Grote, J. (1967) 'Die Sauerstoffdiffusionskonstanten im Lungengewebe und Wasser und ihre Temperatureabhängigkeit', *Pflügers Arch. ges. Physiol.*, *295*, 245-54

Grote, J. (1967) *Hydrodynamik, Elektrolyt- und Säure-Basen-Haushalt im Liquor und Nervensystem*, Thieme, Stuttgart

Grote, J. (1976) *Einführung in die Physiologie des Menschen*, Springer, Berlin, Heidelberg, New York

Grote, J. & Thews, G. (1962) 'Die Bedingungen für die Sauerstoffversorgung des Herzmuskelgewebes', *Pflügers Arch. ges. Physiol.*, *276*, 142-65

Grote, J. & Thews, G. (1973) *Oxygen Transport to Tissue, Instrumentation, Methods and Physiology*, Plenum Press, New York

Grunewald, W. (1968) *Oxygen Transport in Blood and Tissue*, Thieme, Stuttgart

Grunewald, W. (1969) 'Digitale Simulation eines räumlichen Diffusionsmodelles der O_2-Versorgung biologischer Gewebe', *Pflügers Arch.*, *309*, 266-84

Grunewald, W. (1971) 'Einstellzeit der Pt-Elektrode bei Messungen nicht stationärer O_2-Partialdrucke', *Pflügers Arch.*, *322*, 109-30

Harbig, K., Chance, B., Kovách, A.G.B. & Reivich, M. (1976) '*In vivo* Measurement of Pyridine Nucleotide Fluorescence from Cat Brain Cortex', *J. Appl. Physiol.*, *41*, 480-8

Herzog, H. (1977) *Lehrbuch der Inneren Medizin*, Schattauer, Stuttgart, New York

Heyrovsky, J. & Kuta, J. (1965) *Grundlagen der Polarographie*, Akademie Verlag, Berlin

Hirsch, H. (1971) *Lehrbuch der Physiologie in Einzeldarstellungen, Physiologie des Kreislaufs*, Bd. 1, Springer, Berlin, Heidelberg, New York

Honig, A., Schmidt, M., Arndt, H., Kranz, G. & Zapf, C. (1975) 'Über die Regulation des Blutvolumens und der Nierenfunktion im akuten arteriellen Sauerstoffmangel', *Das Deutsche Gesundheitswesen*, *30*, 2257-62, 2353-8

Huch, R., Lübbers, D.W. & Huch, A. (1972) 'Quantitative Continuous Measurement of Partial Oxygen Pressure on the Skin of Adults and New-born Babies', *Pflügers Arch.*, *337*, 185-98

Huch, R., Lübbers, D.W. & Huch, A. (1975) *Oxygen Measurements in Biology and Medicine*, Butterworths, London, Boston

Ingvar, D.H., Lübbers, D.W. & Siesjö, B. (1960) 'Measurement of Oxygen Tension on the Surface of the Cerebral Cortex of the Cat during Hyperoxia and Hypoxia', *Acta Physiol. Scand.*, *48*, 373-81

Ji, S., Chance, B., Bradley, H.S. & Nathan, R. (1977) 'Two-dimensional Analysis of the Redox State of the Rat Cerebral Cortex *in vivo* by NADH Fluorescence Photography', *Brain Res.*, *119*, 357-73

Jöbsis, F.F., O'Connor, M., Vitale, A. & Vreman, H. (1971) 'Intracellular Redox Changes in Functioning Cerebral Cortex. I. Metabolic Effects of Epileptiform Activity', *J. Neurophys.*, *34*, 735-49

Kallman, R.F. (1972) 'The Phenomenon of Reoxygenation and its Implications for Radiotherapy', *Radiology*, *105*, 135-42

Kessler, M. (1974a) *Mitteilungen aus der Max-Planck-Gesellschaft*, Präsidialbüro der Max-Planck-Gesellschaft, München

Kessler, M. (1974b) 'Oxygen Supply to Tissue in Normoxia and in Oxygen Deficiency', *Microvasc. Res.*, *8*, 283-90

Kessler, M., Höper, J., Schäfer, D. & Starlinger, H. (1974) *Klin. Anaesthesiol. u. Intensivtherapie*, Bd. 5, *Midrozirkulation*, Springer, Berlin, Heidelberg, New York

Klingenberg, M. & Lübbers, D.W. (1966) *D-Glucose und verwandte Verbindungen in Medizin und Biologie*, Enke, Stuttgart

Kontos, A.H. (1981) 'Regulation of the Cerebral Circulation', *Ann. Rev. Physiol.*, *43*, 397-407

Kramer, K., Thurau, K. & Deetjen, P. (1960) 'Hämodynamik des Nierenmarks. 1. Mitteilung: Capilläre Passagezeit, Blutvolumen, Durchblutung, Gewebshämatokrit und O_2-Verbrauch des Nierenmarks *in situ*', *Pflügers Arch. ges. Physiol.*, *270*, 251-69

Kreuzer, F. (1950) 'Über die Diffusion von Sauerstoff in Serumeiweißösungen verschiedener Konzentration', *Helv. Physiol. Pharmacol. Acta*, *8*, 505-8

Kreuzer, F. (1967) *Exercise at Altitude*, Excerpta Medica, Amsterdam

Kreuzer, F. (1976) 'Mechanismen beim Gasaustausch im Organismus', *Ergebn. Exp. Med.*, *23*, 121-40

Krogh, A. (1918/1919) 'The Rate of Diffusion of Gases through Animal Tissues with some Remarks on the Coefficient of Invasion', *J. Physiol. (Lond.)*, *52*, 391-408

Krogh, A. (ed.) (1924) *Anatomie und Physiologie der Kapillaren*, Springer, Berlin

Kunze, K. (1966) 'Die lokale kontinuierliche Sauerstoffdruckmessung in der menschlichen Muskulatur', *Pflügers Arch. ges. Physiol.*, *292*, 151-60

Kunze, K. (ed.) (1969) *Das Sauerstoffdruckfeld im normalen und pathologisch veränderten Muskel*, Springer, Berlin, Heidelberg, New York

Lassen, N.A. (1959) 'Cerebral Blood Flow and Oxygen Consumption in Man', *Physiol. Rev.*, *39*, 183

Lehmenkühler, A., Bingmann, D., Lange-Aschenfeldt, H. & Berges, D. (1978) *Oxygen Transport to Tissue III*, Plenum Press, New York, London

Lenfant, C. & Aucutt, C. (1965) 'Oxygen Uptake and Change in Carbon Dioxide Tension in Human Blood Stored at 37 °C', *J. Appl. Physiol.*, *20*, 503

Leniger-Follert, E. & Lübbers, D.W. (1973) 'Determination of Local Myoglobin Concentration in the Guinea Pig Heart', *Pflügers Arch.*, *341*, 271-80

Leniger-Follert, E., Lübbers, D.W. & Wrabetz, W. (1975) 'Regulation of Local Tissue PO_2 of the Brain Cortex at Different Arterial O_2 Pressures', *Pflügers Arch.*, *359*, 81-95

Lochner, W. (1971) *Lehrbuch der Physiologie in Einzeldarstellungen, Physiologie des Kreislaufs*, Bd. 1, Springer, Berlin, Heidelberg, New York

Lübbers, D.W. (1967) *Marburger Jahrbuch 1966/67*, Elwerth, Marburg

Lübbers, D.W. (1968a) *Herzinsuffizienz*, Thieme, Stuttgart

Lübbers, D.W. (1968b) *Oxygen Transport in Blood and Tissue*, Thieme, Stuttgart

Lübbers, D.W. (1969) *Oxygen Pressure Recording in Gases, Fluids and Tissue*, Karger, Basel, New York

Lübbers, D.W. (1970) *Gefäßwand und Blutplasma*, Fischer, Jena

Lübbers, D.W. (1972) *Der Hirnkreislauf*, Thieme, Stuttgart

Lübbers, D.W. (1974a) 'Das O_2-Versorgungssystem der Warmblüterorgane', *Jahrbuch der Max-Planck-Gesellschaft zur Förderung der Wissenschaften e.V.*

Lübbers, D.W. (1977) *Brain and Heart Infarct*, Springer, Berlin, Heidelberg, New York

Lübbers, D.W. (1978a) *Handbuch der allgemeinen Pathologie III/7, Mikrozirkulation/Microcirculation*, Springer, Berlin, Heidelberg, New York

Lübbers, D.W. (1978b) 'Die Sauerstoffversorgung der Warmblüterorgane unter normalen und pathologischen Bedingungen' *RheinischWestfälische Akademie der Wissenschaften Vorträge N 272 Westdeutscher Verlag*, 7-52

Lübbers, D.W. (1981) *Oxygen Transport to Tissue*, vol. 25, Pergamon Press, Budapest

Lübbers, D.W., Baumgärtl, H., Grunewald, W., Leniger-Follert, E & Wodick, R. (1974) 'Micromethods to Monitor Local Tissue Oxygen Supply and Microcirculation', *Proc. Symp. VIIIth Eur. Conf. Microcirculation*, Le Touquet

Lübbers, D.W., Hauck, H. & Weigelt, H. (1976) *Ionic Actions on Vascular Smooth Muscle*, Springer, Berlin, Heidelberg, New York

Lübbers, D.W. & Leniger-Follert, E. (1978) *Cerebral Vascular Smooth Muscle and its Control*, Elsevier/Experta Medica, North-Holland

Lübbers, D.W. & Schuchhardt, S. (1971) 'Supply and Utilization of Oxygen in the Myocardium', *XXVth International Congress of Physiological Sciences*, München

Lutz, J. & Bauereisen, E. (1971) *Lehrbuch der Physiologie in Einzeldarstellungen, Physiologie des Kreislaufs*, Bd. 1, Springer, Berlin, Heidelberg, New York

Messmer, K. & Schmidt-Schönbein, H. (eds) (1972) *Hemodilution*, Karger, Basel, München, Paris, London, New York, Sydney

Messmer, K., Sunder-Plassmann, L. & Kessler, M. (1974) *Mikrozirkulation, Klinische Anästhesiologie und Intensivtherapie*, Springer, Berlin, Heidelberg, New York

Navari, R.M., Gainer, J.L. & Hall, K.R. (1970) *Blood Oxygenation*, Plenum Press, New York

Olsson, A. (1981) 'Local Factors Regulating Cardiac and Skeletal Muscle Blood Flow', *Ann. Rev. Physiol.*, *43*, 385-95

Opitz, E. & Schneider, M. (1950) 'Über die Sauerstoffversorgung des Gehirns und den Mechanismus von Mangelwirkungen', *Ergebn. Physiol.*, *46*, 126-260

Opitz, E. & Thews, G. (1952) 'Einfluß von Frequenz und Faserdicke auf die Sauerstoffversorgung des menschlichen Herzmuskels', *Arch. Kreislaufforsch.*, *18*, 137

Oshino, N., Sugano, T., Oshino, R. & Chance, B. (1974) 'Mitochondrial Function under Hypoxic Conditions: the Steady States of Cytochrome a+a3 and their Relation to Mitochondrial Energy States', *Biochem. Biophys. Acta, 368*, 298-310

Piiper, J. (1972) *Physiologie des Menschen*, Urban & Schwarzenberg, München, Berlin, Wien

Pittmann, R.N. (1981) *Vasodilation*, Raven Press, New York

Pries, A.R., Gaethgens, P. & Kanzow, G. (1981) *Oxygen Transport to Tissue*, vol. 25, Pergamon Press, Budapest

Rodenhäuser, J.H., Baumgärtl, H., Lübbers, D.W. & Briggs, D. (1971) *Ophthalmology*, Excerpta Medica, North Holland

Schmidt-Schönbein, H. (1974) *Mikrozirkulation*, Springer, Berlin, Heidelberg, New York

Schmidt-Schönbein, H. (1981) *Oxygen Transport to Tissue*, vol. 25, Pergamon Press, Budapest

Schuchhardt, S. & Lübbers, D.W. (1969) 'Sauerstoffversorgung und Mikrozirkulation der Herzmuskulatur', *Wiss. Z. d. Univ. Halle-Wittenberg*, R 11

Schuchhardt, S., Schuster, J. & Ryzlewicz, Th. (1978) 'Untersuchungen zur Sauerstoffversorgung des Warmblütermyokards mit Platin-Nadel-Elektroden', *Forschungsberichte des Landes Nordrhein-Westfalen*, Westdeutscher Verlag

Schwickardi, D. (1975) 'Unblutige und kontinuierliche Sauerstoffeüberwachung mit einer transkutanen Sauerstoffelektrode', *Anästh. Prax.*, *11*, 63-6

Silver, I.A. (1965) 'Some Observations on the Cerebral Cortex with an Ultramicro Membrane Covered Oxygen Electrode', *Med. Electron. Biol. Eng.*, *3*, 377-87

Silver, I.A. (1969) *Oxygen Pressure Recording in Gases, Fluids and Tissues*, Karger, Basel, New York

Siésjö, B.K. (1961) 'A Method for Continuous Measurement of the Carbon Dioxide Tension on the Cerebral Cortex', *Acta Physiol. Scand.*, *51*, 297-313

Starlinger, H., Lübbers, D.W. (1973) 'Polarographic Measurements of the Oxygen Pressure Performed Simultaneously with Optical Measurements of the Redox State of the Respiratory Chain in Suspensions of Mitochondria under Steady-state Conditions at Low Oxygen Tensions', *Pflügers Arch.*, *341*, 15-22

Sunder-Plassmann, L. & Messmer, K. (1974) *Mikrozirkulation*, Springer, Berlin, Heidelberg, New York

Thews, G. (1960) 'Die Sauerstoffdiffusion im Gehirn', *Pflügers Arch. ges. Physiol.*, *271*, 197-226

Thews, G. (1960) 'Ein Verfahren zur Bestimmung des O_2-Diffusionskoeffizienten, der O_2-Leitfähigkeit und des O_2-Löslichkeitskoeffizienten im Gehirngewebe',

Pflügers Arch. ges. Physiol., *271*, 227-37

Thews, G. (1976) *Einführung in die Physiologie des Menschen*, Springer, Berlin, Heidelberg, New York

Thurau, K. (1971) *Lehrbuch der Physiologie in Einzeldarstellungen, Physiologie des Kreislaufs*, Bd. 1, Springer, Berlin, Heidelberg, New York

Turek, Z., Kreuzer, F. & Hoofd, L.J.C. (1973) 'Advantage or Disadvantage of a Decrease of Blood Oxygen Affinity for Tissue Oxygen Supply at Hypoxia', *Pflügers Arch.*, *342*, 185-97

Vaupel, P., Brambeck, W. & Thews, G. (1973) *Oxygen Transport to Tissue, Instrumentation, Methods and Physiology*, Plenum Press, New York, London

Vaupel, P., Thews, G. & Wendling, P. (1976) 'Kritische Sauerstoffund Glucoseversorgung maligner Tumoren', *Dtsch. med. Wschr.*, *101*, 1810-6

Weigelt, H., Fujii, T., Lübbers, D.W. & Hauck, G. (1981) 'Specialized Endothelial Cells in the Frog Mesentery — Attempts of an Electrophysiological Characterization', *Bibl. Anat.*, *20*, 89-93

Whalen, W.J., Riley, J. & Nair, P. (1967) 'A Microelectrode for Measuring Intracellular PO_2', *J. Appl. Physiol.*, *23*, 798-801

Wilson, D.F., Erecinska, M., Silver, I.A., Owen, Ch.S. & Nishiki, K. (1981) *Adv. Physiol. Sci.*, vol. 10, *Respiration*, Pergamon Press, Budapest

Zander, R. & Lang, W. (1975) 'Does Oxygen Solubility of Zero Exist in the Biological Field', *Pflügers Arch.*, *355*, R 37

Zander, R. & Schmid-Schönbein, H. (1973) 'Intracellular Mechanism of Oxygen Transport in Flowing Blood', *Resp. Physiol.*, *19*, 279-89

Zweifach, B.W. & Kossmann, C.E. (1937) 'Micro Manipulation of Small Blood Vessels in the Mouse', *Am. J. Physiol.*, *120*, 23-35

6 STUDIES OF RESPIRATORY CONTROL IN MAN

J.M. Patrick

This chapter deals with those aspects of the respiratory control system in which studies in man, both in health and disease, have been particularly revealing. It covers first the special precautions that have to be taken in human experimentation in this field, and then discusses the information that can be obtained from studies involving voluntary control of breathing. The later sections cover the more clinical aspects of respiratory control, dealing with the performance of the system when its inputs or outputs are altered by disease, and when the controller itself is affected by disease, by surgery, or by drugs.

Within each section, the treatment of the material is illustrative rather than exhaustive, and most of the statements made are not supported by citations of the evidence. Many of the papers referred to, however, contain discussions of the relevant observations.

The Problems Arising from the Use of Human Subjects

The Placebo Effect

The main problems associated with respiratory experimentation in man stem from the dual control of the respiratory apparatus by both behavioural (voluntary) and metabolic (automatic) pathways (Mitchell & Berger, 1975). Voluntary control may be helpful (see below) but may interfere. The penalty incurred by recruiting the usual available laboratory personnel for studies in respiratory control is well illustrated by Figure 6.1. Either consciously or unconsciously, these cooperative subjects may produce the outcome expected of them. Similarly, the classical post-hyperventilation apnoea seen by Douglas & Haldane in themselves could not be reproduced by other authors using naive subjects. Fink (1961) even warned that 'respiratory studies based on self-experimentation are of limited objectivity and liable to errors of subjectivity'. Yet the ethical requirement to provide full information to volunteers about the nature and purpose of the proposed experiments, and about the sensations likely to occur during them, may well give clues to any intelligent subject as to what ventilatory responses

203

may be appropriate. Fink argues that 'the best safeguard against the intrusion of volition into the respiratory rhythms is to have subjects who are entirely unaware of the objectives of the experiment'. For reasons of this sort, the practice of using a double-blind protocol when measuring the respiratory responses to drugs is highly recommended.

Figure 6.1: Ventilation Responses in a Subject who became Conditioned to the Information that he was to Receive a Noradrenaline Infusion; breathing 4.5% CO_2 throughout experiment. *A, B*: subject told he was receiving noradrenaline at 3 and 6 μg/min respectively. No noradrenaline was actually administered. *C*: subject told he was no longer receiving noradrenaline. *D-E*: noradrenaline infused at 10 μg/min; subject not told, unaware of infusion.

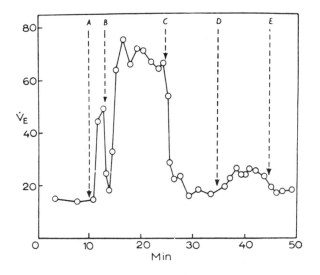

Courtesy of the editors of *The Journal of Physiology*.

Unwanted Conscious Control

Even if the subject cannot anticipate the expected responses, the fact that he knows that his breathing is being measured can superimpose an unwelcome voluntary element onto the automatic control system. Hyperventilation is a common feature of laboratory class experiments, and rhythm disturbances are another characteristic feature of such introspective respiratory behaviour. Severinghaus (1976) lists several precautions to be taken: his advice to counsel the subjects to wait between breaths until they need to breathe is an especially valuable counter-measure against tachypnoea at rest. Light reading, relaxing music, recorded stories and white noise have all been used as distractors: the advantage of reading is that the subject is seen to be neither asleep nor dangerously hypoxic. It may not even be necessary to keep the subject awake if only the metabolic respiratory control is being measured, as the behavioural component is effectively excluded in slow-wave sleep.

The Mouthpiece Effect

Reliable measurement of ventilation and its pattern is fundamental to respiratory physiology; so a method, however precise, that materially alters any component of ventilation may be of less value than one that is less precise but which does not interfere with the spontaneous pattern. Gilbert *et al.* (1972) first drew attention to the possibility that the conventional mouthpiece and noseclip may themselves alter ventilation in man. Resting ventilation, measured with magnetometers across the thorax, was increased by 21 per cent when a mouthpiece and noseclip were applied for 4 minutes: tidal volume was increased by 29 per cent and respiratory frequency slightly reduced. Several authors using a variety of methods have since reported similar findings. The overall picture is that application of respiratory apparatus to the airway increases ventilation with a raised tidal volume but no consistent change in frequency. This implies that our 'normal values' for resting ventilation and tidal volume, based on conventional spirometer methods, may be overestimates by as much as 30 per cent. The use of the mouth-piece and noseclip may alter the pattern of breathing in exercising subjects also, and it is possible that some of the apparent species differences in breathing patterns may be due to distortion of this sort. See also Chapter 8, p. 287.

Variability of Responses

Several authors have commented on the variation in respiratory responses to stimuli even as simple as CO_2 inhalation. Variation between subjects need not be a problem. CO_2-sensitivity is proportional to body size in normal adults: vital capacity is the most useful and appropriate predictor variable reflecting size but one which still accounts for only a small part of the inter-individual variance (Patrick & Cotes, 1974). While age in adults does not alter CO_2-sensitivity (Patrick & Howard, 1972), the spreading of ventilation-perfusion ($\dot{V}/\dot{Q}$) ratios and the fall in the transfer factor of the ageing lung (Cotes, 1979) will lead to hypoxaemia and hyperventilation. This will increase the apparent

CO_2-sensitivity in older subjects unless arterial PO_2 is measured or the contribution of hypoxial drive removed by the administration of oxygen. Ventilation while breathing air in exercise increases with increasing age in older men (Patrick, 1981).

Groups of subjects noted to have unusual values for CO_2-sensitivities are endurance athletes, divers and swimmers, and New Guineans. Although estimates of heritability have been made for some respiratory variables (Arkinstall *et al.*, 1974), there is no consensus as to the relative importance of genetic and environmental influences on the parameters of the respiratory responses to CO_2 and O_2.

Test–retest variation within subjects is also often rather large, even when biological factors like previous eating and drinking, time of day, and posture are controlled. Factors like attentiveness and cooperation, described above, are likely to be major contributors to this type of variation, emphasising the need for careful attention to detail in the experimental protocol. However, Lloyd & Cunningham (1963) report their 'surprise and delight that respiratory output in a trained subject is so highly reproducible'. The irreducible variation means that rather large numbers of subjects are needed in order to demonstrate a weak effect like that of 100 mg of propranolol on CO_2-sensitivity (see below). Placebo controls are particularly important in studies of this sort.

The Advantages of Using Human Subjects

Although, as we have seen, human subjects bring problems to the study of respiratory control, they also make a special contribution both because of the direct link they provide with clinical applications, and because of their ready availability at a time when animal experimentation is becoming increasingly expensive. Furthermore, the volunteer subject can cooperate, where appropriate, in the performance of an experiment and he can report his sensations. Highly quantitative, non-invasive experiments using human subjects have been the hallmark of several schools of respiratory physiology, notably in Copenhagen, Oxford and Buffalo, and these have laid the foundation for our present understanding of the respiratory control system in man.

Lessons from Breath-Holding

A good example of the benefits to be derived from a human subject's ability to cooperate with the experimenter is found in what Rahn called the 'lessons from breath-holding'. After practice, the time for which a voluntary apnoea can be maintained under given circumstances is reasonably constant, and the breaking-point indicates the moment when the rising respiratory drive just cannot be voluntarily resisted. Timing of breath-holds and measurement of the changes in indices of chemical stimuli provide a fascinating picture of this drive. The outcome of breath-holding at different starting lung-volumes and at different partial pressures of CO_2 and O_2 have provided clear illustrations of the interaction of hypoxial and hypercapnic chemical drives, and of the CO_2 threshold for ventilatory stimulation. Non-chemical inputs to the respiratory centre, difficult to study otherwise in man, have also been identified: for example, the reflex effect of lung stretch and shrinkage, and the drive that is proportional to the time elapsed since the last movement of the chest (Figure 6.2). This latter drive appears to resemble von Euler's central inspiratory ramp generator (Clark & von Euler, 1972). In a remarkable series of experiments, Campbell *et al.* (1969) demonstrated the contribution of the efferent innervation of the thoracic muscles to the desire to breathe during a breath-hold. Conscious curarised volunteers, in whom one hand was spared to signal the need for assisted ventilation, were able to tolerate apnoea for very much longer than normal during breath-holds (Figure 6.2). This demonstrates a non-chemical component of the ventilatory drive associated with inspiratory muscle activity, and EMG recordings have supported this hypothesis. This inspiratory muscle activity is probably also important in the sensation of dyspnoea (see below). Godfrey & Campbell (1968) have reviewed some of this work.

Reporting Respiratory Sensation

Unlike experimental animals, human subjects can report their respiratory sensations to the experimenter, and thus provide quantitative psychophysical data. This has particular relevance to the study of breathlessness. Two rather different approaches to the measurement of sensations have been exploited: threshold detection and scaling methods.

The first approach involves adding elastic or resistive loads of various sizes to the breathing circuit, and determining either the load that is detected by the subject on 50 per cent of occasions, or the load that produces a 'just noticeable difference' in sensation. Zechman *et al.* (1981) suggest that this 'difference threshold' should be expressed as a proportion of the baseline stimulus on the grounds that respiratory

Figure 6.2:A Schematic Figure Incorporating the Findings of Several Authors in Breath-holding Experiments. Hypoxia is removed and alveolar PCO_2 is adjusted before the breath-hold by rebreathing or hyperventilation in oxygen. Initial PCO_2 is shown on the ordinate and the dashed diagonal lines show the approximately linear rise in PCO_2 during each breath-hold to the points representing the PCO_2 and the time at the breaking-point. ▲---▲. Fowler's (1954) experiment. At the breaking-point the breath is held again after a short period of rebreathing which does not allow the PCO_2 to fall back to the starting level. Progressively higher breaking-point PCO_2 and shorter durations are found in subsequent breath-holds. The bold line shows the linear relation between breath-hold duration and breaking-point PCO_2. The slope of this line is a function of the lung volume: at lower volumes the tolerance to CO_2 is reduced and the line is steeper (Patrick & Reed, 1969). When the thorax is paralysed (Campbell et al., 1969), the desire to breathe again is delayed until the PCO_2 reaches much higher levels: compare breath-holds under control conditions in two subjects, ●, ■, with those after curare, ○, □.

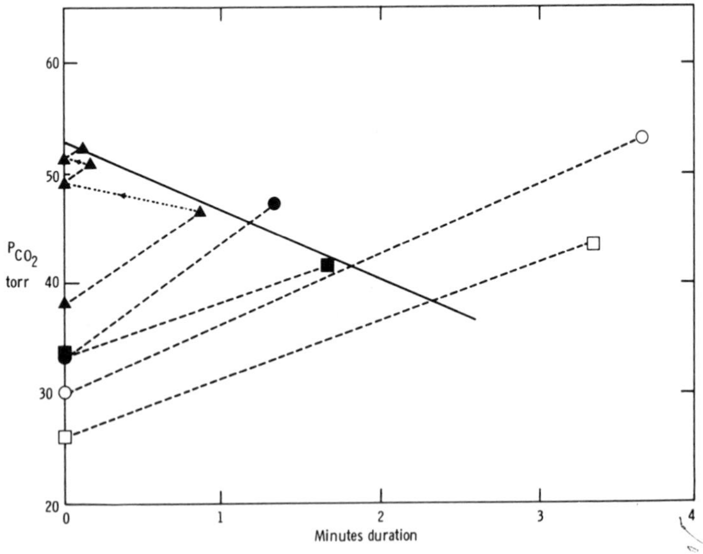

sensations, like other sensory modalities, follow Weber's law and show a constant proportionality between the least detectable increment in stimulus and its initial intensity. Normal subjects can detect an increase of one-third in their baseline resistance and an increase of one-tenth in their baseline elasticity. These fractions appear to be unchanged in small groups of patients with chronic bronchitis or asthma.

The second approach is to scale the sensation. Rating scales for the perception of exertion in exercise studies were pioneered by Borg (1970), and one useful variant of his approach has been to partition the sensation between the thorax and the limbs. More specific respiratory scales that are applicable either to normal sensations or to 'breathlessness' have been described. These are usually based on Stevens' (1957) general demonstration of a linear relation between the logarithms of the intensities of the stimulus and of the sensation. The scale can either be left open-ended or referred to some particular sensation, real or imagined. The slope of the log–log plot, i.e. the exponent of the power function relating sensation to stimulus, is commonly used to characterise respiratory perception. However, this exponent varies from day to day in a given subject, and displacement of the line without alteration in slope also indicates altered sensation which may be overlooked if too much emphasis is placed on the exponent.

To scale the sensation, the subject estimates it numerically or by matching it with another sensory modality. In this latter technique the subject might, for example, produce tension in a hand-grip dynamometer so as to match the magnitude of the respiratory sensation under test. This procedure has the added complication of an extra and different power function between the sensation and stimulus of the new modality. Another method of scaling respiratory sensation is for the subject to alter his breathing so as to produce for himself sensations of stated magnitudes expressed either by a numerical scale or as a given ratio of a standard sensation. Broadly speaking, the conclusions drawn from all these approaches appear to be consistent.

The interaction between the sensations caused by added resistive and elastic loads, and the timing of the sensations produced, suggest that the generation of sensation is related to the changes in intrathoracic pressure or the extra inspiratory work required to overcome the loads. The time-course of the active contraction of the inspiratory muscles appears to be an additional factor influencing respiratory perception (Killian *et al.*, 1980).

Breathlessness is an important and disabling symptom of respiratory disease and can only be studied in man. Adams, Chronos & Guz (1982) have emphasised the distinction between hyperventilation and breathlessness by showing that during progressive hyperventilation the threshold for breathlessness occurs later in hypercapnia than in hypoxia, and later still (i.e. at a higher ventilation) in exercise. This recent interest in the psychophysics of respiratory sensation promises substantial benefits in the investigation of dyspnoea.

Speech and Breathing

Speech is another area in which respiratory studies in man cannot be replaced by animal experimentation. Speech imposes strong constraints on the ventilatory system, limiting mean expiratory airflows to between 0.15 and 0.50 litres/s at rest. Speech is controlled by the voluntary pathway descending from the cortex to the phrenic motor neurones independently of the respiratory centre (Mitchell & Berger, 1975), and it appears to take precedence over the metabolic ventilatory requirements mediated via the autonomic control system. Thus, speech *increases* ventilation at rest, while during the hyperventilation of either CO_2-breathing (Phillipson, 1978) or moderate exercise (Doust & Patrick, 1981) ventilation is profoundly *reduced* by speech, keeping mean expiratory flow rates down below 0.75 litres/s. There appears to be a cross-over point a little above resting ventilations, where the two conflicting requirements are equally satisfied.

The Respiratory Control System in Disease

Figure 6.3 represents a simplified schematic diagram of the respiratory control system. We can regard clinical disturbances of respiration as arising from diseases of the respiratory system itself affecting the afferent or efferent pathways, or the controller, or its central influences; or they may arise because systemic non-respiratory disease alters the inputs to otherwise normal respiratory control mechanisms. We shall consider these first.

Inputs Affected by Disease

J-receptors

A striking example of the latter is the stimulation of ventilation in pulmonary congestion due to the excitation of juxtapulmonary capillary (J-) receptors (Anand, Loeshche, Marek & Paintal, 1982; see also Chapter 3). Although these receptors are situated in the interstitial space between the pulmonary capillaries and the alveoli, they are excited when capillary pressure is raised. This occurs during exercise in normal man, especially at altitude. The evidence concerning the role of J-receptors is largely obtained from animal work and is therefore indirect, but it seems likely that these receptors contribute in man to the sensation of dyspnoea as well as to the hyperventilation in various disorders affecting the pulmonary circulation.

Figure 6.3: A Schematic Diagram of the Respiratory Control System.

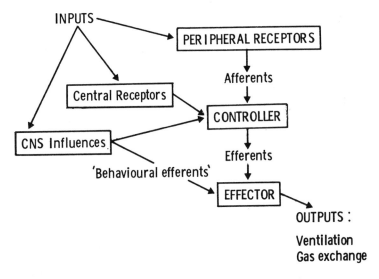

Adaptation to Chronic Hypercapnia

Part at least of the altered respiratory control in patients with chronic bronchitis should be considered as a normal response although the disease has damaged the system's effector-organ, the lung. Here, chronic hypercapnia due to inadequate gas exchange and alveolar hypoventilation may be severe enough to saturate the CO_2-response and depress central respiratory drive directly (see below). Furthermore, the compensatory increase in the bicarbonate concentration of the cerebrospinal fluid (CSF) attenuates the CSF acidosis and reduces the effective (CO_2, H^+) stimulus at the medullary chemoreceptors. (This is an equivalent active transport mechanism to the one which produces the opposite change naturally in the respiratory alkalaemia of high altitude.) For both these reasons, the ventilatory drive in chronic bronchitis with hypercapnia depends predominantly on the hypoxia present, and is therefore at risk from the injudicious use of supplementary oxygen which may take away the drive and widen further the PO_2 gradient between inspired air and arterial blood. See Chapter 2 for a discussion of H^+ sensitivity.

Hyperventilation

Hyperventilation is seen in a variety of diseases. Important examples are diabetic ketosis and during mild exercise in anaemia. These can be accounted for in terms of the normal peripheral chemoreceptor response to the extra (CO_2, H^+) stimulus resulting from the metabolic acidaemia induced: ketoacids and lactic acids are the respective stimuli.

The somewhat paradoxical hypocapnia accompanying hypoxaemia in certain types of respiratory failure is also accounted for in terms of the response of a normal respiratory control system to an altered input. In many diseases of the lung interstitium or alveoli, disturbances of ventilation-perfusion ratios cause a primary hypoxaemia: this stimulates ventilation via the peripheral chemoreceptors which are responding normally. The resulting hyperventilation reduces the arterial PCO_2 while raising the PO_2 back towards normal, but the inevitable removal of part of the normal (CO_2, H^+) stimulus limits the response, just as it does in the classical rebreathing experiments when metabolic CO_2 is absorbed by soda-lime in the spirometer during the ventilatory response to hypoxia. This type of respiratory failure, with hypocapnia concurrent with hypoxaemia, is distinguished from the type in which the output of the control mechanism itself is also compromised by the disease process and *cannot* respond with a hyperventilation. This latter type is termed ventilatory failure and will be considered briefly below.

The 'hyperventilation syndrome' is said to be due to intermittent chronic anxiety affecting the normal respiratory centre via pathways from the forebrain. A variety of signs and symptoms are ascribed to the resulting respiratory alkalosis. Fear and pain have similar effects in the shorter term, and the hyperventilation seen in some patients with forebrain lesions may be due to the lifting of inhibitory influences descending to the medulla from higher centres.

These examples serve as a reminder that some of the respiratory perturbations in systemic and even respiratory disease may result from the intact respiratory system's quite appropriate response to changes in levels of input variables. The effect of the respiratory response, however, may or may not be homoeostatic in terms of the original pathological disturbance. The improvement in oxygenation achieved by hyperventilation, for example, is often at the expense of the systemic sequelae of hypocapnia.

Outputs Affected by Disease

The outputs or effector mechanisms of the control system may be altered by disease while the controller itself is functioning normally and while the inputs, apart from those in the homeostatic feedback loops (which are inevitably altered by any alteration in effective ventilation), remain within the normal range. This is the problem in respiratory failure of peripheral origin which has been thoroughly reviewed by Sykes, McNichol & Campbell (1976). Three illustrations will suffice.

Neurogenic Respiratory Failure
When the neuromuscular link between the control centres and the lungs is impaired, normal central drive results in a ventilation which is lower than normal although the lungs are unaffected. Poliomyelitis affecting respiratory motor neurones in the cervical or thoracic spinal cord provides the classical example, but polyneuritis, myasthenia gravis and spinal cord injury are other causes of the same phenomenon. Management includes mechanically assisted ventilation for part or all of the day to replace the deficit in respiratory drive. Curarisation in routine anaesthesia is an elective equivalent with a short duration, and epidural anaesthesia that is allowed to affect the cervical and thoracic nerve roots has the same effect. The anaesthetist's task is to provide adequate ventilation and gas exchange, and monitoring of arterial blood gases is used to judge its effectiveness.

Altered Lung Mechanics
When the lungs or airways themselves are diseased, the problem of inadequate ventilation is moved further downstream. Normal central drive and normal excitation to the respiratory muscles result in a ventilation that is lower than usual, either because the lungs or thorax are less compliant or because more airway resistance is encountered. In these circumstances the respiratory drive at a given PCO_2 is normal when measured in terms of the work of breathing, but less than normal when measured as ventilation. Recent investigations using the rate of change of airway pressure during the early part of each breath as a convenient index of the inspiratory work have confirmed this finding.

Disturbed Gas Exchange
Third, when gas exchange itself is impaired (usually through distortions by disease of the distribution of $\dot{V}/\dot{Q}$ ratios in different units throughout the lung) a normally operating control system is accompanied by abnormal blood gas tensions, usually a low PO_2. As described in the section on Hyperventilation above, ventilation may be increased in these circumstances as part of the normal homoeostatic response to

such disturbances. There is also a rise in the central ventilatory drive both in the illustration of neurogenic respiratory failure although the lungs may be unable to respond, as well as in the example of pulmonary disease although effective ventilation may be little increased.

The Controller Itself Affected by Disease

Controller Depressed by Hypercapnia or Hypoxia

Several studies have shown that patients with chronic obstructive lung disease are relatively unresponsive to extraneously administered carbon dioxide. These muted ventilatory responses have been determined in several different ways which largely avoid the problem, discussed above that the ventilatory output in lungs whose mechanics are altered by disease does not properly reflect the output of the controller. The cause of this unresponsiveness is unclear. It is usually correlated with the degree of CO_2 retention, but this feature could be an effect rather than a cause of the central disturbance. Schaefer (1963), however, found moderately reduced CO_2-responses in normal subjects after 42 days continuous exposure to 1.5 per cent CO_2 in the inspired air which raised their arterial PCO_2 by only 1.4 Torr: this suggests that hypercapnia is a cause rather than an effect of the central depression. There are no prospective studies to show whether variation in CO_2-sensitivity between patients measured at first referral with chronic bronchitis accounts for later differences in the ultimate course of the disease in terms of CO_2-retention, CO_2-responses and symptoms.

Responses to hypoxia are much harder to measure in patients with obstructive lung disease, and it is difficult to make satisfactory comparisons with normal subjects. The general impression is that such patients have a reduced sensitivity to hypoxia, particularly those with the greatest hypoxaemia. However, some severely hypoxaemic patients on exposure to 100 per cent O_2 have a material reduction in ventilation, indicating at least some preservation of peripheral chemoreceptor drive.

Peripheral Chemoreceptor Denervation (see also Chapter 1)

In intact man, stepwise replacement of the nitrogen in the inspired air by oxygen reduces ventilation in a continuous hyperbolic function without a threshold. This effect, which can be interpreted as a reduction in the CO_2-sensitivity, reveals the interaction between the responses to CO_2 and to hypoxia. At low ventilation, the effect may be obscured by a slower paradoxical stimulation of ventilation resulting from the central sequelae of hyperoxygenation of the blood perfusing the medulla. Nevertheless, for many purposes, the effect of denervation of the peripheral chemoreceptors can be mimicked in intact man by the administration of hyperoxic inspired gases.

Experimental extirpation or denervation of a sensory organ, one of the classical methods in experimental animal physiology, has also been mimicked in clinical studies involving the carotid bodies. These have been resected bilaterally in a vain attempt to alleviate the symptoms of intractable asthma, or denervated incidentally during surgery for carotid arterial stenosis. Their afferent innervation has also been effectively destroyed by the neurological sequelae of diabetes mellitus, tabes dorsalis and familial dysautonomia; and at least two normal subjects have submitted voluntarily to transient anaesthetic blockade of the ninth and tenth cranial nerves in the neck carrying the afferent and efferent innervation of carotid body and the carotid sinus. The consensus is that both the immediate and the steady-state ventilatory response to hypoxaemia are abolished or severely reduced by such denervations. Three factors determine the size of the effect: whether or not the aortic bodies are still able to signal the hypoxia, the degree of concomitant central respiratory depression induced, and the adequacy of the cerebral perfusion in the face of the resultant hypotension. Several authors have reported prolongation of voluntary breath-holding times (see above) with greater apparent tolerance of profound hypoxia. Part of the steady-state response to CO_2 is also lost after bilateral carotid glomectomy.

Patients with chronic surgical or pathological denervations have been reported to have episodic symptoms ascribed to a failure to maintain homoeostasis in the face of hypoxaemia, with cyanosis, nocturnal apnoeas and even respiratory arrests. There is conflicting evidence, however, as to whether normal resting ventilation is affected. Some denervated patients appear no different from control subjects, while others, particularly those whose carotid sinuses were also denervated, have higher resting arterial PCO_2 after surgery. The overall impression provided by those reports of direct and indirect experimentation with a prime sensor in the respiratory control system is that the carotid bodies in man function as they do in cats and dogs, making a major contribution to responses during hypoxaemia but not during air-breathing.

Sleep and Sleep Apnoeas
Phillipson (1978) has provided an excellent review of the control of breathing in sleep, but despite the recent revival of interest in this subject, there are still rather little data available from human subjects.

In normal sleep, the major alteration in respiratory control is the reduction in CO_2-responsiveness which is closely correlated with the depth of sleep as judged by EEG recording. The CO_2-sensitivity is reduced and the threshold is increased both at the onset and at the deepening of sleep. This rise in threshold resembles the withdrawal of the 'wakefulness'-drive during non-REM sleep: ventilation falls and PCO_2 rises. In awake subjects, it is this drive that prevents the apnoea during the hypocapnia that follows hyperventilation (Fink, 1961), and which is mimicked by the modest ventilatory drive accompanying the performance of mental arithmetic. Phillipson & Sullivan (1978) have emphasised the importance of 'arousal' as a response to stimuli like CO_2 and hypoxia during sleep. The arousal restores the full ventilatory response to these stimuli, and also allows an appropriate behavioural response which might expedite the removal of their cause (like airway obstruction). For example, when patients with low ventilatory drive associated with chronic obstructive lung disease (COLD) are asleep and their ventilatory drive is further reduced by the withdrawal of the 'arousal' stimulus, they tend to have transient episodes of hypoxaemia which wake them and, in disturbing their sleep, correct part of the problem.

Apnoea during sleep in adults has received considerable attention recently (cf Guilleminault *et al.*, 1978). Besides the aggravation of COLD mentioned above, three main types of respiratory disturbance have been described (Phillipson, 1978). They are periodic breathing (see below) which is also seen during wakefulness, central apnoea, and upper airway obstructive apnoea. The last two are not always clearly delimited, as mixed apnoeas are often seen.

Central apnoeas encompass several types of disturbance of the metabolic controller and occur particularly during non-REM sleep when the alternative 'behavioural' efferent output to the respiratory muscles is relatively quiescent. In infants some such disorder of 'metabolic' control may account for those cases of 'sudden infant death syndrome' which occur during non-REM sleep with no evidence of struggling or arousal.

The nasopharynx or the laryngeal airway may become partially obstructed during sleep, even in the absence of any demonstrable anatomical abnormality. Such obstruction is a characteristic feature of the Pickwickian syndrome in which obesity and daytime drowsiness are associated with episodes of respiratory failure during sleep. This pattern may occur in REM sleep, as part of a general relaxation of the muscles concerned with breathing, as well as during non-REM sleep when the 'behavioural' output stimulates muscular activities such as swallowing or phonation that might prejudice regular patterns of airflow (see above). In this situation the metabolic controller is still sensitive to hypoxia though not to CO_2, and arousal can still occur, thus enabling the patient to break out of the chain.

Periodic Breathing

Cheyne-Stokes respiration is characterised by waxing and waning of ventilation (mainly tidal volume) with a frequency of around 0.02 Hz, interspersed with periods of apnoea. It occurs in normal subjects at high altitude, particularly during sleep, and in two main groups of patients: those with heart failure and those with central nervous system disease, particularly supramedullary lesions of the motor pathways. The control system itself is probably intact and the periodicity represents an instability of the feedback loops due to one or more of the following abnormalities: (a) a fluctuating set-point for ventilation in light sleep as rising PCO_2 raises the wakefulness-drive and the resultant normocapnia allows it to fall again; (b) a dependence on the non-linear poorly damped responses to hypoxia in the absence of CO_2-drive; (c) prolonged time-delays between the controller and controlled elements, and (d) lifting of the supra-medullary inhibition of the CO_2-response.

Some Drugs Affecting the Respiratory Control System

Several groups of drugs have effects on the respiratory control system: three are discussed to illustrate different aspects of their use. Stimulation of ventilation may be the primary desired effect: this is illustrated here by the use of progesterone. Drugs used for non-respiratory disease may have unwanted side-effects on the respiratory system, and the patient's ventilation may be compromised. This may be particularly hazardous for patients whose ventilatory function is already impaired. This problem is illustrated by the use of beta-adrenergic antagonists. Finally, drugs have been used directly as experimental tools in the investigation of the respiratory control system. The use of naloxone is discussed in relation to the possible involvement of endogenous opiates in respiratory control.

Progesterone

Progesterone, given intramuscularly or orally to obese men with

Pickwickian syndrome, stimulates ventilation and improves gas exchange, correcting the respiratory acidosis. It is also effective in controlling the polycythaemia of high altitude by improving the arterial PO_2. A central site of action has been suggested on the grounds that progesterone has a time-course that parallels the appearance in the CSF of the hormone itself and its metabolites. However, it appears that the predominant respiratory effect is an increase in the ventilatory response to hypoxia rather than to carbon dioxide, suggesting an action at the peripheral chemoreceptor.

Adrenergic Mechanisms and Beta-blockade

Beta-adrenoreceptor antagonist drugs like propranolol are widely used in clinical practice, particularly for the treatment of patients with angina pectoris or hypertension. Bronchoconstriction is among the unwanted effects which may make beta-blockade hazardous in those whose airways are maintained by beta-adrenergic sympathetic tone, particularly asthmatic patients. A problem of more direct concern here is whether beta-adrenergic mechanisms have any role in respiratory control, and whether there are patients whose ventilatory drive might be dangerously impaired by beta-blockade.

Ventilation in man is stimulated by infusions of adrenaline, noradrenaline, isoprenaline and salbutamol: this points to the involvement of at least one adrenergic link in the control system. The effect of noradrenaline at least is likely to be at the peripheral chemoreceptors: when these are silenced by hyperoxia, noradrenaline affects neither ventilation nor carotid body discharge (measured in dogs). These effects might have been attributable to stimulation of alpha-adrenoreceptors but they are blocked by propranolol and not by phentolamine, and the alpha-agonist phenylephrine has a slightly *depressant* action on ventilation, even in hypoxia. The pure beta-adrenoceptor agonist isoprenaline and salbutamol stimulate ventilation by enhancing the responses of the carotid body to hypoxia and hypercapnia, and these effects are blocked by propranolol. Much of this work has been reviewed by Cunningham (1974).

Several authors have attempted to assess the effect of propranolol on the ventilatory response to CO_2 in man. The rather poor reproducibility of the rebreathing technique has meant that few have found significant changes. A review of six studies (Patrick & Pearson, 1980) suggested that propranolol caused a modest depression of the CO_2 drive in the absence of hypoxia. Two further studies support this contention. The response to hypoxia at controlled levels of CO_2 drive is not affected by propranolol. In exercising man propranolol causes a transitory depression of ventilation when given intravenously, and it impairs the early rise in ventilation at the onset of exercise. In the steady-state however, doses of propranolol that reduce exercise tolerance appear not to affect ventilation, except briefly following intravenous administration.

The overall impressions provided by these rather conflicting data are that some beta-adrenergic mechanisms *are* involved in the respiratory control system, probably at the peripheral chemoreceptors and possibly centrally as well, but that beta-blockade causes minimal reductions in ventilatory drive in man. There is little danger, therefore, that patients will come to harm because of the respiratory side-effects of beta-adrenergic antagonist drugs, as opposed to their effects on the airways.

Opiates

The respiratory depression caused by morphine and other narcotics is well known. The CO_2-response line is shifted to the right with or without a reduction in slope, and the effect is distinct from that of mere sleepiness. A central site of action is likely. Nalorphine and pentazocine reverse the respiratory effects of large doses of morphine, but their own agonist-actions may mask this property when used against smaller doses of morphine.

Naloxone is a newer narcotic antagonist without partial-agonist properties, and is thought to be a specific blocker of central opiate-receptors that are plentiful in the medulla. It blocks the depressant effects both of morphine and of the recently discovered endogenous opiate, beta-endorphin.

A potentially exciting new field is the possible involvement of the endogenous opiates in respiratory control. Moss & Friedman (1978) found that intracisternal injection of beta-endorphin in lightly anaesthetised dogs markedly depressed their respiratory drive (measured as occlusion pressure or ventilation). This was reversed or delayed by systemic administration of naloxone. In decerebrate or anaesthetised cats, naloxone increases respiratory drive. However, the results of naloxone administration in man have so far been disappointing. Fleetham *et al.* (1980) showed that hypoxial responses were increased, but the change was no more than with placebo injections. Santiago *et al.* (1981) found no effect of naloxone in normal subjects nor in those patients with COLD who responded normally to a resistive load. In other patients, naloxone prevented the depression of ventilation experienced during inspiratory loading, but no placebo controls were performed and the effect needs confirmation.

However, endogenous opiates are probably released in a variety of circumstances, and might be responsible for some of the variation in respiratory responses to other agents in different studies.

Some of the conclusions drawn in this chapter about respiratory control in man could be or have been confirmed or extended in animal experiments. There is considerable interaction between the two approaches to the topic, and species differences are not much of a problem.

However, the clinical importance of disorders of respiratory control and the serious consequences of its involvement in diseases of other systems, make human studies of respiratory control mandatory. Furthermore, human experimentation with normal subjects is of value in its own right, with the aim of providing an account of the operation of the respiratory control system as an integrated whole. While human subjects do raise certain problems for the experimenter, they also provide a rich source of unique material for the physiologist, the anaesthetist, and the physician.

References

Adams, L., Chronos, N. & Guz, A. (1982) 'Breathlessness during Different Ventilatory Drives in Normal Subjects', *J. Physiol.*, *324*, 33P

Anand, A., Loeschcke, H.H., Marek, W. & Paintal, A.S. (1982) 'Significance of the Respiratory Drive by Impulses from J-receptors', *J. Physiol.*, *325*, 14P

Arkinstall, W.W., Nirmel, K., Klissouras, V. & Milic-Emili, J. (1974) 'Genetic Differences in Ventilatory Responses to Inhaled CO_2', *J. Appl. Physiol.*, *36*, 6-11

Borg, G. (1970) 'Perceived Exertion as an Indicator of Somatic Stress', *Scand. J. Rehabil. Med.*, *2*, 92-8

Campbell, E.J.M., Godfrey, S., Clark, T.J.H., Freedman, S. & Norman, J. (1969) 'The Effect of Muscular Paralysis Induced by Tubocurarine on the Duration and Sensation of Breath-holding during Hypercapnia', *Clin. Sci.*, *36*, 323-8

Clark, F.J. & Euler, C. von. (1972) 'On the Regulation of Depth and Rate of Breathing', *J. Physiol.*, *222*, 267-95

Cotes, J.E. (1979) *Lung Function*, 4th ed, Blackwell, Oxford

Cunningham, D.J.C. (1974) 'Integrative Aspects of the Regulation of Breathing: a Personal View', in J.G. Widdicombe (ed.), *Respiratory Physiology*, MTP International Review of Science, Physiology Series One, vol. 2, 303-69

Doust, J.H. & Patrick, J.M. (1981) 'The Limitation of Exercise Ventilation during Speech', *Resp. Physiol.*, *46*, 137-47

Fink, B.R. (1961) 'Influence of Cerebral Activity in Wakefulness on Regulation of Breathing', *J. Appl. Physiol.*, *16*, 15-20

Fleetham, J.A., Clark, H., Dhingra, S., Chernick, V. & Anthonisen, N.R. (1980) 'Endogenous Opiates and Chemical Control of Breathing in Humans', *Am. Rev. Resp. Dis.*, *121*, 1045-9

Fowler, W.S. (1954) 'The Breaking-Point of Breath-Holding', *Resp. Physiol.*, *6*, 529-45

Gilbert, R., Auchincloss, J.H., Brodsky, J. & Boden, W. (1972) 'Changes in Tidal Volume, Frequency, and Ventilation Induced by their Measurement', *J. Appl. Physiol.*, *33*, 252-4

Godfrey, S. & Campbell, E.J.M. (1968) 'The Control of Breath-holding', *Resp. Physiol.*, *5*, 385-400

Guilleminault, C., Hoed, J. van den & Mitler, M. (1978) 'Clinical Overview of Sleep Apnea Syndromes', in C. Guilleminault & W. Dement (eds), *Sleep apnea syndromes*, Liss, New York, pp. 1-12

Killian, K.J., Mahutte, C.K., Howell, J.B.L. & Campbell, E.J.M. (1980) 'The Effect of Timing, Flow, Lung Volume and Threshold Pressures on Resistive Load Detection', *J. Appl. Physiol.*, *49*, 958-63

Lloyd, B.B. & Cunningham, D.J.C. (1963) 'A Quantitative Approach to the Regulation of Human Respiration', in D.J.C. Cunningham & B.B. Lloyd (eds), *The Regulation of Human Respiration*, Blackwell, Oxford, pp. 331-50

Mitchell, R.A. & Berger, A.J. (1975) 'Neural Regulation of Respiration', *Am. Rev. Resp. Dis.*, *111*, 206-24

Moss, I.R. & Friedman, E. (1978) 'Beta-endorphin: Effects on Respiratory Regulation', *Life Sciences*, *23*, 1271-76

Patrick, J.M. (1981) 'Longitudinal Changes in the Ventilatory Cost of Exercise in Normal 65-year-old Men', *J. Physiol.*, *320*, 98-9P

Patrick, J.M. & Cotes, J.E. (1974) 'Anthropometric and Other Factors Affecting Respiratory Responses to CO_2 in New Guineans', *Phil. Trans. R. Soc. (Series B)*, *268*, 363-73

Patrick, J.M. & Howard, A. (1972) 'The Influence of Age, Sex, Body Size and Lung Size on the Control and Pattern of Breathing During CO_2 Inhalation in Caucasians', *Resp. Physiol.*, *16*, 337-50

Patrick, J.M. & Pearson, S.B. (1980) 'Beta-adrenoreceptor Blockade and Ventilation in Man', *Br. J. Clin. Pharmacol.*, *10*, 624-5

Patrick, J.M. & Reed, J.W. (1969) 'The Influence of Lung Volume and CO_2 on Breath-Holding', *J. Physiol.*, *204*, 91-2P

Phillipson, E.A. (1978) 'Control of Breathing During Sleep', *Am. Rev. Resp. Dis.*, *118*, 909-39

Phillipson, E.A., McClean, P.A., Sullivan, C.E. & Zamel, N. (1978) 'Interaction of Metabolic and Behavioural Respiratory Control During Hypercapnia and Speech', *Am. Rev. Resp. Dis.*, *117*, 903-9

Phillipson, E.A. & Sullivan, C.E. (1978) 'Arousal: the Forgotten Stimulus', *Am. Rev. Resp. Dis.*, *118*, 807-9

Santiago, T.V., Remolina, C., Scoles, V. & Edelman, N.H. (1981) 'Endorphins and the Control of Breathing', *N. Engl. J. Med.*, *304*, 1190-5

Schaefer, K.E., Hastings, B.J., Carey, C.R. & Nichols, G. (1963) 'Respiratory Acclimatization to Carbon Dioxide', *J. Appl. Physiol.*, *18*, 1071-8

Severinghaus, J.W. (1976) 'Proposed Standard Determination of Ventilatory Responses to Hypoxia and Hypercapnia in Man', *Chest*, *70*, 129-31

Stevens, S.S. (1957) 'On the Psychophysical Law', *Psycholog. Rev.*, *64*, 153-81

Sykes, M.K., McNicol, M.W. & Campbell, E.J.M. (1976) *Respiratory Failure*, 2nd edn, Blackwell, Oxford

Zechman, F.W., Wiley, R.L. & Davenport, P.W. (1981) 'Ability of Healthy Men to Discriminate Between added Inspiratory Resistive and Elastic Loads', *Resp. Physiol.*, *45*, 111-20

7 RESPIRATORY OSCILLATIONS IN HEALTH AND DISEASE

Christopher B. Wolff

Introduction

It has become apparent, over the years, that the stability of arterial PCO_2 ($PaCO_2$) is closely linked to the level of hydrogen ion activity (H^+) somewhere in the medullary region of the brain (Pappenheimer, Fencl, Heisey & Held, 1965). That the chosen $PaCO_2$ might be altered in normal man is clear from the lowered alveolar PCO_2 ($PACO_2$) and $PaCO_2$ in the chronically hypoxic state of high altitude acclimatisation (Fitzgerald, 1913; Rahn & Otis, 1949). Although Lahiri has shown that the $PACO_2$ is not as low in high altitude natives as in altitude-acclimatised lowlanders it was lower than at sea level (Lahiri, 1968). Wolff (1980) calculated the likely relationship between arterial PCO_2 and PO_2 (PaO_2) in acclimatised normal subjects, and proposed that the same relationship might apply to chronically hypoxic patients at sea level, so long as they had a normally functioning respiratory control system. In ambulant patients with chronic stable asthma the same basic relationship between $PaCO_2$ and PaO_2 was found to be present as in normal subjects acclimatised to the hypoxia of high altitude (Cochrane, Prior & Wolff, 1980b). A close approximation to both the relationship calculated for normal subjects at altitude and that measured in the patients with chronic stable asthma is:

$$PaCO_2, \text{expected} = 0.25 \times PaO_2 + 15 \text{ mm Hg} \tag{1a}$$

or

$$PaCO_2, \text{expected} = 0.25 \times PaO_2 + 2 \text{ kPa} \tag{1b}$$

Since the sensitivity of the peripheral arterial chemoreceptors to hypoxia plays a part in determining mean arterial PCO_2 in the chronic hypoxia of altitude acclimatisation, it is apparent that a peripheral arterial chemoreceptor input can, in the long term, affect the 'chosen' value of $PaCO_2$. The chosen $PaCO_2$ is then held stable by the central chemoreceptors.

In chronic acid-base disorders of the metabolic type there is a slow readjustment of ventilation, *pari-passu* with changes in the bicarbonate

222

production rate by the choroid plexus (Herrera & Kazemi, 1980). These act in much the same way as in high-altitude acclimatisation to chronic hypoxia. Ventilation in metabolic acidosis, as in acclimatisation, gradually increases as CSF bicarbonate falls. The drop in CSF PCO_2, resulting from increased ventilation, can then occur without the restraining effect on ventilation of decreased CSF pH. Here the respiratory control mechanism is involved in the control of arterial pH.

It can be seen that insights into respiratory control mechanisms can be gained from normal animals and man, and also from studying the adjustments which occur in response to the abnormalities in function imposed by disease processes.

Respiratory Oscillations in $PACO_2$ and $PaCO_2$ as a Possible Key to the Understanding of Ventilatory Adjustments in Exercise

The increase in ventilation which occurs in exercise is relatively linear with respect to both the oxygen consumption rate ($\dot{V}O_2$) and the carbon dioxide production rate ($\dot{V}CO_2$), up to slightly differing work rates. Figure 7.1 shows the relationships for expired ventilation ($\dot{V}E$) against $\dot{V}O_2$ (a) and $\dot{V}CO_2$ (b). The metabolic rate at which the $\dot{V}E/\dot{V}CO_2$ curve becomes non-linear has been referred to as the anaerobic threshold by Wasserman *et al.* (1967, 1977). These authors found a sharp increase in circulating lactic acid above the 'anaerobic threshold', as others had done. It has been suggested that hydrogen ion excess adds an extra increment to the ventilatory response to exercise above this critical work load.

Below the anaerobic threshold ventilation appears to match CO_2 production rate sufficiently well for there to be little change in the mean level of the alveolar PCO_2. If we consider alveolar ventilation ($\dot{V}A$), measured under the same conditions as $\dot{V}CO_2$ and $PACO_2$, we have the following simple relationship, in air-breathing subjects:

$$\dot{V}A = PB \times \dot{V}CO_2 / PACO_2 \tag{2}$$

where PB is the ambient barometric pressure (see Wolff, 1980 for this equation expressing $\dot{V}CO_2$ in moles).

In Figure 7.2 we have $\dot{V}E$ and $\dot{V}A$ plotted against $\dot{V}CO_2$ in the upper two panels and end-tidal PCO_2 ($PetCO_2$) plotted against $\dot{V}CO_2$ in the lower panel. $PetCO_2$ is taken to represent $PACO_2$ (though the precise relationship between these two requires further elucidation). Results were obtained in normal subjects breathing through an external dead space of 140 ml.

Figure 7.1: Minute Ventilation ($\dot{V}_E$) related to the rate of oxygen consumption ($\dot{V}O_2$) in the steady state (a); and $\dot{V}_E$ related to carbon dioxide production rate ($\dot{V}CO_2$) in (b).

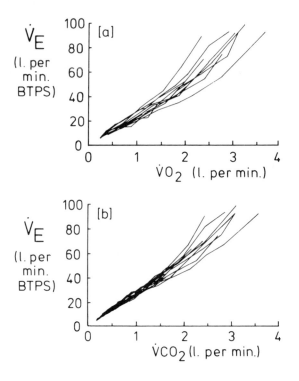

Source: after Wasserman *et al.*, 1967, 1977.

This meant that the total ventilation required to achieve (near) constancy of $PetCO_2$ and $PACO_2$ mean value, was a little larger than would have been needed normally (without added dead space). It can be seen from equation 2 that a constant PCO_2 means that increases in alveolar ventilation precisely match increases in CO_2 production rate. In other words the respiratory control system somehow causes normal man to increase his alveolar ventilation in direct proportion to his increase in CO_2 production rate, as illustrated for the subjects of Figure 7.2. Similar results were obtained for total ventilation (Asmussen,

Nielsen & Wieth-Pedersen, 1943b) in normal man, though in their study $\dot{V}_E$ and $PACO_2$ were plotted against $\dot{V}O_2$. Asmussen *et al.* (1943b) found more convincing constancy of $PACO_2$, around 38.5 mm Hg, than is seen in Figure 7.2, though results in the literature vary on this point.

Figure 7.2: $\dot{V}_E$ Plotted against $\dot{V}CO_2$ (upper panel); Alveolar Ventilation ($\dot{V}A$) against $\dot{V}CO_2$ (middle panel); end-tidal PCO_2 ($PetCO_2$) against $\dot{V}CO_2$ (lower panel). ●, rest and mild exercise results from five medical students; o, rest to moderately heavy exercise results from seven further medical students. All points, steady state.

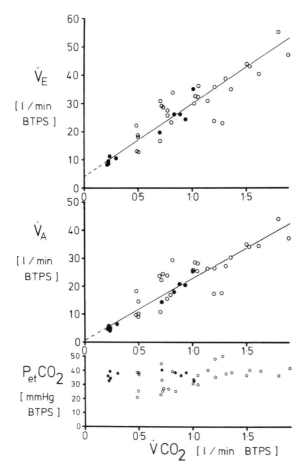

Source: Openshaw, Wolff & Woodroof, unpublished.

In a companion study to Asmussen *et al.* (1943b) Asmussen, Christensen & Nielsen (1943a), showed that ventilation was little, if at all, reduced in exercising subjects in whom the circulation to the legs was supposedly cut off. Despite some uncertainty as to how completely circulation was cut off, there was a reduction in $\dot{V}CO_2$ and $\dot{V}O_2$ and hence an increase in the ratio of ventilation to both $\dot{V}O_2$ and $\dot{V}CO_2$, with a consequent drop in $PACO_2$. Asmussen *et al.* (1943a) took this to mean that ventilation during the occlusion period was driven by neural afferent impulses. Asmussen *et al.* (1943a) also studied a subject with tabes dorsalis who had a virtually normal steady-state ventilation to $\dot{V}O_2$ relationship. They pointed out that the neural drive in this subject could not be carried in the dorsal tracts of the spinal cord. However, a neural drive to ventilation in exercise seems likely to be an important component of respiratory control.

In a paper by Kao, Michel, Mei & Li (1963) ventilation and the arterial PCO_2 were each plotted against $\dot{V}O_2$, for one of a triad of cross perfused dogs. One dog received blood of constant composition from a donor animal to perfuse its cerebral vessels, and in turn donated venous blood from its lower limbs to a third animal. When its legs were exercised its peripheral and central chemoreceptors were unlikely to have received any change in their chemical environment. This animal was referred to as 'the neural dog'. The $\dot{V}_E/\dot{V}O_2$ and $PACO_2/\dot{V}O_2$ plots were very similar to those shown in the middle and lower panels of Figure 7.2. This strongly suggested that the ventilatory changes could be accomplished without involvement of peripheral or central chemoreceptors. Though the results appear to be flawless, it is difficult to understand how ventilation can be precisely adjusted by the respiratory control system in exercise to give homeostasis for $PaCO_2$ without the involvement of chemical changes related to $\dot{V}CO_2$. Related experimental evidence also appears in the studies by Michel & Kao (1964) and Kao, Michel & Mei (1964), and have all been outlined by Kao (1963), and discussed by Mahler (1979). If we assume that chemical control is likely to be necessary for the normal adjustment of ventilation in the face of varying CO_2 loads, we face the problem as to how a normal $PaCO_2$ value is sustained, since the mean value of PCO_2 itself cannot be the signal. The respiratory system is capable of making ventilatory adjustments on the basis of the mean PCO_2 where it is altered from its normal value; as seen with CO_2 response curves. However, there appears to be insufficient PCO_2 change in exercise for the sensitivity obtained from CO_2 response curves in man to account for the ventilatory changes, even in severe exercise (Bannister, Cunningham & Douglas, 1954).

Figure 7.3: The Time Course of Alveolar PCO_2 (or FCO_2 — dry fraction) shown above for One Respiratory Cycle at Rest. Below, is shown the corresponding pattern of airway flow with an indication of the times at which dead-space gas is cleared.

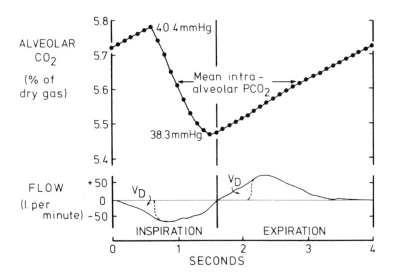

Source: after DuBois *et al.*, 1952.

Yamamoto (1960) pointed out that alveolar PCO_2 varies within each respiratory cycle. Krogh & Lindhard (1914) had demonstrated increases in airway and (theoretical) rate of alveolar PCO_2 increase during expiration, and DuBois, Britt & Fenn (1952) illustrated the likely profile of $PACO_2$ oscillations within the respiratory cycle at rest (Figure 7.3). Yamamoto (1960) considered the possibility that the increased oscillations of exercise could provide a stimulus to ventilation, over and above that provided by the mean level of the PCO_2. He pointed out (Yamamoto, 1962) that the oscillations in alveolar PCO_2 could act as a time variant signal which might then be transmitted, via the intervening circulatory pathway, to the systemic arterial blood stream. In order for the information in the oscillations to be available to the respiratory control system there would have to be a suitable receptor for such rapid oscillations in arterial PCO_2. At that time arterial pH electrodes, as indirect monitors of rapid arterial PCO_2 changes, only had sufficient speed of response to detect oscillations of respiratory frequency in the arterial blood at low respiratory rates.

We now know that pH oscillations are present at normal and high respiratory rates, and have evidence that they reflect PCO_2 oscillations. Furthermore, it was not known, in the early 1960s that fast $PaCO_2$ changes are detected at the carotid body in addition to its well-known sensitivity to hypoxia.

Nevertheless, Yamamoto & Edwards (1960) undertook experiments, in lightly anaesthetised rats, in which the increased $\dot{V}CO_2$ of exercise was simulated by infusion of CO_2-enriched blood to the venous side of the lungs. This extra volume infusion was balanced by an outflow of blood to a reservoir from the arterial side of the circulation. There was sufficient constancy of mean arterial PCO_2 to support their suggestion that a chemical signal related to $\dot{V}CO_2$ acted to maintain PCO_2 homeostasis. Appropriate ventilatory adjustments therefore appeared not to require any peripheral neural signal from exercising muscle.

The suggestion had therefore been made by 1960 that respiratory oscillations in the arterial PCO_2 play a crucial part in respiratory control, particularly with regard to the appropriate matching of ventilation to the rate of whole body CO_2 production.

The Carotid Body as a Detector of Rapid $PaCO_2$ Changes and Receptor for Reflex Respiratory Response to Dynamic $PaCO_2$ Change

For the theory of Yamamoto (1960, 1962) to work there needs to be a suitable receptor which will detect rapid changes in $PaCO_2$ such as respiratory oscillations. Until the 1960s the carotid body was thought of mainly as a detector of steady-state hypoxia. However, Hornbein, Griffo & Roos (1961) first suggested that there were respiratory oscillations in the afferent chemoreceptor discharge. They studied the afferent sinus nerve discharge, after elimination of baroreceptor impulses. They recorded chemoreceptor discharge from the circumference of the whole nerve. Figure 7.4 shows a respiratory oscillation at relatively normal arterial oxygen tension (PaO_2). Hornbein (1965) suspected that arterial oscillations in PCO_2 caused this oscillation in discharge frequency. He passed arterial blood through a mixing chamber, before it reached the carotid body, to eliminate blood gas oscillations, and thereby removed respiratory oscillations in chemoreceptor discharge frequency. He could not pin the chemical stimulus down explicitly to $PaCO_2$ rather than PaO_2 or pH.

Biscoe and Purves (1965, 1967) showed that arterial PO_2 oscillations were present in cats and in lambs. Band & Semple (1966, 1967) showed that arterial pH oscillations were present, at normal and high respiratory rates, both in anaesthetised cats and in resting man. Their pH electrode system had sufficient speed of response fully to outline the arterial pH waveform. Band, Cameron & Semple (1969a, b) were able to demonstrate that the PCO_2 changes which occur at the lung within each breath were detected indirectly by their fast response pH electrode as arterial pH oscillations.

Figure 7.4: A Ratemeter Count of Sinus Nerve Afferent Chemoreceptor Discharge (arbitrary units) at Normal Oxygen Tension Shows an Oscillation of Respiratory Frequency.

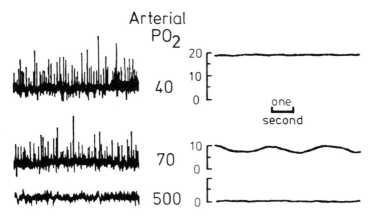

Source: after Hornbein *et al.*, 1961.

Fitzgerald, Lietner & Liaubet (1969) were able to show that PCO_2 and/or pH oscillations, rather than PO_2 oscillations were the likely cause of respiratory oscillations in sinus nerve chemoreceptor discharge frequency. These authors infused blood with altered levels of PCO_2 or PO_2 in an intermittent manner into the carotid arterial blood stream of anaesthetised cats, such that either PCO_2 or PO_2 fluctuated between the value in the infusate and the value in the animals' systemic arterial blood. $PaCO_2$ fluctuations resulted in fluctuations in chemoreceptor discharge frequency between low rates and rates well above normal respiratory rates, whereas PaO_2 fluctuations in the blood were only effective at very low rates. The precise role of pH change was not explored.

Further evidence that rapid changes in PCO_2 or pH were detected at the carotid body was provided by Black, McCloskey & Torrance (1971). While recording single unit afferent chemoreceptor discharge, blood from a rubber sac connected to the headward end of the carotid artery was forced retrogradely past the offshoot of the carotid body blood supply. When the administered blood had a high PCO_2 value there was a very abrupt response with a high peak in the discharge frequency, followed by a lower plateau. The plateau persisted while the high PCO_2 perfusion continued. The overshoot was not seen with PO_2 changes.

This response may represent a phasic response, causing the over-shoot, followed by a static response to the mean pH and/or PCO_2 change. Only the plateau was obtained if the perfusate contained a carbonic anhydrase inhibitor, which presumably prevented rapid chemical equilibration from taking place in the carotid body. Although discharge frequency altered quite quickly with abrupt changes in PaO_2 in these experiments, the PaO_2 changes were probably very much larger than occur naturally in respiratory PaO_2 oscillations.

The specificity of the carotid body to rapid changes in $PaCO_2$, rather than to arterial pH (pHa) was demonstrated by Band & Phillips (1973). Their study was outlined in detail by Band *et al.* (1978). Sinus nerve afferent chemoreceptor discharge was monitored and small boluses of isotonic saline solutions were injected through a catheter with its tip in the aortic root. The solutions contained high CO_2 and high hydrogen ion (a buffered acid was used), high hydrogen ion and active carbonic anhydrase, or carbonic anhydrase alone as a control. Only the high CO_2 solution and mixed acid and carbonic anhydrase solution were effective in stimulating the carotid body to give an increase in discharge. These findings have been interpreted to mean that rapid changes in PCO_2 cause rapid carotid body responses, whereas rapid hydrogen ion changes will only stimulate the carotid body when equilibration with arterial blood is sufficiently accelerated to release CO_2 before the blood reaches the carotid body. Co_2 is therefore a specific fast-acting stimulus.

Band, Cameron & Semple (1970) showed that there were reflex effects of rapid PCO_2 change in the carotid arterial blood stream even when these only caused pH changes comparable to naturally occuring oscillations.

This type of reflex, first demonstrated by Black & Torrance (1967a), was also outlined in some detail by Black & Torrance (1971) and Eldridge (1972a, b). It consisted of an increase in the ongoing inspiratory tidal volume, so long as the stimulus was timed to occur at the right moment during inspiration (Figure 7.5). Black & Torrance (1967, 1971) and Eldridge (1972a) demonstrated that very similar phenomena to those occurring with CO_2 stimuli, also occurred if brief electrical stimuli were applied to the sinus nerve, showing that the carotid body was the receptor for this reflex. In the study by Band *et al.* (1970) acid and hypoxic injectates failed to give rise to the reflex tidal volume changes which were seen with carbon dioxide, again pointing to PCO_2 as a dynamic stimulus rather than PO_2 or pH.

Figure 7.5: Shows how Brief CO_2 'Stimuli' have Differing Effects on Tidal Volume (inspiration upward) According to their Timing in the Respiratory Cycle. A stimulus in mid to late inspiration increases and prolongs that inspiration.

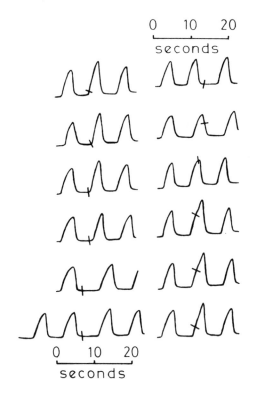

Overall reflex effects, of a slightly more prolonged $PaCO_2$ change, were shown to depend on an intact carotid body in dogs by Bouverot, Flandrois, Puccinelli & Dejours (1965). Conscious dogs were given two breaths of inspirate with increased CO_2 concentration. The animals with intact carotid bodies showed a rapid ventilatory change, whereas those without carotid bodies showed an attenuated and delayed response (Figure 7.6).

Figure 7.6: The Average Time Courses of Changes in Alveolar PCO_2 ($\Delta PACO_2$) and Ventilation ($\Delta \dot{V}_E$) from Two Series of Conscious Dogs. In both series (32 control animals and 16 in which the carotid body function had been eliminated), two breaths of 7% CO_2 ($FICO_2$, dry = 0.07; FIO_2, dry = 0.21%) were given after a suitable control period. Apart from these two ('CO_2') breaths the animals breathed air. Ventilatory changes were delayed and attenuated in the chemodenervated dogs.

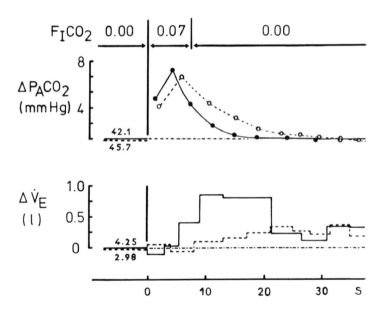

Source: after Bouverot *et al.*, 1965; also illustrated in Cunningham, 1972.

The $\dot{V}CO_2$ Information Transmission Line to the Central Nervous System

We can now consider how well information as to the CO_2 production rate, represented by alveolar PCO_2 oscillations, is relayed to the central nervous system.

Specific Properties of Respiratory $PACO_2$ Oscillations

An increase in amplitude of alveolar PCO_2 ($PACO_2$) oscillations with exercise was known to occur by Haldane & Priestley in 1905. Krogh & Lindhard (1914) used a Haldane (CO_2 absorbtion) apparatus to measure samples of expired gas (dry fraction), whereas DuBois, Fowler, Soffer & Fenn (1951) used an infra-red CO_2 gas analyser (wet or 'total' fraction). Both showed that there was an increase in the steepness of the expiratory airway PCO_2 slope in exercise. DuBois *et al.* (1951) monitored airway PCO_2 on the rapid CO_2 gas analyser which had recently been described by Fowler (1949).

Figure 7.7 shows recordings made by Dubois *et al.* (1951) which illustrate the increased steepness of airway CO_2 in exercise. The record taken immediately post-hyperventilation is also steepened. The reasons behind these effects have led to methods by which part of the expiratory slope of alveolar PCO_2 can be examined. It has been shown that alveolar PCO_2 during expiration rises at a rate ($dPACO_2/dt$) which is directly proportional to the rate of CO_2 production in the steady state (Cochrane & Wolff, 1979; Newstead, Nowell & Wolff, 1980). Saunders (1980) like Yamamoto (1960) has predicted, from theoretical analysis (Saunders, Bali & Carson, 1980), that $dPACO_2/dt$ in expiration ($dPCO_2/dt$, max. of Saunders) will increase in exercise. The theoretical relationship given by Saunders *et al.* (1980) has also been expressed by Wolff (1981) as:

$$dPACO_2/dt = PB \cdot \dot{V}CO_2/ELV \qquad (3)$$

where ELV is the effective volume of gas in the lung into which CO_2 is lost from the blood – a term used originally by Du Bois *et al.* (1952), the initials standing for Equivalent Lung Volume. Saunders specifically estimated the relationship:

$$dPACO_2/dt,max. = 4.5 \times \dot{V}CO_2 \qquad (4A)$$

where $dPCO_2/dtmax.$ is in mm Hg per second and $\dot{V}CO_2$ in litres per minute. A convenient figure to remember is that for a $\dot{V}CO_2$ of 222 ml per minute Saunders' equation predicts that alveolar PCO_2 will rise at a rate of 1 mm Hg per second. Equation 4A is very similar to the average result of Newstead *et al.* (1980), which was that $dPACO_2/dt$ in expiration equals 4.45 mm Hg per second for each litre per minute of CO_2 production rate. A rate of rise of $PACO_2$ in expiration of 1 mm Hg per second then corresponds to a CO_2 production rate of 225 ml per minute. The resting CO_2 production rates were higher than this so that a normal resting expiratory alveolar PCO_2 rate of rise may be greater than 1 mm Hg per second. Cochrane & Wolff (1979) experimentally obtained a smaller constant relating $dPACO_2/dt$ to $\dot{V}CO_2$ such that

$$dPACO_2/dt,exp = 2.58 \times \dot{V}CO_2 \qquad (4B)$$

where $dPACO_2/dt,exp$ was rate of rise of $PACO_2$ during expiration in mm Hg per second and $\dot{V}CO_2$ was in litres per minute (see also Cochrane *et al.*, 1982).

Figure 7.7: Expiratory Airway CO_2 Recordings are Shown with a Control Profile (A), and Profiles from Exercise (B), and Brief Hyperventilation at Rest (C). The hyperventilation reduces end-tidal PCO_2, whereas in exercise it remains relatively normal. The 'plateau' is steepened in both (B) and (C) due to increased CO_2 delivery to the alveolar compartment from the pulmonary blood supply.

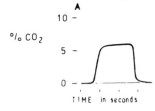

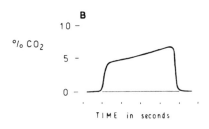

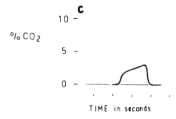

Source: after DuBois *et al.*, 1951.

If we go back to the plot of DuBois *et al.* (1952) illustrated in Figure 7.3, the expiratory $PACO_2$ rose in a remarkably linear manner, but at only 0.68 mm Hg per second. $\dot{V}CO_2$ was 316 ml per minute, corrected to standard temperature and pressure dry (STPD), or 383 ml per minute if we convert into body temperature and pressure saturated with water vapour (BTPS). This resting $\dot{V}CO_2$ is closer to the values obtained by Newstead *et al.* (1980) at rest than the hypothetical values, 222 and 225 ml per minute given above. If we scale up the resting value for $dPACO_2/dt$ obtained from Figure 7.3 (i.e. 383 ml per minute) in direct proportion to $\dot{V}CO_2$ we obtain only 1.77 mm Hg per second at one litre per minute $\dot{V}CO_2$. This is less than the other three values given for a $\dot{V}CO_2$ value of one litre per minute (Cochrane & Wolff, 2.58 mm Hg s^{-1}; Newstead *et al.*, 4.45 mm Hg s^{-1}; Saunders, 4.5 mm Hg s^{-1}). As seen from Equation 3, these differences may result from differences in lung gas volume (ELV).

The proportional relationship between steady-state (mean) values of $dPACO_2/dt$ in expiration and $\dot{V}CO_2$ is well illustrated by the individual graphs for each subject of Newstead *et al.* (1980) as seen in Figure 7.8. Such a proportional relationship was suggested first in 1979 by Cochrane and Wolff (experimentally) and by Saunders (theoretically). Its importance is that, in the steady state, alveolar PCO_2 oscillations contain information (Yamamoto, 1962) which is related to $\dot{V}CO_2$ in a remarkably simple manner. It is strongly reminiscent of the directly proportional relationship between alveolar ventilation ($\dot{V}A$) and $\dot{V}CO_2$ seen in Figure 7.2. If the $dPACO_2/dt$ signal reaches the nervous system the transmission line carrying information as to $\dot{V}CO_2$ will be complete. This will not prove, of course, that the information is utilised by the brain in respiratory control, merely that it reaches the brain and is available to the control system. It is necessary to show whether the nervous system is capable of using such information.

Carriage of Respiratory PCO_2 Oscillations in the Circulation, from Lung to Carotid Body

The changes in $PACO_2$ oscillations with increases in CO_2 production rate are illustrated diagramatically in Figure 7.9. When the CO_2 production increases, from its basal value to three times its basal value, the expiratory $PACO_2$ rate of rise (expiratory slope, i.e. $dPACO_2/dt,exp$) increases threefold.

The greater expiratory rate of $PACO_2$ rise will be present as a result of a CO_2 production rate three times that associated with the lesser expiratory $PACO_2$ rate of rise. There will be no mean $PACO_2$ change if

Figure 7.8: Individual Plots for each of Seven Subjects, Showing Mean Steady-State Values of $dPACO_2/dt$ in Expiration (ordinate) and $\dot{V}CO_2$ (abcissa). Lines equivalent to direct proportionality are shown.

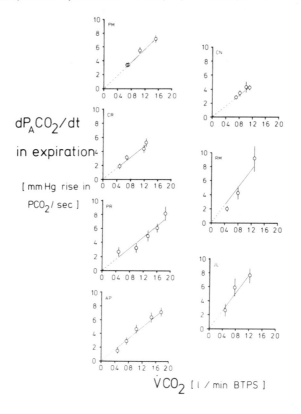

dP_ACO_2/dt

in expiration

[mm Hg rise in

PCO_2/ sec]

$\dot{V}CO_2$ [l / min BTPS]

Source: After Cochrane *et al.*, 1982.

alveolar ventilation also increases to three times the basal value. (If ventilation failed to increase, there would have to be an increase in mean $PACO_2$ to three times its initial value before all the CO_2 made at the tissues could be cleared at the lungs. This would not be compatible with life if the original $PACO_2$ mean value had been normal and the subject breathed air, since the alveolar oxygen tension would fall to only 10 to 20 mm Hg.)

The effect of an increase in respiratory rate is illustrated in Figure 7.10. Keeping $\dot{V}CO_2$ constant and comparing the oscillation at the lower respiratory rate with the faster one, we see that oscillation

Figure 7.9: The Lines T1–P1, and T2–P2 Represent Expiratory Alveolar PCO_2 in Two Different States, in which CO_2 Production Rate is Basal and Three Times Basal. The basal value of the $PACO_2$ rate of rise (or 'slope') is $X_1 P_1 / T_1 X_1$, and the 'slope' when $\dot{V}CO_2 = 3 \times$ basal is $X_2 P_2 / T_2 X_2$. However, in this figure, the intervals $T_1 X_1$ and $T_2 X_2$ are the same. The 'slope' therefore changes by a factor $X_2 P_2 / X_1 P_1$, i.e. by a ratio of three.

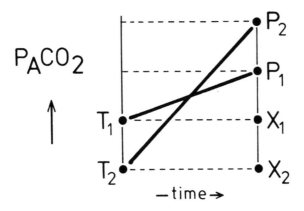

Figure 7.10: Two Diagrammatic Alveolar PCO_2 Oscillations Illustrating the Effects of Steady-State Changes in Respiratory Rate at Constant $\dot{V}CO_2$. The rising and falling ramps ('expiratory $PaCO_2$') each have the same rates of change at the two respiratory rates, so by similar triangles, the amplitudes are theoretically related in proportion to the respiratory periods.

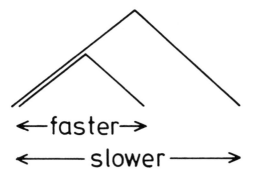

amplitude is proportional to respiratory period (where there is a constant $\dot{V}CO_2$).

Figure 7.11 shows that the directly proportional relationship between amplitude and respiratory period has been confirmed experimentally for human subjects in the resting state.

Figure 7.11: Mean Amplitude of pH Oscillation at Rest Plotted Against Respiratory (oscillation) Period in each of Five Normal Subjects. With similar mean PCO_2 values these changes in pH oscillations indicate changes in $PaCO_2$ oscillations. Hence, there is a directly proportional relationship between $PaCO_2$ oscillation amplitude and respiratory period in resting subjects ($\triangle$pH is $10^{-3} \times$ y axis value).

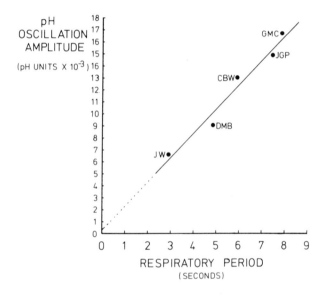

Source: Band, Wolff, Ward, Cochrane and Prior, unpublished.

When the resting alveolar PCO_2 oscillations are transmitted to the systemic arterial circulation there is an equivalent oscillation in arterial pH. The precise pH oscillation will depend upon the way in which CO_2 and carbonic acid are buffered. The oscillation in the arterial blood occurs later than the original event at the lung, because of the time taken for blood to travel from lung to electrode. Inspiratory dilution of alveolar PCO_2 gives rise to a fall in $PACO_2$ and when this blood arrives at the pH electrode we see an alkaline shift in pH. The acid change seen in each pH oscillation reflects the rise in PCO_2 which was originally generated at the lung during expiration.

Figure 7.12: Photographic Record of Arterial pH Oscillations Before and After the Onset of Exercise (40 watts) in Normal Man. The records below the arterial pH are: inspiratory tidal volume (insp. $\dot{V}_T$), airway carbon dioxide concentration (airway CO_2 %), and vertical lines which show the passage of time in seconds. $\dot{V}CO_2$ changed from 304 ml min^{-1} BTPS (barometric pressure 760 mm Hg) to 827 ml min^{-1}; and dpH/dt (amplitude/half period) from 4.41×10^{-3} to 11.95×10^{-3} pH units per second.

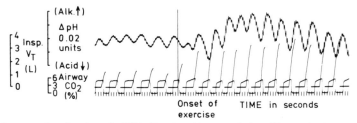

Source: after Band *et al.*, 1980. Reprinted by permission. (Copyright Macmillan Journals Ltd.)

The recording in Figure 7.12 was obtained by Band *et al.* (1980), and shows arterial pH, *in vivo*, initially at rest and later in exercise. It was possible to assess the way in which the slope of the downward (or acid going) limb of the pH oscillations changed from the resting to the exercising state, and to compare this with the way in which the CO_2 production rate changed. Both increased by a factor of 2.7. Hence, this study confirmed that arterial PCO_2 oscillations may behave much as shown for alveolar PCO_2 oscillations. That is, with a simple directly proportional relationship between PCO_2 oscillation upslope (pH oscillation downslope in the systemic arterial blood) and the rate of CO_2 production. The values obtained by Band *et al.* (1980) for pH oscillation downslope (see legend to Figure 7.12) were obtained by dividing the amplitude by half the period. Values obtained by this method, with moderately smoothed sinusoidal waveforms can be scaled up by a factor of $\pi/2$ to obtain the maximum slope. This scaled-up version has been taken as a reasonable, objective, estimate of the downslope of the original pH oscillation. Relating the pH oscillation downslope to $\dot{V}CO_2$ we obtain an equivalence of 22.7×10^{-3} pH units s^{-1} to one litre min^{-1} CO_2 production rate (at body temperature). Taking the buffer slope values of Siggaard Anderson (1962), 1963), the equivalent PCO_2 upslope is 3.28 mm Hg per second, (at a mean PCO_2 of 40 mm Hg, haemoglobin 15 g/100 ml and buffer slope $\Delta log PCO_2 = \Delta pH/0.637$). This value falls between those quoted for Cochrane & Wolff (1979) and Newstead *et al.* (1980) in alveolar gas. There may, therefore, be little attenuation of the respiratory PCO_2 oscillations in the systemic arterial blood stream.

Our information transmission line for $\dot{V}CO_2$ has so far travelled from the pulmonary gas compartment as far as the systemic arterial blood stream. We will now consider whether the peripheral arterial chemoreceptors are capable of detecting $PaCO_2$ oscillations.

Properties of the Carotid Body as a Receptor for Rapid $PaCO_2$ Changes

The oscillations in firing frequency in chemoreceptor afferent sinus nerve discharge, found by Hornbein *et al.* (1961) (Figure 7.4) resulted specifically from $PaCO_2$ oscillations according to the evidence presented here (and normal PaO_2 and pH oscillations are not dynamic stimuli at respiratory frequencies). We come now to studies of single or few fibre afferent chemoreceptor preparations of the sinus nerve (see Chapter 1 of this volume).

Figure 7.13: Sinus Nerve Discharge in an Afferent Chemoreceptor Single Unit in the Anaesthetised Cat, and One-second Time Marks Beneath. The carotid body was naturally perfused. The discharge shows the characteristic apparently random (Poisson) pattern.

Source: after Band *et al.*, 1978.

Biscoe & Taylor (1963, see also Taylor, 1968), examined the pattern of single unit afferent chemoreceptor fibre discharge in cats, and showed that it closely resembled a Poisson distributed (random) train of impulses. A recording of sinus nerve afferent chemoreceptor discharge (made in an air-breathing cat at normal PaO_2), is shown in

Figure 7.13. Under conditions of artificial perfusion the pattern of firing showed a narrower frequency range over which discharge appeared random, than natural perfusion. Under resting conditions breathing air the naturally perfused carotid body of the cat yields a mean discharge frequency in single chemoreceptor fibres of, approximately, one to two impulses per second. Since each respiratory period varies from about two to four seconds, the number of impulses per respiratory period is too small to see any oscillation in discharge frequency. However, Leitner & Dejours (1968) showed that, by taking the average discharge frequency in small time intervals over a number of respiratory cycles, a strong tendency was revealed for discharge frequency to vary in a cyclical manner with respiration. Their illustration has been re-drawn as Figure 7.14. The excursion of the oscillation in discharge frequency is large when we compare it with the oscillation expected from a truly random 'Poisson' discharge (Saunders & Taylor, 1974).

Figure 7.14: This Shows the Average Discharge Rate, over 75 Respiratory Cycles in Six Portions of the Respiratory Cycle, in an Afferent Chemoreceptor Few Fibre Preparation.

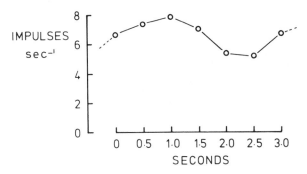

Source: after Leitner & Dejours, 1968.

The randomness is more apparent than real. Furthermore, the oscillation is far greater in amplitude than would be expected from plots of mean discharge rate against $PaCO_2$ in the steady state. These plots may be referred to as CO_2 response curves; they describe the static component of carotid body responsiveness to CO_2. We can profit from examining the dynamic responsiveness separately, since interactions between oxygen and CO_2 which are present for the static response do not occur with dynamic CO_2 stimulation (Band & Wolff,

1978). The CO_2 response curve for afferent chemoreceptor discharge
(static relationship) as obtained by Lahiri & Delaney (1975) is shown
diagrammatically in Figure 7.15.

Figure 7.15: Discharge Rate of a Single Sinus Nerve Chemoreceptor
Unit, Plotted against Arterial PCO_2 (shallow sloping line). This static
curve was re-drawn from Lahiri & Delaney, (1975). The short steeper
line is a dynamic curve. It was drawn on the assumption that the peak
and trough of the oscillation in discharge frequency obtained by
Leitner & Dejours (1968) correspond to the peak and trough of the
$PaCO_2$ oscillation (probably about 2 mm Hg amplitude).

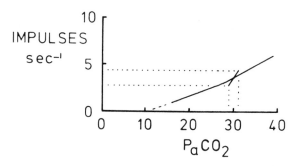

If the arterial oscillation amplitude in cats is 2 mm Hg and mean
$PaCO_2$ is 30 mm Hg (see Herbert & Mitchell, 1971) we would expect
the chemoreceptor discharge rate to oscillate far less than is the case,
in Figure 7.14, if the response was the same as that of Lahiri & Delaney
(1975). The studies of Leitner & Dejours (1968), Goodman, Nail &
Torrance (1974) and Band *et al.* (1978) all show that the amplitude
of the respiratory oscillation in chemoreceptor discharge is far too
great, simply to reflect sensitivity on the same basis as the static CO_2
relationship.

One can also understand the difference between the static and
dynamic responses to arterial PCO_2 by considering the relation between
the oscillation amplitude and the mean values, for both $PaCO_2$ and
discharge frequency. For $PaCO_2$ (in the cat), the amplitude is approximately 7 per cent of the mean value at normal PaO_2. For afferent
chemoreceptor discharge frequency under the same conditions the
oscillation amplitude was 40 per cent of the mean value in the study
of Leitner & Dejours (1968), and up to 150 per cent in studies by
Goodman *et al.* (1974) and Band *et al.* (1978).

It has been pointed out by Band *et al.* (1978) that the afferent chemoreceptor discharge oscillations are not simply proportional responses to the variations in $PaCO_2$ in the arterial blood perfusing the carotid bodies. Band *et al.* (1978) suggested that the carotid body was sensitive at least in part, to the rate of change of $PaCO_2$. The oscillations then represent the dynamic response to the naturally occurring oscillations in $PaCO_2$ whereas the mean discharge frequency represents the static component of the carotid body sensitivity (response to the mean value of the $PaCO_2$ stimulus).

The study by Black *et al.* (1971), which showed an overshoot discharge frequency in single units in response to a step change in $PaCO_2$, is also indicative of carotid body sensitivity to $PaCO_2$ rate of change. It is possible that there may have been some reduction in the frequency response in the preparation of Black *et al.* (1971) due to effects of residual blood products on carotid body function.

It is reasonably certain that the mean value of the afferent chemoreceptor discharge frequency is determined principally by the mean PaO_2, pH and $PaCO_2$. That the oscillations in discharge are due to oscillations in arterial blood chemistry was confirmed by Band *et al.* (1978), in the same manner as described by Hornbein (1965). A mixing chamber was incorporated into the arterial pathway upstream from the carotid body in the study by Band *et al.* (1978). When this was used to eliminate oscillations in $PaCO_2$, the oscillations in chemoreceptor discharge were also eliminated. The mean firing rate, in afferent chemoreceptor discharge, was unaltered when the oscillations were removed by means of the mixing chamber, in the experiments of Band *et al.* (1978).

We can refer to the responses of the carotid body to mean $PaCO_2$, pH and $PaCO_2$ as static responses and the response to rapid $PaCO_2$ change, in particular $PaCO_2$ oscillations, as a dynamic response. It would appear then, that the static responses are proportional in the case of pHa and $PaCO_2$, (hyperbolic in the case of oxygen) whereas the dynamic response of the carotid body probably involves considerable sensitivity to the rate of change of $PaCO_2$. Sensitivity to rate of change is referred to in engineering terminology as differential sensitivity. A simple description of the carotid body sensitivity to $PaCO_2$ is therefore 'differential plus proportional'; and this description includes both dynamic and static $PaCO_2$ sensitivities.

The difference between static and dynamic responses of the carotid body can be illustrated in relation to oxygen sensitivity. In the study of mean chemoreceptor discharge in response to various steady state values of $PaCO_2$, performed by Lahiri & Delaney (1975), hypoxia had a multiplicative effect on the CO_2 response (closely analogous to the findings for ventilatory responses to CO_2 in man — Lloyd & Cunningham, 1963; Lloyd, 1966). Band & Wolff (1978) studied responses to $PaCO_2$ oscillations in the anaesthetised cat, at both normal and low steady-state oxygen tensions. If there had been a multiplicative effect of the hypoxia on the dynamic response to CO_2 the oscillations in discharge frequency would have been increased. There was, however, no amplification (Figure 7.16).

Figure 7.16: Afferent Chemoreceptor Discharge at Normal and at Low Oxygen Tensions. Horizontal bars show peak and trough values of the averaged oscillations in discharge frequency. Record from sinus nerve filament. Nerve sectioned cephalad to site of recording electrodes.

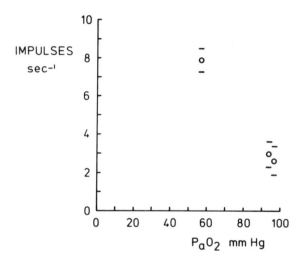

Source: after Band & Wolff, 1978.

It can be seen that oscillation amplitude is unchanged in hypoxia though mean firing rate is increased. The unchanged oscillation amplitude is to be expected since it depends on the dynamic response of the carotid body to carbon dioxide, whereas the mean firing rate depends upon the static response to hypoxia. It is apparent that dynamic sensitivity of the carotid body to CO_2 is not potentiated by hypoxia, in contrast to the static sensitivity. Thus, potentiating effects of hypoxia on reflex ventilatory effects of rapid changes in arterial PCO_2 are unlikely to depend upon multiplicative effects at the carotid body, and must therefore be mediated in the brain.

Dynamic responses, of the carotid body to $PaCO_2$ changes are likely to be the basis of breath-by-breath alternation in tidal volume magnitude, which has been demonstrated in conscious man in response to inspiration of high and low CO_2 gas mixtures (Ward & Cunningham, 1977), and in the anaesthetised cat in response to CO_2 free and CO_2 rich gas mixtures (Wolff, 1977; Wolff & Pearson, 1978).

The carotid body response to $PaCO_2$ is in engineering terminology linear; i.e. the dynamic stimulus to the carotid chemoreceptors does not affect the static response. Thus, any effects of oscillations in determining mean ventilation (Yamamoto, 1960, 1962) cannot be a simple result of $PaCO_2$ oscillations having a direct effect on the mean chemoreceptor discharge. The question therefore remains, as to how, if at all, oscillations in chemoreceptor discharge frequency affect ventilation. This question is re-enforced by the fact that the oscillatory response to $PaCO_2$ oscillations did not, *per se*, affect the mean firing rate (Band *et al.*, 1978). Similarly, oscillations of two respiratory periods duration did not affect mean ventilation, in experiments where alternate breaths of high and low CO_2 mixtures were given; either in man (Ward & Cunningham, 1977) or cat (Wolff, 1977).

We can see that the available evidence suggests that the information transmission line is complete, between the generation of alveolar PCO_2 oscillations, (containing information as to $\dot{V}CO_2$), and the receipt, by the brain, of a related message. This information travels via the circulatory pathway from the lung to the peripheral arterial chemoreceptor, and then via the sinus nerve afferent chemoreceptor fibres to the brain. It is not known whether the aortic chemoreceptors are sensitive to the oscillatory component of the $PaCO_2$ signal. The mechanisms whereby information, in the form of oscillations in chemoreceptor discharge frequency, are processed to participate in the control of breathing remain obscure. Some speculation seems appropriate, however, regarding some of the mechanisms involved in transmission of the signal from the carotid body to the brain via the sinus nerve.

Speculation Concerning Sinus Nerve Discharge at Rest and in Exercise

We have seen that the carotid body function as a peripheral arterial chemoreceptor includes some factors, which we have referred to as stimuli, giving rise to a static component of the response. These determine the mean rate of firing in sinus nerve fibres. There is also the dynamic sensitivity to rapid changes in $PaCO_2$ which appears to have no proven effect on mean discharge. In general terms, the thesis of Yamamoto (1960, 1962) amounts to the idea that $PACO_2$ oscillations, dependent on $\dot{V}CO_2$, may act as a source of information (as to $\dot{V}CO_2$) to the part of the central neuronal mechanism which acts as controller (Saunders *et al.*, 1980). We may therefore ask what form the oscillations in afferent chemoreceptor discharge frequency are likely to take when $\dot{V}CO_2$ increases with exercise. Measurement of respiratory oscillations in sinus nerve afferent chemoreceptor discharge frequency at increased rates of CO_2 production have yet to be published, though Marsh, Nye and Torrance have some preliminary results (Nye, personal communication). We can make some predictions, however, from presently available knowledge.

Figure 7.17: Resting and Exercising $PACO_2$ Time Course in Expiration, Shown Diagrammatically, where Exercise $\dot{V}CO_2$ is Three Times Resting $\dot{V}CO_2$. The expiratory slope of the alveolar (and the correspondingly delayed arterial) PCO_2 will steepen accordingly. With carotid body sensitivity to rate of $PaCO_2$ change we may expect the peak chemoreceptor oscillation discharge frequency (SN discharge rate) to exceed the mean chemoreceptor discharge frequency by three times as much in the exercising state (B) as it does in the resting state (A).

(N.B. $PaCO_2$ will be delayed in its effect on sinus nerve discharge by the time taken for blood to travel from lung to carotid body in the intervening circulatory pathway.)

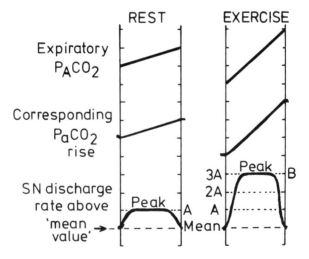

Figure 7.17 illustrates how an increase in $\dot{V}CO_2$ by a ratio of three is likely to be accompanied by a proportional increase in alveolar and arterial PCO_2 rate of change during expiration. In chemoreceptor fibres sensitive to $PaCO_2$ rate of change this would result, during expiration, in an increased height of the peak discharge frequency above the mean value. The height of the peak above the mean firing rate would therefore increase to three times its original value. It can therefore be seen that the averaged chemoreceptor afferent discharge oscillation amplitude may be expected to change in direct proportion to steady-state $\dot{V}CO_2$. This will, however, only be the case if the mean discharge is great enough to accommodate the deeper troughs which will occur, (on the basis of sensitivity to $PaCO_2$ rate of change) when $\dot{V}CO_2$ increases in exercise (Figure 7.18). A relatively simple way in which this might happen would be if the central respiratory control mechanism kept the trough value at zero. Both peak discharge rate, and mean discharge rate would then vary in direct proportion to $\dot{V}CO_2$ (Wolff, 1975). Thus, in the case shown in Figure 7.18, the peak and mean chemoreceptor firing rates would each be adjusted to values in exercise which were three times their resting values (Figure 7.19).

These suggestions are pure speculation. We remain ignorant of important aspects of carotid body sensitivity to rapid $PaCO_2$ changes (dynamic CO_2 sensitivity) – for example:

(i) Unmedullated chemoreceptor fibres may have properties differing from the medullated fibres so far studied.

(ii) The proportion of chemoreceptor fibres showing directly proportional sensitivity, (to $PaCO_2$,) on the one hand, and sensitivity to $PaCO_2$ rate of change, on the other, is not yet clearly established. Band, Willshaw & Wolff (1976) pointed out that there should be very similar expiratory $PaCO_2$ rates of change in $PaCO_2$ oscillations at high and low respiratory rates, in anaesthetised cats which were artificially ventilated at constant $PaCO_2$. In chemoreceptor units exhibiting pure 'rate of change' sensitivity to $PaCO_2$ these very similar expiratory PCO_2 rates of change would give rise to similar differences between mean afferent chemoreceptor discharge frequency and peak discharge frequency. On this basis one would expect similar neuronal oscillation amplitudes at different, steady-state, respiratory rates. On the other hand, in chemoreceptor units with purely proportional sensitivity, the amplitude of the oscillation would decrease as the respiratory period decreased (just as $PaCO_2$ oscillations at constant $\dot{V}CO_2$ change in direct proportion to the respiratory period, see Figure 7.11).

Figure 7.18: The Same Two Steady States of Resting $\dot{V}CO_2$ (A) and Three Times Resting $\dot{V}CO_2$ (B) are Shown as in Figure 7.17. A simplified version of the oscillation waveform in averaged chemoreceptor afferent discharge frequency is shown, with the trough as far below mean discharge as the peak is above it. The increased depth of the trough can only be present if there is a sufficient mean discharge to accommodate it.

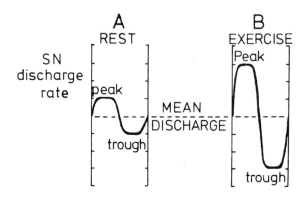

Figure 7.19: Hypothetical Oscillations in Chemoreceptor Discharge Frequency in the Resting and Exercising State, (as in Figures 7.17 and 7.18) with Rate of Change Sensitivity and Efferent Control Maintaining the Oscillation Trough at Zero. The mean and peak firing rates now change in direct proportion to $\dot{V}CO_2$.

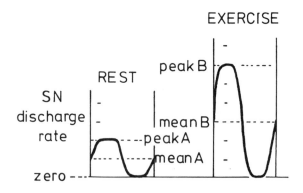

Figure 7.20 shows that the amplitude of the averaged oscillation in neuronal frequency in the cat experiment of Band *et al.* (1976) was virtually unchanged when the respiratory rate was increased (respiratory period decreased) in the unit illustrated. In other units a reduction in neuronal oscillation amplitude was seen with increasing respiratory rate. A sustained amplitude with increased respiratory rate (illustrated in Figure 7.20), is expected with 'rate of change' sensitivity. Further investigations along these lines would clarify what proportion of units show such 'rate of change' sensitivity.

Figure 7.20: Averaged Afferent Chemoreceptor Discharge Frequency from a Single Unit in the Cat Sinus Nerve. Averaging was undertaken in relation to the respiratory period. In the plot on the left the respiratory rate was half that in the middle plot (two cycles shown), and the same as that of the plot on the right. Note the very similar amplitudes despite the change in oscillation (and respiratory) period. The similarity of the amplitudes depends on (a) constant $PaCO_2$ oscillation (maximum) slope, and (b) carotid body sensitivity to $PaCO_2$ rate of change, at least for this unit.

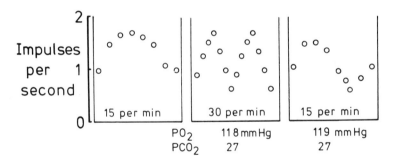

Source: after Band, Willshaw & Wolff, 1976.

(iii) Mean (or even dynamic) afferent chemoreceptor discharge frequency may be affected by both efferent carotid body innervation (Neil & O'Regan, 1971; O'Regan, 1976) and by its sympathetic nerve supply (McCloskey, 1975; O'Regan, 1981).

Advances in technique may eventually make it possible to monitor afferent sinus nerve chemoreceptor discharge in the intact conscious animal during exercise. This seems unlikely to be achieved in man in the foreseeable future. We remain ignorant as to the precise form in which the information in intact carotid sinus nerve chemoreceptor fibres travels to the central nervous sytem. A recent study, briefly reported by Donoghue, Felder, Jordan & Spyer (1982a) suggests that we may more readily record from the chemoreceptor cell bodies in the petrosal ganglion than from sinus nerve filaments. Studies of baroreceptor units have already been made from the nodose ganglion in this manner, and have recently been reported in detail with evidence as to the sites of their central projections (Donoghue, Garcia, Jordan & Spyer, 1982b). With further advances it may become possible to interpret activity recorded centrally, after information processing of chemoreceptor afferent activity has taken place.

Respiratory Phenomena Due to Fast $PaCO_2$ Changes

Linearity of Carotid Body $PaCO_2$ Transient Sensitivity and Non-linearity of the Central Nervous Communications between Arterial Chemoreceptor Projections and Efferent Respiratory Motor Neurones

Brief $PaCO_2$ transients which each caused an increased inspiratory tidal volume (Figure 7.5), were found to have the greatest effect at a particular time during inspiration (Black & Torrance, 1967a, 1971; Band *et al.*, 1970, 1978). The particularly marked effect of $PaCO_2$ transient stimuli, given at a critical time during inspiration, has been accepted as representing a physiological reflex. This precise timing is apparently not mediated at the carotid body chemoreceptor. The evidence for this assertion is that chemoreceptor afferent discharge shows a burst of impulses at whatever time in the respiratory cycle a $PaCO_2$ transient arrives at the carotid body. Furthermore, the timing of the peak of the averaged oscillation in discharge frequency may be found at any part of the respiratory cycle (Band *et al.*, 1978). If the carotid body were only sensitive to arterial PCO_2 oscillations at a particular short interval during inspiration there would only be a brief peak in the averaged discharge frequency, and this peak would always occur at the specific timing at which sensitivity was present during inspiration. Therefore it must be the central respiratory mechanism which determines the specific effectiveness of $PaCO_2$ transients at a particular part of the respiratory cycle (during mid- or late inspiration). Further studies of the arterial chemoreceptor central projections are required before we can explain this graded effect of a burst of afferent chemoreceptor discharge.

A variation in the reflex effectiveness of a given afferent discharge, would not be expected in a system which had a simple linear response. It is therefore referred to as a 'non-linearity' of the respiratory system. The mechanism appears to reside in the central nervous connections between the afferent chemoreceptor projections and those efferent motor neurones which have axons descending in the spinal cord, to link with the peripheral motor output to respiratory muscles.

Effects of Tube Breathing upon Ventilation; a Second Non-linear Phenomenon

We have seen that, in exercise, steepened $PaCO_2$ oscillations and increased ventilation occur together, with little change in the mean $PaCO_2$ value. The increased steepness of the rising and falling limbs of the $PaCO_2$ oscillations seen in exercise is due to the increase in $\dot{V}CO_2$ (Wolff, 1981). Since this increased steepness of $PaCO_2$ oscillations in exercise is not a result of the increase in ventilation, we may reasonably ask whether the steeper oscillations cause the ventilatory change.

If $PaCO_2$ oscillations, in the steady state, contribute to the ventilatory increase in exercise the problem remains as to how their information content (as to $\dot{V}CO_2$) is utilised by the respiratory control system.

We should consider some of the outstanding problems as to how $PaCO_2$ oscillations may function as a part of the respiratory control mechanism in exercise. Firstly, we do not yet know what happens to the afferent chemoreceptor oscillations in exercise. Secondly, we saw in (a) above that, there is a 'non-linearity' in the central linkage between the sensory afferent neuronal 'input' to the central nervous system and the motor efferent 'output' via the spinal cord to the respiratory muscles. Thirdly, although the studies by Pappenheimer *et al.* (1965) showed that increased ventilation accompanies decreased extracellular pH somewhere in the tissues of the central nervous system, this may well remain unchanged in exercise.

In order to see whether the time course of oscillations could have effects in addition to the mean arterial PCO_2, Cunningham, Howson & Pearson (1973) studied tube breathing and a similar reverse manoeuvre. To achieve the tube breathing profile in alveolar PCO_2 they gave a raised CO_2 mixture early in inspiration and CO_2-free gas for the remainder of the inspiratory period. The alveolar PCO_2 oscillation resulting from this manoeuvre would be expected to show a sharp fall late in inspiration. Apart from testing their subjects with a CO_2 mixture throughout inspiration (chosen to give a similar increase in $PACO_2$ mean value above normal), they reversed the sequence of gases used in the tube breathing manoeuvre, so that CO_2-free gas was taken by the subject early in inspiration and the raised CO_2 mixture was breathed for the rest of the inspiratory period. This reverse manoeuvre should give rise to a sharp fall in $PACO_2$ early in inspiration (rather than late as in tube breathing). Precisely the expected arterial pH profile in tube breathing was shown to occur in cats by Band *et al.* (1969b). It was also compared with the profiles obtained when the cats breathed either air or a CO_2 mixture. Though the differences in the oscillation profiles were detectable they were small. In spite of the small differences between the likely $PACO_2$ oscillations in the human experiments of Cunningham *et al.* (1973) there were significant differences in ventilation. Ventilation in the tube manoeuvre was significantly greater than ventilation in the reverse manoeuvre, particularly when there was background hypoxia.

Figure 7.21: Airway and Alveolar PCO_2 Shown Diagrammatically (several breaths). A. 'Tube' breathing, in which a raised CO_2 mixture is breathed by the subject during early inspiration. B. 'Ordinary' CO_2 breathing, in which a lower concentration of CO_2 is given throughout inspiration. Different alveolar PCO_2 oscillation profiles are achieved in A and B despite very similar mean $PACO_2$ values. PA, Oscillation peak A; PB, Oscillation peak B; ET, End tidal PCO_2; t, oscillation trough; FIA, Inspiratory PCO_2 A; FIB, Inspiratory PCO_2 B; 1, time for expiration of dead space; 2, time for inspiration of dead space. (Airway CO_2 profiles drawn as for 'constant flow' expiration.)

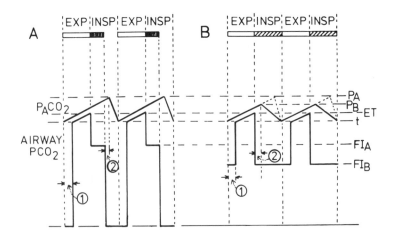

Arterial blood was taken during test and control runs in these studies, to ensure correct interpretation of the 'CO$_2$ response' resulting from each manoeuvre.

The main difference in the $PaCO_2$ stimulus to ventilation in the tube breathing and the reverse manoeuvre appears to have been the small difference in the time course of the $PaCO_2$ oscillations. In other words a difference in the dynamic stimulus ($PaCO_2$ oscillation) gave rise to a difference in the static response (mean ventilation). This is another example of a non-linear response.

It is possible that the raised mean $PaCO_2$ in the tube experiments, which can be interpreted as a failure of $PaCO_2$ homeostasis, acted to unmask the intrinsic properties of the respiratory control system response to oscillations. This cannot be confirmed until we know more about the respiratory control system. For example: (a) whether respiratory $PaCO_2$ oscillations are fundamental to normal respiratory control and, (b) if they are fundamental, more about (i), the precise effects of the oscillations, and (ii), the mechanisms involved in their action.

We have virtually no idea how effects on ventilation might be produced by the respiratory system in response to $PaCO_2$ oscillations. The difficulties are compounded by the necessarily non-linear nature of such a response. One particularly obvious difficulty is the temporary increase in oscillation slope in expiration during hyperventilation. In a feedback loop with the expiratory $PaCO_2$ slope acting positively this would tend to give progressively increasing ventilation. There may either be a phase relationship link putting the inspiratory slope into the controller during inspiration, or else central chemoreceptor sensitivity to lowered mean $PaCO_2$ may counteract peripheral chemoreceptor sensitivity in these circumstances (see Cunningham, 1972).

It seems that we will have to understand much more about the individual components of the relevant pathways and their interactions before the mechanism of action of $PaCO_2$ oscillations in respiratory control can be fully understood.

Arterial PCO_2 Oscillations, Reflexes Mediated via the Carotid Body, and Ventilatory Adjustments to Chronic Hypoxia and Increased Rates of CO_2 Production

Evidence of Reflex Effects of Respiratory Oscillations in Arterial Blood Gas Tensions or pH

In two types of study Black & Torrance (1967b, 1971) gave experimental evidence that respiratory oscillations were present in arterial blood, and could reflexly affect phrenic motor discharge and ventilation. In the first type of study the thorax of an anaesthetised cat was widely opened while the animal was artificially ventilated. Phrenic motor discharge frequency was monitored so that the artifical (ventilator) and spontaneous (phrenic burst) rhythms could both be recorded. When the artificial rate was made slower than the spontaneous rate, intermittent increases in the intensity of the phrenic discharge bursts could be seen. These were at the interval which would be expected from intermittent re-enforcement of the phrenic activity by the second (slower) rhythm of artificial ventilation. It was thought that there must therefore be a blood-borne signal resulting from the artificial ventilation, since all afferent inputs from lung and chest wall had been eliminated.

The second type of experiment of Black & Torrance (1971) also supported the idea that arterial chemical oscillations, generated by pulmonary ventilation, had caused the beats in phrenic nerve discharge in the first experiment. In the second type of experiment a spontaneously breathing cat received blood from an artificially ventilated donor cat. Arterial blood from the donor was made to perfuse the carotid arteries of the recipient. A decrease in the respiratory rate of the donor cat below that of the recipient either caused slowing of the respiratory rhythm of the recipient, or slow periodic changes in ventilation (respiratory rate unchanged), at the frequency expected for re-enforcement of two periodic functions. This suggested transmission of an oscillatory stimulus from the donor cat. In an experiment, modelled on the first of those of Black & Torrance (1967b, 1971) Cross *et al.* (1979) confirmed the effects on phrenic motor neuronal discharge, and were able to examine the phase relation between the ventilator and the arterial pH oscillations in the carotid arteries close to the carotid bodies.

A rather different experiment was undertaken by Linton, Miller & Cameron (1976). This compared the ventilation which occurred in anaesthetised rabbits when they received an intravenous infusion of CO_2-rich blood while breathing air, with the ventilation which occurred during inspiration of a CO_2-enriched gas mixture (1.5–2 per cent CO_2, 21 per cent O_2, balance N_2). The results showed a greater increase in ventilation per mm Hg rise in arterial PCO_2 with CO_2 infusion than with inhalation of the CO_2 mixture. This experiment resembles that of Yamamoto & Edwards (1960) in rats, (see above) except that no comparison was made in the rats between CO_2 infusion and CO_2 inhalation. Yamamoto & Edwards (1960) simply concluded that constancy of $PaCO_2$ in the very lightly anaesthetised rats indicated homeostasis for $PaCO_2$ in the face of a venous CO_2 load. Linton, Miller & Cameron (1977) repeated their investigation to see whether their earlier result was due to oscillations in arterial PCO_2. They introduced a mixing chamber into the carotid arterial circulatory pathway to remove arterial oscillations. Mean $PaCO_2$ then rose when CO_2-rich blood was infused, to a significantly higher value than had occurred when the arterial oscillations were allowed access to the carotid bodies. A similar result was obtained with oscillations present if the carotid sinus nerves had been sectioned.

These results support the role of $PaCO_2$ oscillations and also of the carotid bodies, in maintaining constancy of mean $PaCO_2$ in the face of an intravenous CO_2 load.

Dependence of Alveolar Ventilation on Circulatory Delivery of Carbon Dioxide to the Lungs

Phillipson, Duffin & Cooper (1981c) examined the effect on ventilation in conscious sheep of removing carbon dioxide from venous blood en route to the lungs, via a new type of gas exchanger (Kolobow *et al.*, 1977).

When CO_2 was removed from venous blood at a rate equal to the animal's own, spontaneous, CO_2 production rate breathing ceased, with no apparent harm to the animal. This was reproducible. Others have found that in patients undergoing renal haemodialysis, loss of CO_2 from the patient to the dialysate can reduce ventilation (Patterson *et al.*, 1981; Dolan *et al.*, 1981). When CO_2 has been removed in the course of renal haemodialysis hypoxia has occurred, due to the underventilation. In the experiment of Phillipson *et al.* (1981c) oxygenation was maintained via the gas exchanger. When extra CO_2 was given to the sheep, via the gas exchanger alveolar ventilation increased in direct proportion to steady-state $\dot{V}CO_2$ at the mouth, up to a mouth $\dot{V}CO_2$ value which was twice the animal's own resting CO_2 production rate. This meant that $PaCO_2$ remained constant (see above for explanation). The same was found to apply in a second study (Phillipson *et al.*, 1981a) in which the rate of CO_2 production at the mouth (and hence also $\dot{V}CO_2$ at the mixed venous blood to pulmonary gas interface), was varied, either via the gas exchanger or by exercising the animal (Figure 7.22). The $\dot{V}A/\dot{V}CO_2$ relationship was the same for natural exercise engendered and exchanger produced increases in $\dot{V}CO_2$. $PaCO_2$ was isocapnic with both methods of CO_2 loading. In a third series of experiments Phillipson *et al.* (1981b) showed that arterial

Figure 7.22: Graphs Showing Alveolar Ventilation ($\dot{V}A$) Linearly Related to CO_2 Production Rate ($\dot{V}CO_2$) at the Mouth, Whether due to CO_2 Added via a Venous Exchanger (o) or as a Result of Exercising the Animal (Conscious Sheep) (●). Large filled circle represents initial state. Symbol for exercise plus venous CO_2 loading, △. It can be seen that $\dot{V}A$ changes in direct proportion to $\dot{V}CO_2$. Arterial PCO_2 was constant in three of the four animals. It was also constant, from zero $\dot{V}CO_2$ to 1.5 times resting $\dot{V}CO_2$, in the fourth animal, only starting to rise as $\dot{V}CO_2$ increased beyond this range.

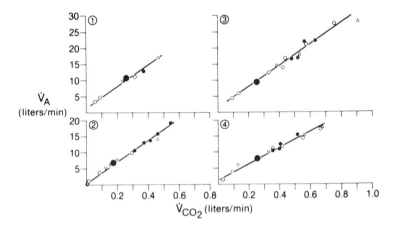

Source: Phillipson *et al.*, 1981a, with permission.

oxygenation affected the result. $PaCO_2$ rose if the sheep was hyperoxic (high PaO_2), whereas it fell with exercise in hypoxia (low PaO_2). $PaCO_2$ remained constant ($\dot{V}A$ directly proportional to $\dot{V}CO_2$) when oxygenation was normal. Finally, in this third paper, the effect of carotid body resection was investigated. Instead of an isocapnic response to increased steady-state CO_2 delivery to the lung (as occurred in intact animal with normal PaO_2), $PaCO_2$ rose with increasing $\dot{V}CO_2$, presumably this was the simplified response of the central chemoreceptor mechanism acting alone (Figure 7.23).

The result of carotid body resection in Phillipson *et al.* (1981b) is similar to that of Linton *et al.* (1977) with anaesthetised rabbits, in which they found that CO_2 infusion after sinus nerve section gave a greater $PaCO_2$ rise than was seen prior to removal of the carotid body arterial chemoreceptor mechanism.

It can be seen that the studies by Phillipson *et al.* (1981a, b, c) and

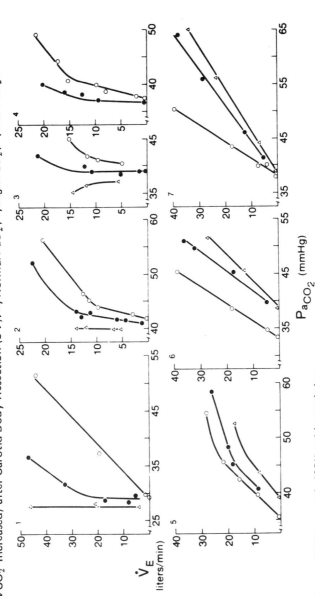

Figure 7.23: Near 'Vertical' CO_2 Responses to Exercise and CO_2 Loading (intravenously) seen in Conscious Sheep Prior to Carotid Body Resection (1-4), and Sloping Responses (Indicating an increasing $PACO_2$, and $PaCO_2$, as $\dot{V}CO_2$ Increased) after Carotid Body Resection (5-7). ●, Normal PaO_2; ○, high PaO_2; △, low PaO_2.

Source: Phillipson *et al.*, 1981b, with permission.

those of Linton *et al.* (1976, 1977) are consistent with the idea that ventilation is normally matched to the CO_2 delivery to the lung, in precisely the manner outlined above. In other words, as $\dot{V}CO_2$ increases, $\dot{V}A$ increases in direct proportion and thereby maintains isocapnia. It can be seen that the ventilatory increases in exercise, or venous CO_2 loading, in which $\dot{V}A$ is matched to $\dot{V}CO_2$, and hence $PaCO_2$ is sustained at the correct value, are precisely appropriate to that CO_2 load. It is important to recognise that such ventilation is normal and reserve the term 'hyperpnoea' (Wasserman *et al.*, 1977) for excessive ventilation, where mean PCO_2 falls.

We can now see that the appropriate ventilatory adjustments to increased CO_2 flux to the lung ($\dot{V}CO_2$, lung) are dependent for their precision on the presence of the carotid bodies (Phillipson *et al.*, 1981b), as well as there being a very strong suspicion that carotid arterial blood chemical oscillations (probably $PaCO_2$) are the necessary stimulus, or signal (Linton *et al.*, 1977).

Dependence of the Normal Resting Mean Arterial PCO_2 *upon an Intact Carotid Body Chemoreceptor*

There are now numerous reports available of studies in which the carotid body chemoreceptor mechanism has been removed, both in man and in animals. Several studies illustrate the fact that the arterial PCO_2 tends to be higher in animals without carotid bodies than in intact animals. The study by Bouverot *et al.* (1965) (see above, Figure 7.6) showed an average increase in the arterial PCO_2 of 3.6 mm Hg in carotid body resected dogs, and a diminution in resting ventilation. The mean $PaCO_2$ in the carotid body resected (CBR) sheep in the study of Phillipson *et al.* (1981b) similarly was 6.1 mm Hg higher than that of the intact animals. The increased $PaCO_2$ reflected a primary change in alveolar ventilation ($\dot{V}A$ in CBR sheep was 65.5 per cent of $\dot{V}A$ in normal sheep).

The study by Forster, Bisgard & Klein (1981) – like those by Bouverot *et al.* (1965) and Phillipson *et al.* (1981b) showed a higher $PaCO_2$ at normal ambient oxygen tension in carotid body resected goats than was found in normal ones. In Figure 7.24 it can be seen that acclimatisation in normal animals to a lowered ambient oxygen tension, (after 72 hours exposure) led to an arterial PCO_2/PO_2 relationship which was very similar to that found by Cochrane *et al.* (1980b) in patients with chronic stable asthma (see also above). This contrasts with the failure of the carotid body-resected goats to acclimatise to chronic hypoxia, where $PaCO_2$ if anything, rose slightly. The difference in

$PaCO_2$ (mean value) between carotid body-resected and normal goats can therefore be seen to be much more striking in chronic hypoxia than it is at normal PaO_2. Presumably, if both normal and CBR goats were exposed to a small degree of chronic hyperoxia $PaCO_2$ in the two groups of animals would become indistinguishable. With a greater degree of hyperoxia, such as 1.5 atmospheres ambient, one might expect $PaCO_2$ in the normal goats to rise higher than the unchanged value in the CBR goats. If this were to happen it would show that the stimulus from the carotid bodies could, in chronic acclimatisation, inhibit ventilation, as well as stimulate it as in exercise.

Figure 7.24: Steady-State (Acclimatised) Arterial CO_2 Tensions ($PaCO_2$) Related to the Corresponding Arterial Oxygen Tension (PaO_2) in conscious goats (Forster *et al.*, 1981). ●, normal animals (mean values at normal and at low ambient oxygen pressure. These points lie close to the line found by Cochrane *et al.* (1980b) for ambulant patients with chronic stable asthma at sea level. ○, carotid body-resected goats, showing, if anything, a rise in $PaCO_2$ in hypoxia. Line for patients with asthma: $PaCO_2 = 0.23 PaO_2 + 16.6$ mm Hg (or $PaCO_2 = 0.23 PaO_2 + 2.2$ kPa). I, ±1 SEM.

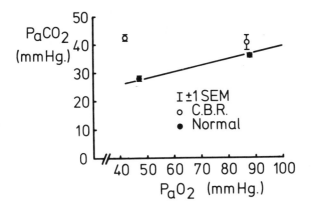

In man acute hypoxic sensitivity was lost in a series of patients in whom carotid enderectomy had been undertaken for atheromatous disease of the common carotid artery (Wade *et al.*, 1970). Absence of such acute hypoxic sensitivity may be taken to indicate that these patients had lost their carotid body function. $PaCO_2$ values rose, on average, 5.8 mm Hg in these patients.

It is not yet known whether it is purely the lack of hypoxic stimulation which causes the inappropriately reduced alveolar ventilation (and hence the increase in $PaCO_2$ mean value) in man and animals who lack carotid arterial chemoreceptors. It now seems likely that $PaCO_2$ oscillations help us to adjust $\dot{V}A$ correctly in relation to $\dot{V}CO_2$, and hence to sustain a normal, constant, $PaCO_2$. There have been suggestions that exercise in hyperoxia can lead to a rise in $PaCO_2$ in normal man (Conway, personal communication − in control subjects for experiments on beta blockade in exercise − Conway, 1982). This would be expected from the $PaCO_2$ rise in hyperoxic, CO_2-loaded, sheep (Phillipson *et al.*, 1981c). There may, as for altitude acclimatisation, be a purely oxygen-dependent mechanism underlying a $PaCO_2$ rise in hyperoxic exercise, or alternatively, the normal adjustments which are found where oxygenation is normal (and their failure in hyperoxia) may also depend upon sensitivity to and the presence, or absence of $PaCO_2$ oscillations.

The Role of $PaCO_2$ Oscillations and Pulmonary Afferent Innervation, in Determining Mean $PaCO_2$ at Rest and Ventilation in Exercise, in Patients with Obstructive Airways Disease

Resting Mean $PaCO_2$

Patients with chronic lung disease who produce sputum and have damaged pulmonary airways but no attenuation of lung markings on X-ray are said to have chronic obstructive bronchitis. Respiratory function tests show such patients to have mechanical limitation of expiratory gas flow during maximal expiratory effort and relatively normal gas transfer tests (in particular, transfer factor for carbon monoxide per unit lung volume, i.e. transfer coefficient, expressed as a percentage of the normal KCO per cent normal, Cotes, 1975). Cochrane, Prior & Wolff (1981a) showed that these patients have a range of attenuation of their arterial pH oscillations *in vivo*. Since the pH oscillations will have reflected the arterial PCO_2 oscillations as a logarithmic function it was important to calculate the likely $PaCO_2$ oscillations; especially as the mean $PaCO_2$ values differed by a significant ratio between patients. This was done by using (*in vitro*) buffer slope values from the studies of Siggaard Anderson (1962, 1963). The simplest expression of $PaCO_2$ oscillation magnitude is the rate of change of its rising limb (upslope) or, more crudely, the mean of both the upslope and downslope values (see above). The mean upslope, in normal subjects in the steady state, is dependent on $\dot{V}CO_2$. So, in resting normal subjects with similar values of $\dot{V}CO_2$ the mean oscillation upslope values should all be similar, as shown in Figure 7.10, where amplitudes are consequently directly proportional to oscillation period. Since the $PaCO_2$ oscillation amplitude will be affected by the respiratory period the magnitudes of $PaCO_2$ oscillations, and hence the degree of attenuation, are most appropriately assessed by comparison of the mean rate of change values (as outlined in Cochrane *et al.*, 1981a).

Chronic obstructive bronchitis represents one end of a spectrum of clinical varieties of obstructive airways disease. At the other end of this spectrum are the patients in whom cough and sputum production are much less conspicuous, and dyspnoea is the presenting symptom. This second type of patient is probably diagnosed later in his disease than the 'bronchitic' patient. This conspicuously dyspnoeic group of patients are commonly referred to as emphysematous, or better clinic-ally emphysematous (since the term emphysema is strictly a pathological one). In these clincally emphysematous patients radiology of the lung fields shows greater translucency than normal with attenuation of the vascular markings, unless there has been an overlay from a bronchitic component. There is considerable argument about pathological corre-lates of these two types of chronic obstructive airways disease though it is accepted that they exhibit conspicuous physiological differences. These two types of patient have been referred to as 'pink and puffing' and 'blue and bloated'; since the patients presenting with dyspnoea usually have virtually normal arterial oxygen tensions and are obviously short of breath – 'pink and puffing', whereas, the 'bronchitic' patients eventually progress to a late stage in which they are cyanosed (due mainly to hypoxia) and tend to have dependent oedema – 'blue and bloated'. These picturesque categories are useful insofar as they pro-mote a comparison of the latest stages of the two more extreme varieties of obstructive lung disease, but detract from useful comparisons between patients in the two groups at earlier stages of the disease processes. In the earlier stage of the bronchitic type of illness there may not be cyanosis, despite these patients being as disabled as some of the patients with clinical emphysema. Burrows, Fletcher, Heard, Jones & Wootliffe (1966) referred to the two clinical groups as 'type B' (bronchitic) and 'type A' (clinically emphysematous) with an intermediate group ('type X') which manifested clinical features of both the other groups. Burrows *et al.* (1966) found certain pathological correlations, but the purpose of mentioning their study is to use their clinical terminology, i.e. to,

Figure 7.25: Arterial pH (Bold Trace), Inspiratory Tidal Volume, and Airway CO_2 Records in (a) a Patient with Chronic Stable Asthma, (b) a Patient with Type B Obstructive Airways Disease, and (c) a Patient with Type A Obstructive Airways Disease. Oscillations were universally shallow in type A patients, and covered the range from zero to normal mean slope in type B patients.

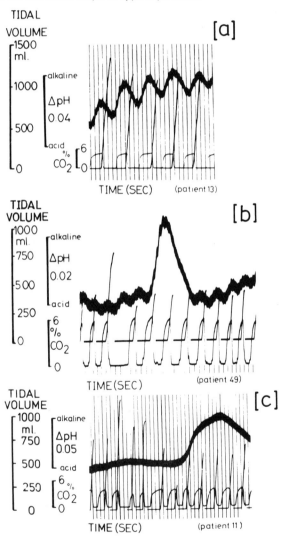

Source: after Cochrane *et al.*, 1981a.

from here on, use the terms type B and type A in regard to the two types of patient studied by Cochrane *et al.* (1981a), and referred to by them as having bronchitis and clinical emphysema respectively.

Comparison of type A and type B patients by Cochrane *et al.* (1981a) showed distinct physiological differences between the two groups. Apart from type A patients showing a significantly lower KCO per cent normal than type B patients, the arterial pH and tidal volume records showed there to be a significant prolongation of the circulatory transport time (transit time) from lung to peripheral artery. This was confirmed by Cochrane (1980a) studying single pass radionuclide transit time from lung to aorta in a selection of the same patients. The two measurements of transit time gave a line of unit slope, which would be expected if the time intervals were correctly measured by both methods.

Figure 7.25 shows typical arterial pH records from patients with (a) chronic stable asthma, (b) type B disease (chronic obstructive bronchitis) and, (c) type A disease ('clinical emphysema'). The patients with type B disease, had pH (and calculated $PaCO_2$) oscillations with mean slopes ranging from zero to normal. In contrast, patients with type A disease had universally shallow oscillations.

Figure 7.26: Mean $PaCO_2$ Plotted against the Average Slope of $PaCO_2$ Oscillations. Each mark on the x axis is 0.1 mm Hg s^{-1}. Slope was obtained as amplitude/half the respiratory period ($\Delta PaCO_2 /\Delta t$). (If this is multiplied by $\pi/2$ an estimate of the maximal slope is obtained.) On the left are results from patients with type B disease; on the right from patients with type A disease. The least squares regression line for type B patients had a significant slope, $P < 0.01$.

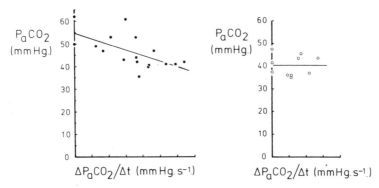

Source: after Cochrane *et al.*, 1981a.

In the patients with type B disease the shallower the oscillation the higher the mean $PaCO_2$. In the patients with type A disease the mean $PaCO_2$ values were virtually normal, despite the universally shallow pH and (calculated) $PaCO_2$ oscillations. These findings are illustrated in Figure 7.26.

Reduced alveolar ventilation (relative to $\dot{V}CO_2$) reflected by high $PaCO_2$ values, can be expressed by a comparison with normal alveolar ventilation. This is done by calculating the ratio of the patient's $PaCO_2$ to that expected from the normal line illustrated in Figure 7.24 (Cochrane *et al.* 1980b). This $PaCO_2$ ratio is the inverse of the alveolar ventilation ratio, so that a high $PaCO_2$ relative to normal (indicating underventilation) is reflected by a high value of the $PaCO_2$ ratio. If the ratio is given the symbol VR the highest values for VR represent the greatest degree of underventilation.

$$\text{i.e. } VR = \text{Patient's } PaCO_2 / \text{Normal } PaCO_2.$$

The relation between VR, as an index of alveolar hypoventilation, and the $PaCO_2$ oscillation slope value in the two types of patient is shown in Figure 7.27. There is an even more significant relation (than in Figure 7.26) in the type B patients ($P < 0.001$), and again, there is no significant correlation in type A patients. The type A patients showed only a slight degree of underventilation ($VR > 1$). This result has been briefly reported by Cochrane, Prior & Wolff (1981b).

The results from patients plotted in Figure 7.26 and 7.27 are ambiguous. Either shallow oscillations lead to impaired alveolar ventilation in some patients (type B) and yet not in others (type A) or, the correlation in type B patients is spurious. Before considering the problem further let us briefly examine another aspect of respiratory control in these two types of patient.

Upon Ventilatory Changes Following the Onset of Exercise

Wasserman, Whipp, Koyal & Cleary (1975) showed that asthmatic patients, in whom the carotid bodies had been resected, increased their ventilation more slowly than did a comparable group of normal subjects. The asthmatic patients without carotid bodies were asymptomatic at the time of their exercise study. By examining the ventilatory time courses obtained in these studies, (Figure 7.28) it is apparent that, in the normal subjects, (in Figure 7.28 (i)) ventilation had increased from the resting value to 80 per cent of the change between rest and steady-state exercise value after two minutes exercise had been performed. In contrast, the subjects without carotid bodies had only increased their

Figure 7.27: V_R, an Index of Alveolar Hypoventilation, Plotted against $PaCO_2$ Oscillation mean Slope (Calculated as for Figure 7.26). Type B patients' results plotted on the left, Type A patients' results plotted on the right. $\Delta PaCO_2/\Delta t$ represents mean rate of change, or slope, of the rising and falling limbs of the $PaCO_2$ oscillations.

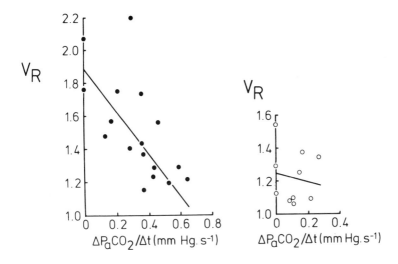

Figure 7.28: Average Breath by Breath Ventilation at Rest and after the Onset of Exercise, Expressed as a Percentage of Total Change between Resting and Steady-State Exercise Ventilation. (i) Below the 'anaerobic threshold', and (ii) above it. ●, Normal subjects; ○, asymptomatic carotid body resected asthmatic patients.

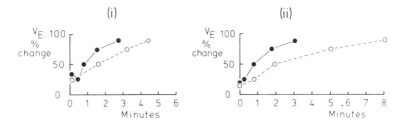

Source: after Wasserman *et al.*, 1975.

ventilation 55 per cent of the way from the resting to the steady-state exercise value after two minutes exercise. Cochrane, Powlson, Prior & Wolff (1979, and in preparation) assessed the rate of ventilatory change early in exercise in patients with chronic obstructive airways disease. The increment in ventilation at two minutes (mean ventilation between one and three minutes) above the resting value, was expressed as a percentage of the rest to exercise (mean ventilation between four and six minutes) change. It was necessary to limit the exercise period to six minutes. Average $\dot{V}CO_2$ in exercise (between four and six minutes) was 2.7 times the resting value. The index of the rate of ventilatory change described here was referred to by Cochrane *et al.* (1979) as α_{1-3}.

$$\text{i.e. } \alpha_{1-3} = (\dot{V},_{1-3} - \dot{V}, \text{rest}) / (\dot{V},_{4-6} - \dot{V}, \text{rest})$$

Figure 7.29 shows the results which were obtained by Cochrane *et al.* (1979) on the time course of exercise ventilation in patients with chronic obstructive airways disease. In Figure 7.29 (i) α_{1-3} is shown for patients with asthma, and type B lung disease. In patients with type B obstructive airways disease the slowest ventilatory change occurs in the patients with the shallowest resting oscillations.

In Figure 7.29 (ii) we see that α_{1-3}, in the type B patients plots against the rate of change of resting $PaCO_2$ oscillations whereas in patients with type A disease there is no significant slowing effect despite universally shallow resting $PaCO_2$ oscillations.

Conclusions from Patient Studies

The appropriateness of the alveolar ventilation to CO_2 production rate (reflected by $PaCO_2$ or VR), and the rate of ventilatory change in exercise (e.g. α_{1-3}), may be regarded as aspects of the chemical control of breathing.

The problem comes down to there being significant relationships between indices of respiratory control (VR or $PaCO_2$ and α_{1-3}) and the mean slope of the resting $PaCO_2$ oscillations in the bronchitic or type B patients, with complete absence of such relationships in type A or clinically emphysematous patients. It seems necessary to suggest the possibility that there is impaired chemical control in type B patients who have lost their respiratory $PaCO_2$ oscillations; whereas the patients with type A obstructive airways disease, with universally attenuated oscillations, must then have a non-chemical component of respiratory control, overriding the inhibitory effect of losing their $PaCO_2$ oscillations.

Figure 7.29: (i) Proportion of the Rest to Exercise Change in Ventilation (alpha) Reached at a given Time. Dashed lines show average time course for the (faster) normal subjects and (slower) carotid body resected subjects of Wasserman *et al.* (1975) — below anaerobic threshold — Figure 7.28 (i). Δ, Patients of Cochrane *et al.* (1979) with asthma; ●, patients with type B obstructive airways disease; ○, patients with type A obstructive airways disease. h, high (normal); m, medium; l, low values of $PaCO_2$ oscillation slope. (ii) α_{1-3} (as plotted in (i) at two minutes — for calculation see text) plotted against the (mean) upslope of $PaCO_2$ oscillations in the same patients ($\Delta PaCO_2/\Delta t$). Compatible with effects of oscillations in type B disease (and asthma) but incompatible with a simple inhibitory effect of oscillation loss in type A disease. α_{1-3} and $\Delta PaCO_2/\Delta t$ are plotted in groups, each ± 2 SEM.

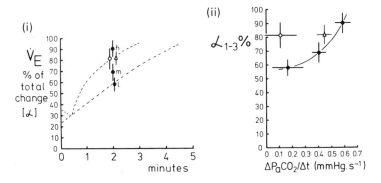

It is proposed that the positive effect on ventilation overriding loss of $PaCO_2$ oscillations in patients with type A obstructive airways disease may be a result of functional loss of pulmonary vagal afferent inhibition. That this is possible is emphasised by the findings of Thompson, Marchak, Bryan & Froese (1981). These authors found that while high-frequency oscillatory ventilation (15 Hertz) caused apnoea in dogs, this was reversed by vagotomy. Prior to vagotomy the high-frequency oscillation presumably eliminated $PACO_2$ (and $PaCO_2$) oscillations, while also stimulating the pulmonary stretch receptors. Removal of the stretch receptor input would leave an animal with adequate induced alveolar ventilation but no $PaCO_2$ oscillations. This should closely resemble the situation in the sheep of Phillipson *et al.* (1981c) in which tissue CO_2 production rate was matched by venous CO_2 removal. In these sheep there would again be no oscillations but, unlike the dogs of Thompson *et al.* (1981) there would not have been vagal inhibition (see Table 7.1).

Table 7.1: Ventilation, Oscillations, Pulmonary Hyperinflation and the Vagi

	$PaCO_2$ oscillations	Hyper- inflation	Vagi intact (i) cut (c)	Ventilation
Normal man	+	o	i	+
Sheep — normal $\dot{V}CO_2$	+	o	i	+
— zero $\dot{V}CO_2$	o	o	i	o
HFV dogs −1	o	+	i	o
−2	o	+	C	+
Hering-Breuer	+	+	i	o
Type A patients	o	+	?	+

Note: HFV dogs — High Frequency Ventilation in dogs; 1 — vagi intact; 2 — vagi cut; +, present; o, absent.

Head (1889) pointed out that the precise effects of vagal section depend on whether or not there is a conducting pathway between the severed ends of the afferent filaments. Usually, in his hands, the final result of vagal section involved a strong inspiratory stimulation. However, where there was no conducting path this inspiratory stimulation, with large inspiratory tidal volumes, was coupled with slow breathing and underventilation. However, with a conducting pathway present between the severed vagus nerve endings actual ventilatory stimulation occurred, at least in the early stages.

In relation to the two clinical studies reported here, it is suggested that there may be loss of vagal inhibition, with greater ventilation than expected from loss of $PaCO_2$ oscillations in patients with type A obstructive airways disease. It is possible that there may be less vagal inhibitory tone than would be expected for the large lung volumes found in type A patients.

A striking clinical feature of the type A patients is their dyspnoea. The cause of this dyspnoea has been as obscure as the alveolar hypoventilation of patients in the later stages of type B obstructive airways disease. It is only speculation, but it seems possible that the dyspnoea in these patients is a result of a selective loss of vagal afferent tone, due to the nature of the pulmonary disorder.

Here we can see that there may be a very important link between the respiratory phenomena in disease and the understanding of normal respiratory control. If no type A patients had been studied, we might reasonably have thought that respiratory control depended solely on $PaCO_2$ oscillations with central chemoreceptors as a background stabilising influence. While the $PaCO_2$ oscillations seem likely to be important, the findings in the patients with type A obstructive airways disease have alerted us to the likelihood that other factors, such as pulmonary afferent innervation, are important in respiratory control. In the absence of patient studies other factors might not have been suspected for some time.

Conclusion

Arterial pH oscillations have been shown indirectly to monitor arterial PCO_2 oscillations. The $PaCO_2$ oscillations arise from alveolar PCO_2 oscillations at the lungs in normal man and animals. In turn, the expiratory slope of $PACO_2$ oscillations depends upon the rate at which CO_2 is delivered to the lungs (see equation p. 233).

There can be seen to be a source of information as to $\dot{V}CO_2$ in the $PACO_2$ oscillations which, with probably little attenuation, pass into the systemic arterial circulation. The carotid body has been shown to detect the oscillations in $PaCO_2$ in its blood supply with, at least partial, rate of change sensitivity. After this the information transmission line becomes less well understood, since there are uncertainties about sinus nerve discharge characteristics in the intact animal, and non-linearities in the transition from arterial PCO_2 to ventilatory changes via the respiratory control system.

Certain properties of the system as a whole appear to be strong pointers to a role of $PaCO_2$ oscillations in respiratory control. Examples are (i) the precise relation between alveolar ventilation and the CO_2 production rate in exercise, which is upset by abnormal oxygenation or carotid body resection (Yamamoto & Edwards, 1960; Linton *et al.*, 1976, 1977; Phillipson *et al.*, 1981a, b, c); (ii) the failure of mean $PaCO_2$ at rest to remain normal in the face of carotid body resection, or (in type B patients with obstructive airways disease) when $PaCO_2$ oscillations are attenuated; and (iii) the slowing of the ventilatory increase early in exercise in the carotid body resected subjects of Wasserman *et al.* (1975) and in type B patients with attenuated $PaCO_2$ oscillations (Cochrane *et al.*, 1979, and in preparation).

Though many questions remain unanswered it appears that $PaCO_2$ oscillations of respiratory frequency are likely to play an important role in respiratory control, with the reservation that pulmonary afferent and other receptor information may normally be integrated with, or in disease interfere with, the chemical control of breathing. It may only be necessary for the central chemoreceptors to be actively involved in the control of breathing when homeostasis of the cerebral environment is threatened. Experiments in which $\dot{V}CO_2$ is increased, but where $PaCO_2$ oscillations are eliminated, should shortly show whether $PaCO_2$ oscillations normally sustain homeostasis for mean $PaCO_2$ (Phillipson, personal communication). Once this effective proof of the importance of $PaCO_2$ oscillations is available details of their role in the control of breathing should be easier to elucidate than has been the case to date.

Acknowledgements

Thanks are due to T.J.H. Clark and the Medical Research Council for extra laboratory facilities, and to D.M. Band, G.M. Cochrane, S.J.G. Semple, R.W. Torrance, D.J.C. Cunningham, B.B. Lloyd, E.A. Phillipson, J.G. Prior, R.V. Nowell and C.G. Newstead for many points found here which arose in discussion.

References

Asmussen, E., Christensen, E.H. & Nielsen, M. (1943a) 'Humoral or Nervous Control of Respiration during Muscular Work?' *Acta Physiol. Scand.*, 6, 160-7

Asmussen, E., Nielsen, M. & Wieth-Pedersen, G. (1943b) 'Cortical or Reflex Control of Respiration during Muscular Work?' *Acta Physiol. Scand.*, 6, 168-75

Band, D.M., Cameron, I.R. & Semple, S.J.G. (1969a) 'Oscillations in Arterial pH with Breathing in the Cat', *J. Appl. Physiol.*, 26, 261-7

Band, D.M., Cameron, I.R. & Semple, S.J.G. (1969b) 'Effect of Different Methods of CO_2 Administration on Oscillations of Arterial pH in the Cat', *J. Appl. Physiol.*, 26, 268-73

Band, D.M., Cameron, I.R. & Semple, S.J.G. (1970) 'The Effect on Respiration of Abrupt Changes in Carotid Artery pH and PCO_2 in the Cat', *J. Physiol.*, 211, 479-94

Band, D.M., McClelland, M., Phillips, D.L., Saunders, K.B. & Wolff, C.B. (1978) 'Sensitivity of the Carotid Body to Within-breath Changes in Arterial PCO_2', *J. Appl. Physiol.*, 45, 768-77

Band, D.M. & Phillips, D.L. (1973) 'The Modification, by the Addition of Carbonic Anhydrase, of the Effect of Non-gaseous Acids on the Carotid Body Chemoreceptor Discharge of the Cat', *J. Physiol.*, 229, 11-12P

Band, D.M. & Semple, S.J.G. (1966) 'Intravascular Electrode for Continuous Measurement of Arterial pH in Man and Animals', *J. Physiol.*, 184, 58-9P

Band, D.M. & Semple, S.J.G. (1967) 'Continuous Measurement of Blood pH with an Indwelling Arterial Glass Electrode', *J. Appl. Physiol.*, 22, 854-7

Band, D.M., Wilshaw, P. & Wolff, C.B. (1976) 'The Speed of Response of the Carotid Body Chemoreceptor' in A.S. Paintal (ed.), *Morphology and Mechanisms of Chemoreceptors*, Vallabhbhai Patel Chest Institute, University of Delhi, Delhi, pp. 197-208

Band, D.M. & Wolff, C.B. (1978) 'Respiratory Oscillations in Discharge Frequency of Chemoreceptor Afferents in Sinus Nerve of Anaesthetised Cats at Normal and Low Arterial Oxygen Tensions', *J. Physiol., 282*, 1-6

Band, D.M., Wolff, C.B., Ward, J., Cochrane, G.M. & Prior, J.G. (1980) 'Respiratory Oscillations in Arterial Carbon Dioxide Tension as a Control Signal in Exercise, *Nature, 283*, 84-5

Bannister, R.G., Cunningham, D.J.C. & Douglas, C.G. (1954) 'The Carbon Dioxide Stimulus to Breathing in Severe Exercise', *J. Physiol., 125*, 90-117

Biscoe, T.J. (1971) 'Carotid Body: Structure and Function', *Physiol. Rev., 51*, 437-95

Biscoe, T.J. & Purves, M.J. (1965) 'Rhythmical Variation in the Carotid Body Chemoreceptors', *J. Physiol., 177*, 72-3P

Biscoe, T.J. & Purves, M.J. (1967) 'Observations on the Rhythmic Variation in the Cat Carotid Body Chemoreceptor Activity which has the Same Period as Respiration', *J. Physiol., 190*, 389-412

Biscoe, T.J. & Taylor, A. (1963) 'The Discharge Pattern Recorded in Chemoreceptor Afferent Fibres from the Cat Carotid Body with Normal Circulation and During Perfusion', *J. Physiol., 168*, 332-44

Black, A.M.S., McCloskey, D.I & Torrance, R.W. (1971) 'The Responses of Carotid Body Chemoreceptors in the Cat to Sudden Changes of Hypercapnic and Hypoxic Stimuli', *Resp. Physiol., 13*, 36-49

Black, A.M.S. & Torrance, R.W. (1967a) 'Chemoreceptor Effects in the Respiratory Cycle', *J. Physiol., 189*, 59-61P

Black, A.M.S. & Torrance, R.W. (1967b) 'The Effect of Respiratory Variations in Chemoreceptor discharge upon Inspiratory Activity', *J. Physiol., 191*, 114-5P

Black, A.M.S. & Torrance, R.W. (1971) 'Respiratory Oscillations in Chemoreceptor Discharge in the Control of Breathing', *Resp. Physiol., 13*, 221-37

Bouverot, P., Flandrois, R., Puccinelli, R. & Dejours, P. (1965) 'Etude du Role des Chémorécepteurs Artériels dans la Régulation de Respiration Pulmonaire chez le Chien Eveillé', *Arch. Int. Pharmacodyn., 157*, 253-71

Burrows, B., Fletcher, C.M., Heard, B.E., Jones, N.L. & Wootliff, J.S. (1966) 'The Emphysematous and Bronchial Types of Airways Obstruction', *Lancet, i*, 930-5

Cochrane, G.M., Hilson, A.J., Maisey, M.N., Nowell, R., Prior, J.G. & Wolff, C.B. (1980a) 'Circulatory Transit Times and their Prolongation in Clinical Emphysema', *Clinical Science, 59*, 12P

Cochrane, G.M., Newstead, C.G., Nowell, R.V., Openshaw, P. & Wolff, C.B. (1982) 'The Rate of Rise of Alveolar Carbon Dioxide Pressure during Expiration in Man', *J. Physiol., 333*, 17-27

Cochrane, G.M., Powlson, M., Prior, J. & Wolff, C.B. (1979) 'Respiratory pH Oscillations and the Timecourse of Exercise Ventilation', *Clin. Sci., 57*, 14P

Cochrane, G.M., Prior, J.G. & Wolff, C.B. (1980b) 'Chronic Stable Asthma and the Normal Arterial Pressure of Carbon Dioxide in Hypoxia', *Br. Med. J., 281*, 705-7

Cochrane, G.M., Prior, J.G. & Wolff, C.B. (1981a) 'Respiratory Arterial pH and PCO_2 Oscillations in Patients with Chronic Obstructive Airways Disease', *Clin. Sci., 61*, 693-702

Cochrane, G.M., Prior, J.G. & Wolff, C.B. (1981b) 'Control of Breathing and Respiratory PCO_2 Oscillations in Patients with Airways Obstruction', *Thorax, 36*, 708-9

Cochrane, G.M. & Wolff, C.B. (1979) 'The Partial Pressure of Carbon Dioxide in Alveolar Gas during Expiration and the Rate of Carbon Dioxide Production in Man', *J. Physiol.*, *289*, 69-70P

Conway, M.A. (1982) 'Effects of beta-Adrenergic Blockade on the Ventilatory Response to Hypoxic and Hyperoxic Exercise in Man', *J. Physiol.*, *325*, 18P

Cotes, J.E. (1975) *Lung Function: Assessment and Application in Medicine.* IIIrd edn, Blackwells, Oxford, pp. 380-91

Cross, B.A., Grant, B.J.B., Guz, A., Jones, P.W., Semple, S.J.G. & Stidwell, R.P. (1979) 'Dependence of Phrenic Motor Neurone Output on the Oscillatory Component of Arterial Blood Gas Composition', *J. Physiol.*, *290*, 163-84

Cunningham, D.J.C. (1972) 'Time Patterns of Alveolar Carbon Dioxide and Oxygen: the Effects of Various Patterns of Oscillations on Breathing in Man', in I. Gilliland and Jill Francis (eds), *The Scientific Basis of Medicine, Annual Reviews*, The Athlone Press, University of London, pp. 333-62

Cunningham, D.J.C., Howson, M.G. & Pearson, S.B. (1973) 'The Respiratory Effects of Altering the Time Profile of Alveolar Carbon Dioxide and Oxygen within each Respiratory Cycle', *J. Physiol.*, *243*, 1-28

Dolan, M.J., Whipp, B.J., Davidson, W.D., Weitzman, R.E. & Wasserman, K. (1981) 'Hypopnea Associated with Acetate Hemodialysis: Carbon Dioxide-Flow-Dependent Ventilation', *N. Engl. J. Med.*, *305*, 72-5

Donoghue, S., Felder, R.B., Jordan, D. & Spyer, K.M. (1982a) 'Brain Stem Projections of Carotid Baroreceptors and Chemoreceptors in the Cat', *J. Physiol.*, *330*, 53-4P

Donoghue, S., Garcia, M., Jordan, D. & Spyer, K.M. (1982b) 'Identification and Brain-Stem Projections of Aortic Baroreceptor Afferent Neurones in Nodose Ganglia of Cats and Rabbits', *J. Physiol.*, *322*, 337-52

DuBois, A.B., Britt, A.G. & Fenn, W.O. (1952) 'Alveolar CO_2 during the Respiratory Cycle', *J. Appl. Physiol.*, *4*, 535-48

DuBois, A.B., Fowler, R.C., Soffer, A. & Fenn, W.O. (1951) 'Alveolar CO_2 Measured by Expiration into the Rapid Infrared Gas Analyser', *J. Appl. Physiol.*, *4*, 526-34

Eldridge, F.L. (1972a) 'The Importance of Timing on the Respiratory Effects of Intermittent Carotid Sinus Nerve Stimulation', *J. Physiol.*, *222*, 297-318

Eldridge, F.L. (1972b) 'The Importance of Timing on the Respiratory Effects of Intermittent Carotid Body Chemoreceptor Stimulation', *J. Physiol.*, *222*, 319-33

Fitzgerald, M.P. (1913) 'The Changes in the Breathing and the Blood at Various High Altitudes', *Phil. Trans. B*, *203*, 351-7

Fitzgerald, R.S., Leitner, L.-M. & Liaubet, M.-J. (1969) 'Carotid Chemoreceptor Response to Intermittent or Sustained Stimulation in the Cat', *Resp. Physiol.*, *6*, 395-402

Forster, H.V., Bisgard, G.E. & Klein, J.P. (1981) 'Effect of Peripheral Chemoreceptor Denervation on Acclimatization of Goats during Hypoxia', *J. Appl. Physiol.*, *50*, 392-8

Fowler, R.C. (1949) 'A Rapid Infra-red Gas Analyser', *Rev. Scient. Instruments*, *20*, 175-8

Goodman, N.W., Nail, B.S. & Torrance, R.W. (1974) 'Oscillations in the Discharge of Single Carotid Chemoreceptor Fibres of the Cat', *Resp. Physiol.*, *20*, 251-69

Head, H. (1889) 'On the Regulation of Respiration. Part I. Experimental', *J. Physiol.*, *10*, 1-70

Haldane, J.S. & Priestley, J.G. (1905) 'The Regulation of the Lung-ventilation', *J. Physiol.*, *32*, 225-66

Herbert, Dorothy A. & Mitchell, R.A. (1971) 'Blood Gas Tensions and Acid Base Balance in Awake Cats', *J. Appl. Physiol.*, *30*, 434-6

Herrera, L. & Kazemi, H. (1980) 'CSF Bicarbonate Regulation in Metabolic

Acidosis: Role of HCO_3^- Formation in CNS', *J. Appl. Physiol.*, *49*, 778-83

Hornbein, T.F., Griffo, Z.J. & Roos, A. (1961) 'Quantitation of Chemoreceptor Activity: Interrelation of Hypoxia and Hypercapnia', *J. Neurophysiol.*, *24*, 561-8

Hornbein, T.F. (1965) 'Effect of Respiratory Oscillation of Arterial PO_2 and PCO_2 on Carotid Chemoreceptor Activity and Phrenic Nerve Activity', *Physiologist*, *8*, 197

Kao, F.F. (1963) 'An Experimental Study of the Pathways Involved in Exercise Hyperpnoea Employing Cross-circulation Techniques', in D.J.C. Cunningham & B.B. Lloyd (eds), *The Regulation of Human Respiration (Haldane Centenary Symposium)*, Blackwell, Oxford, pp. 461-502

Kao, F.F., Michel, C.C. & Mei, S.S. (1964) 'Carbon Dioxide and Pulmonary Ventilation in Muscular Exercise', *J. Appl. Physiol.*, *19*, 1075-80

Kao, F.F., Michel, C.C., Mei, S.S. & Li, W.K. (1963) 'Somatic Afferent Influence on Respiration', *Ann. N.Y. Acad. Sci.*, *109*, 696-710

Kolobow, T., Gattinoni, L., Tomlinson, T., White, D., Pierce, J. & Iapichino, G. (1977) 'The Carbon Dioxide Membrane Lung (CDML): a New Concept', *Trans. Am. Soc. Artif. Int. Organs*, *23*, 17-21

Krogh, A. & Lindhard, J. (1914) 'On the Average Composition of the Alveolar Air and its Variation during the Respiratory Cycle', *J. Physiol.*, *47*, 431-45

Lahiri, S. (1968) 'Alveolar Gas Pressures in Man with Life-time Hypoxia', *Resp. Physiol.*, *4*, 376-86

Lahiri, S. & Delaney, R.G. (1975) 'Stimulus Interaction in the Responses of Carotid Body Chemoreceptor Single Afferent Fibres', *Resp. Physiol.*, *24*, 249-66

Leitner, L.-M. & Dejours, P. (1968) 'The Speed of Response of the Arterial Chemoreceptors', in R.W. Torrance (ed.), *The Arterial Chemoreceptors*, Blackwell, Oxford, pp. 79-88

Linton, R.A.F., Miller, R. & Cameron, I.R. (1976) 'Ventilatory Response to CO_2 Inhalation and Intravenous Infusion of Hypercapnic Blood', *Resp. Physiol.*, *26*, 383-94

Linton, R.A.F., Miller, R. & Cameron, I.R. (1977) 'Role of PCO_2 Oscillations and Chemoreceptors in Ventilatory Response to Inhaled and Infused CO_2', *Resp. Physiol.*, *29*, 201-10

Lloyd, B.B. (1966) 'The Interactions between Hypoxia and other Ventilatory Stimuli', in *Proc. Int. Symp. Cardiovasc. Respir. Effects Hypoxia*, Karger, Basel, New York, pp. 146-65

Lloyd, B.B. & Cunningham, D.J.C. (1963) 'A Quantitative Approach to the Regulation of Human Respiration', in D.J.C. Cunningham & B.B. Lloyd (eds), *The Regulation of Human Respiration*, Blackwell, Oxford, pp. 331-49

Mahler, M. (1979) 'Neural and Humoral Signals for Pulmonary Ventilation arising in Exercising Muscle', *Medicine and Science in Sports*, *11*, (No. 2), 191-7

McCloskey, D.I. (1975) 'Mechanisms of Autonomic Control of Carotid Chemoreceptor Activity', *Resp. Physiol.*, *25*, 53-61

Michel, C.C. & Kao, F.F. (1964) 'Use of a Cross Circulation Technique in Studying Respiratory Responses to CO_2', *J. Appl. Physiol.*, *19*, 1070-4

Neil, E. & O'Regan, R.G. (1971) 'Efferent and Afferent Impulse Activity Recorded from Few-fibre Preparations of Otherwise Intact Sinus and Aortic Nerves', *J. Physiol.*, *215*, 33-47

Newstead, C.G., Nowell, R.V. & Wolff, C.B. (1980) 'The Rate of Rise of Alveolar Carbon Dioxide Tension during Expiration', *J. Physiol.*, *307*, 42P

O'Regan, R.G. (1976) 'Efferent Control of Chemoreceptors', in A.S. Paintal (ed.), *Morphology and Mechanisms of Chemoreceptors*, Vallabhbhai Patel Chest

Institute, University of Delhi, Delhi, pp. 228-47

O'Regan, R.G. (1981) 'Responses of Carotid Body Chemosensory Activity and Blood Flow to Stimulation of Sympathetic Nerves in the Cat', *J. Physiol.*, *315*, 81-98

Pappenheimer, J.R., Fencl, V., Heisey, S.R. & Held, D. (1965) 'Role of Cerebral Fluids in Control of Respiration as Studied in Unanaesthetised goats', *Am. J. Physiol.*, *208*, 436-50

Patterson, R.W., Nissenson, A.R., Miller, J., Smith, R.T., Narins, R.G. & Sullivan, S.F. (1981) 'Hypoxemia and Pulmonary Gas Exchange during Hemodialysis', *J. Appl. Physiol.*, *50*, 259-64

Phillipson, E.A., Bowes, G., Townsend, E.R., Duffin, J. & Cooper, J.D. (1981a) 'Role of Metabolic CO_2 Production in Ventilatory Response to Steady-state Exercise', *J. Clin. Invest.*, *68*, 768-74

Phillipson, E.A., Bowes, G., Townsend, E.R., Duffin, J. & Cooper, J.D. (1981b) 'Carotid Chemoreceptors in Ventilatory Responses to Changes in Venous CO_2 Load', *J. Appl. Physiol.*, *51*, 1398-403

Phillipson, E.A., Duffin, J. & Cooper, J.D. (1981c) 'Critical Dependence of Respiratory Rhythmicity on Metabolic CO_2 Load', *J. Appl. Physiol.*, *50*, 45-54

Rahn, H. & Otis, A.B. (1949) 'Man's Respiratory Response During and after Acclimatisation to High Altitudes', *Am. J. Physiol.*, *157*, 445-62

Saunders, K.B. (1979) 'Isocapnic 'Exercise' in a Breathing Model of the Respiratory System', *Clin. Sci.*, *56*, 9P

Saunders, K.B. (1980) 'Oscillations of Arterial CO_2 Tension in a Respiratory Model: Some Implications for the Control of Breathing in Exercise', *J. Theor. Biol.*, *84*, 163-79

Saunders, K.B., Bali, H.N. & Carson, E.R. (1980) 'A Breathing Model of the Respiratory System: the Controlled System', *J. Theor. Biol.*, *84*, 135-61

Saunders, K.B. & Taylor, A.(1974) 'A Statistical Model of Chemoreceptor Afferent Discharge, The Detection of Modulation by Bin Averaging', *Med. Biol. Eng.*, *12*, 280-6

Siggaard Anderson, O. (1962) 'The pH-logPCO_2 Blood Acid-Base Nomogram Revised', *Scand. J. Clin. Lab. Invest.*, *14*, 598-604

Siggaard Anderson, O. (1963) 'Blood Acid-Base Alignment Nomogram', *Scand. J. Clin. Lab. Invest.*, *15*, 211-7

Taylor, A. (1968) 'The Discharge Pattern of Single Carotid Body Chemoreceptor Units in Relation to Possible Sensory Mechanisms', in R.W. Torrance (ed.), *The Arterial Chemoreceptors*, Blackwell, Oxford, pp. 195-204

Thompson, W.K., Marchak, B.E., Bryan, A.C. & Froese, A.B. (1981) 'Vagotomy Reverses Apnea Induced by High-frequency Oscillatory Ventilation', *J. Appl. Physiol.*, *51*, 1484-7

Wade, J.G., Larson, C.P., Hickey, R.F., Ehrenfeld, W.K. & Severinghaus, J.W. (1970) 'Effect of Carotid Endartectomy on Carotid Chemoreceptor and Baroreceptor Function in Man', *N. Engl. J. Med.*, *282*, 823-9

Ward, S.A. & Cunningham, D.J.C. (1977) 'The Relation between Hypoxia and CO_2-induced Reflex Effects of Breathing in Man', *Resp. Physiol.*, *29*, 379-90

Wasserman, K., Van Kessel, A.L. & Burton, G.G. (1967) 'Interaction of Physiological Mechanisms during Exercise', *J. Appl. Physiol.*, *22*, 71-5

Wasserman, K., Whipp, B.J., Casaburi, R. & Beaver, W.L. (1977) 'Carbon Dioxide Flow and Exercise Hyperpnea: Cause and Effect', *Am. Rev. Resp. Dis.*, *115*, (No. 6 part 2), 225-7

Wasserman, K., Whipp, B.J., Koyal, S.N. & Cleary, M.G. (1975) 'Effect of Carotid Body Resection on Ventilatory and Acid Base Control During Exercise', *J. App. Physiol.*, *39*, 354-8

Wolff, C.B. (1975) 'Breath by Breath Chemoreceptor Responses of the Carotid Body in the Anaesthetised Cat', Ph.D. Thesis, University of London

Wolff, C.B. (1977) 'The Effects on Breathing of Alternate Breaths of Air and a Carbon Dioxide Rich Gas Mixture in Anaesthetised Cats', *J. Physiol.*, *268*, 483-91

Wolff, C.B. (1980) 'Normal Ventilation in Chronic Hypoxia', *J. Physiol.*, *308*, 118-9P

Wolff, C.B. (1981) 'Respiratory Oscillations in Alveolar Carbon Dioxide Partial Pressure in the Steady State', *J. Physiol.*, *319*, 87-8P

Wolff, C.B. & Pearson, S.B. (1978) 'The Influence of Low, Normal and High PaO_2 on the Respiratory Effects of $PaCO_2$ Oscillations in Anaesthetised Cats', *Resp. Physiol.*, *32*, 281-92

Yamamoto, W.S. (1960) 'Mathematical Analysis of the Timecourse of Alveolar Carbon Dioxide', *J. Appl. Physiol.*, *15*, 215-9

Yamamoto, W.S. (1962) 'Transmission of Information by the Arterial Blood Stream with Particular Reference to Carbon Dioxide', *Biophys. J.*, *2*, 132-59

Yamamoto, W.S. & Edwards, McI.W. (1960) 'Homeostasis of Carbon Dioxide during Intravenous Infusion of Carbon Dioxide', *J. Appl. Physiol.*, *15*, 807-8

8 CONTROL OF RESPIRATION IN THE FETUS AND NEWBORN

Peter J. Fleming and Jose Ponte

Introduction

The onset of air-breathing at the time of birth is arguably the most important adaptation which the fetus must make. Under stable conditions prior to birth, respiratory exchange via the placenta is principally determined by the mother (Dawes *et al.*, 1969). Immediately after birth, however, the newborn must establish and maintain pulmonary gas exchange precisely matched to metabolic demands.

Breathing is the only major autonomic function which is under a significant degree of voluntary control. Plum (1970) and Bryan & Bryan (1978) have drawn attention to the limitations of the traditional approach of considering only the metabolic control system and its feed-back loops. The behavioural or voluntary and the metabolic or automatic control of breathing are mediated by several neural structures and have separate descending pathways (Mitchell & Berger, 1975). The relative contribution made by each of these two systems is variable. When awake and quietly resting the metabolic system probably predominates. At other times whilst awake the metabolic system is subordinated to the behavioural, for example, when talking (Phillipson, McClean, Sullivan & Zanel, 1978). During sleep the predominant control system is determined by the type of sleep. In quiet sleep metabolic control predominates, whilst in rapid eye movement (R.E.M.) sleep, there is a substantial 'behavioural' input (Phillipson, 1977; Bryan & Bryan, 1978).

The importance of this state-dependence of respiratory control is exemplified by the fetus and newborn. The fetus makes breathing movements mainly in R.E.M. sleep (Dawes *et al.*, 1972). In quiet sleep such activity is reduced or absent, though the site and nature of the inhibition is uncertain. Whatever the site of inhibition, it must clearly be reversed at birth for regular respiration to be established. Information on sleep state is thus essential for any detailed discussion of respiratory control in the fetus and newborn.

In the next section of this chapter we review the available information on fetal breathing movements and the changes occurring at the time of birth. In the final section we review the postnatal development of respiratory control and its clinical significance.

Breathing Movements in the Fetus

Evidence of Respiratory Movements

Breathing movements are present in fetal life, and this has been documented in various species, including man, by a variety of methods which have included direct observation through the abdominal wall, or of the exposed fetus, transillumination of the pregnant uterus, injection of X-ray contrast into the amniotic fluid, the use of ultrasound, and the insertion of tracheal catheters in chronic animal preparations (Snyder & Rosenfeld, 1937; Barcroft & Barron, 1936; Steele & Windle, 1939; Davies & Potter, 1946; Boddy & Mantell, 1972; Martin *et al.*, 1974; Towell & Salvador, 1974; Mantell 1976). Earlier work, in which observations were made in 'acute' preparations of animal fetuses, exteriorised under some form of general or regional anaesthesia, led to controversy over the presence or not of respiratory movements in the fetus. However little doubt is left of their presence, since the introduction of the 'chronic' preparation by Meschia, Cotter, Breathnach & Barron (1965), in which monitoring is made in undisturbed fetuses, fully recovered from the surgery carried out to implant the necessary devices. Using this technique in the sheep, Dawes *et al.* (1972) found that in the last third of gestation respiratory movements are present for about 40 per cent of the time; they are irregular, rapid in rate (1-4 Hz), occur rather unpredictably and are unrelated to the natural variations in maternal and fetal blood gases (Figure 8.1).

Bowes *et al.* (1981) have recently documented the changes in pattern of these breathing movements with gestational age in the sheep fetus, from 100 days to term. As the fetus matures, respiratory movements become increasingly periodic and their rate is lower. Periods of breathing are interspersed with apneoic periods lasting from 20 per cent of the time at 110 days to 60 per cent at 140 days. Cutaneous stimulation (Scarpelli, Condorelli & Cosmi, 1977) or peripheral nerve stimulation of the fetus *in utero* (Chapman *et al.*, 1977) have negligible effect upon the occurrence or the characteristics of these movements, but a rise of 15 mmHg in the $PaCO_2$ of the ewe, accompanied by a rise of 11 mmHg in the $PaCO_2$ of the fetus, usually initiated breathing movements when they were not present, and caused them to be deeper and more regular (Chapman *et al.*, 1980). Hypoxia always causes these movements to disappear, and, if extreme, will give rise to gasps (Snyder & Rosenfeld, 1937; Boddy & Dawes, 1975). However, the occasional spontaneous gasp is also found in the undisturbed fetus at a rate of 1 to 3 per minute (Dawes *et al.*, 1972).

This apparent randomness of the fetal respiratory movements has raised doubts in many investigators as to their nature. However, the accumulated evidence leaves little doubt about their respiratory origin, since they are accompanied by movements of the amniotic fluid in the trachea, and electromyographic activity in the diaphragm (Maloney *et al.*, 1975). In addition, these movements are synchronous with phasic neuronal activity in medullary and phrenic motor nerones (Bystrzycka, Nail & Purves, 1975).

Respiratory Movements and Electrical Activity in the Brain

Recent studies of the fetal electroencephalogram (EEG) by Jost *et al.* (1972), and of the ocular movements by Dawes *et al.* (1972) in the sheep, have shown that there is a clear association between the periods of rapid, irregular breathing and a pattern of low voltage high frequency EEG, in the last third of gestation. This is accompanied by rapid eye movements (REM), a fall in heart rate and blood pressure, and a decrease in the evoked cortical auditory response, and seems to correspond to similar patterns of rapid and irregular respiration in the adult during REM sleep. In this state the chemo- and mechanoreceptor afferents are partially disconnected from the generator of the respiratory rhythm, as opposed to slow-wave sleep in which respiration is closely regulated by these inputs (Phillipson, 1978). Detailed electrophysiological studies carried out by Persson (1973) throughout maturation of the sheep fetus have shown evidence of central inhibitory mechanisms which appear by the 90th day, and evoked cortical responses similar to those of the adult from 100 days onwards, which are consistent with the EEG findings mentioned above.

Inhibitory Mechanisms

What therefore remains to be understood is why the respiratory movements are not present during slow-wave EEG which occupies 40-60 per cent of the time, in spite of the powerful chemical drive, i.e. arterial PO_2 around 28 mmHg and PCO_2 of 54 mmHg (Dawes *et al.*, 1972).

Firstly we may rule out the question of inadequate maturity since the newborn, even if premature, sustains adequate ventilation immediately after birth, irrespective of the sleep rate.

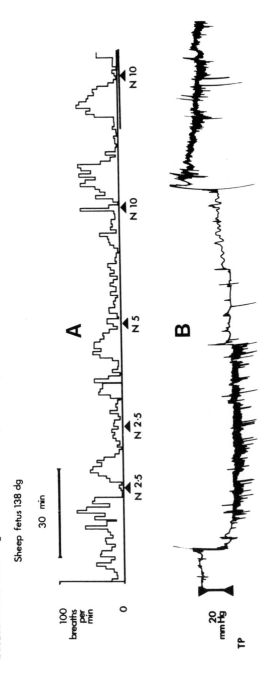

Figure 8.1: Tracings Showing from Above: A, Computed Rate of Breathing over a 3 hour Period, Derived from Bursts of Diaphragmatic Electromyographic Activity (EMG) Recorded in C; B is the Tracheal Pressure and D the Electroencephalogram, in a Sheep Fetus of 138 Days Gestation. N stands for *naloxone* injected i.v. and the numbers are doses in mg. B, C and D tracings correspond to the underlined part at the end of A. Large shifts of baseline in B are due to the ewe standing up or lying down. Tracing C is included as a rough guide to the presence of EMG activity; there is no calibration bar because most of the signal is above the frequency response of the pen recorder.

Figure 8.1: Continued

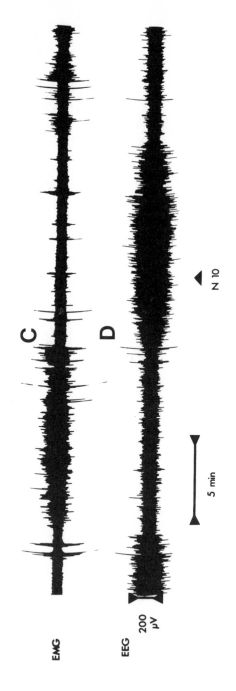

Note: we are thankful to Ms B. Carruthers, Dr M.A. Day and Dr A.R. Milne of Bristol University for this picture.

So, there must either be some form of inhibition, as originally proposed by Barcroft & Barron (1936), or a lack of the excitation derived from peripheral and central chemoreceptors. We shall examine these two possibilities in turn.

Two types of inhibition have been proposed, first an active neuronal discharge arising either in the suprapontine structures of the nervous system, or in the periphery, most likely in receptors located within the lungs or upper airways. Evidence for descending inhibition was not observed in the transections performed by Barcroft & Barron (1937) and has had little experimental support since that earlier work. Afferent activity measured in the vagus (Ponte & Purves, 1973) has shown that stretch receptors are active in the fetus, but they are unlikely to be the cause of inhibition since the section of these nerves does not lead to sustained respiratory activity (Boddy *et al.*, 1974). Secondly, a form of chemical inhibition has been proposed, namely the release of morphine-like peptides, possibly by the placenta. Evidence for this mechanism was obtained in acute preparations involving the exteriorisation of the fetus, in which naloxone, a specific morphine antagonist, was found to initiate breathing and increase the response to CO_2 (Moss & Scarpelli, 1979). However, in more recent experiments (Carruthers *et al.*, unpublished data) carried out in the chronically catheterised fetus, naloxone had an insignificant effect upon breathing patterns (Figure 8.1) lending little support to the endorphin mediated inhibition.

The second possible explanation for the lack of respiratory movements is centred upon the absence of a peripheral chemoreceptor drive, since it was found by Biscoe, Purves & Sampson (1969) that the carotid body chemoreceptors of the sheep fetus do not respond to chemical stimuli in the same manner as in the new-born or the adult; little activity is found in the afferent nerves supplying the receptors, and the few receptors found lack their normal adult sensitivity to chemical stimuli. However, this insensitivity can be readily reversed by exposing the receptors to higher oxygen pressure *in vitro* (Jansen, Purves & Tan, 1980) pointing to a possible underlying vascular mechanism, of a similar nature to that occurring in the ductus arteriosus and in the pulmonary vessels.

In contrast, the aortic chemoreceptors are active and responsive to hypoxia in fetal life (Ponte & Purves, 1973) but their central projections seem to be confined to cardiovascular reflexes (Dawes *et al.*, 1969), namely the redistribution of cardiac output and adjustment of umbilical blood flow (Reuss & Rudolph, 1980).

However, the absence of chemoreceptor drive may not be the sole cause of the absence of breathing movements, since the central chemo-receptor does show some activity when the carbon dioxide concentration is raised (Boddy *et al.*, 1974), though the threshold of activation is raised and the sensitivity reduced compared to the mature individual.

At present the main problem is in finding a cause for the reduced sensitivity of the peripheral and central chemoreceptors in the fetus. With respect to the central chemoreceptors, cerebrospinal fluid in the fetus is of identical composition to the newborn or the adult (Hodson *et al.*, 1968), ruling out an inhibition caused by low concentration of hydrogen ion.

Catecholamines

Our lack of understanding of the nature of the chemoreceptor process, and of the role of transmitters within the carotid body has been the main difficulty in interpreting the changes in sensitivity of these receptors occurring at birth. However, recent evidence has shown that there is an important β-adrenergic mechanism in their response to hypoxia and hypercapnia (Folgering, Ponte & Sadig, 1982); the administration of propranolol, for example, causes a marked inhibition of the chemoreceptor response to hypoxia, particularly noticeable at low oxygen pressures.

The interest of this finding lies in the fact that the circulating catecholamines of the fetus rise many fold just prior to birth (Jones, 1980), from levels of the order of less than 1 μg/ml to values between 10 and 20 μg/ml in the day preceding delivery. The possibility then arises that, in the fetus, the synthetic pathway leading to the formation of noradrenaline from tyrosine is inhibited at some stage, being activated at birth, perhaps by the rise in oxygen pressure. This offers a plausible explanation for the lack of sensitivity of the carotid chemoreceptors of the fetus which may also apply to the central chemoreceptors, but requires experimental verification.

Clinical Implications

The clinical benefits of these findings are important in providing a basis for the prevention and management of the respiratory problems of the newborn, particularly common in the premature. Several drugs, which may be administered to women in labour, e.g. β-adrenergic blocking agents in the management of hypertension, may adversely affect the adaptation of the respiratory control system to the first hours of extra-uterine life.

Respiratory movements only occur in the undisturbed, well-oxygenated, fetus; this has led some authors to believe that they should be taken as an index of the state of health of the fetus (Boddy & Dawes, 1975).

Control of Respiration in the Newborn

Introduction

Immediately after birth the newborn animal's most important action is to take the first breath of air; the next most important action is the second breath. As discussed in the preceding section, the newborn animal is in possession of a mechanically well-tried respiratory apparatus. One major section of the control system, the metabolic control system, has been inhibited prior to birth, and is thus essentially unused at the time of birth. Perhaps not surprisingly, therefore, there are major changes in the metabolic control of respiration in the newborn period, and these are described in the following pages.

The Peripheral Chemoreceptors

In the mature mammal the peripheral chemoreceptors (carotid bodies ± aortic chemoreceptor) respond to changes in arterial oxygen and CO_2 tension. In the fetal lamb close to term, Biscoe *et al.* (1969) found minimal or absent electrical activity in the carotid sinus nerve despite blood gas tensions which would be expected to produce vigorous activity in the adult. Within minutes of the onset of respiration, however, chemoreceptor activity was present and responded to changes in arterial PO_2 and PCO_2. The nature of this activation of the peripheral chemoreceptors at birth is unknown, but it appears to be irreversible and to depend on the rise in arterial PO_2 which occurs at the onset of air-breathing.

Whatever the mechanism involved, the establishment of normal respiration after birth in lambs requires intact chemoreceptors. Although chemodenervated lambs breathe, their respiration is irregular and reduced (Harned *et al.*, 1967).

The role of the peripheral chemoreceptors in human infants has been difficult to define. In both pre-term and term infants hypoxia leads to a transient rise in ventilation, followed by a marked fall. The adult response of sustained hyperventilation does not appear until one to three weeks after birth (Brady & Ceruti, 1966; Rigatto, Brady & Torre, 1975a). Recent work by Rigatto *et al.* (1979) suggests that the response to hypoxia may depend on sleep state. They studied six infants at post-conceptional ages of 33–35 weeks, and showed comparable immediate increases in ventilation in rapid eye movement sleep (REM), quiet sleep (QS) and awake state. In REM or when awake this was followed by the expected respiratory depression, but in QS the hyperventilation was sustained.

An alternative way to quantify respiratory drive is to produce hyperoxia, as the inhalation of 100 per cent oxygen in the newborn infant leads to a transient fall in ventilation followed by a marked rebound hyperventilation (Rigatto *et al.*, 1975a). Numerous authors have demonstrated that this response to 100 per cent O_2 is present and equal in REM and QS in infants at post-conceptional ages of 33–52 weeks (Rigatto *et al.*, 1979; Bolton & Herman, 1974; Fagenholz, O'Connell & Shannon, 1976).

In the adult, hypoxia enhances and hyperoxia depresses the ventilatory response to CO_2. In the newborn pre-term infant the reverse is true – hypoxia depresses CO_2 responsiveness (Rigatto, Torre & Cates, 1975b). In a cool environment hypoxia produces ventilatory depression in pre-term infants, with no initial hyperpnoea (Ceruti, 1966).

Occlusion of the airway at end-expiration is a technique widely used in the study of peripheral chemoreceptor activity. The negative pressure generated by the attempt to inspire during airway occlusion is a measure of total 'inspiratory drive' unaffected by inhibitory vagal stretch receptor feedback. Airway occlusion results in a rise in $PaCO_2$ and fall in PaO_2 leading to a mixed stimulus to the peripheral chemoreceptors. The rate of rise of inspiratory pressure in successive breaths over a 5 to 10 second period has been used as an index of overall peripheral chemoreceptor effects. In the newborn infant especially if pre-term, this method has the disadvantage that vigorous inspiratory efforts against an occluded airway lead to inward distortion of the very compliant rib cage, particularly in REM sleep, in which there is decreased inspiratory activity of the intercostal muscles (Bryan, 1978). Inspiratory rib cage distortion may result in inhibition of diaphragmatic contraction via the intercostal-phrenic reflex (see section on vagal trigeminal and proprioceptive reflexes). Frantz *et al.* (1976) used the technique of airway occlusion to investigate the effects of gestational age, postnatal age and sleep state on peripheral chemoreceptor effects. The presence of rib cage distortion in many of their infants, particularly in the more pre-term ones and those in REM sleep, make their results difficult to interpret. When controlled for the degree of rib cage distortion, however, they show increasing chemoreceptor effects with increasing gestational and post-conceptional age, but no difference between REM and QS.

The initial rise in ventilation in hypoxic infants is due to carotid body activity and the late fall is probably due to central depression. This biphasic response may be due to a fall in chemoreceptor output with prolonged hypoxia or to increasing central depression or to a combination of both. As discussed in the section on fetal breathing movements, the newborn has very low levels of noradrenaline just before birth. If synthesis of noradrenaline in the peripheral chemoreceptors is delayed until after birth, then stores will be low in the first few days and may perhaps be depleted with prolonged hypoxia leading to decreased chemoreceptor activity.

Tenney & Ou (1977) described separate neural pathways from the cerebral cortex and from the hypothalamus to the medulla. Hypoxia leads to respiratory inhibition via the former and facilitation via the latter pathways. Thus in the mature animal there is a balanced effect. The very low levels of CNS noradrenaline (a facilitatory neurotransmitter) in the newborn period may reduce the facilitatory effect of the hypothalamic pathway, leading to an inhibitory bias in the output from higher centres in hypoxia. The greater importance of cortical inputs in respiratory control in REM than in QS might explain the more marked hypoxic respiratory depression in REM (Bryan, 1978).

The paradoxical response to hypoxia may also be partly related to the complicated cardiovascular adaptations of the perinatal period. Purves (1966) showed that because of right to left shunts in newborn lambs, administration of a few breaths of 100 per cent oxygen produced only very small changes in carotid PaO_2. Hypoxia produces pulmonary vasoconstriction and an increase in right to left shunting in newborn infants (Klaus, Fanaroff & Martin, 1979). In the presence of hypoxia and right to left shunting at the ductus arteriosus the carotid PaO_2 will be higher and the $PaCO_2$ lower than in the descending aorta. The carotid body will thus see a 'mixed' stimulus of a fall in both PaO_2 and $PaCO_2$, leading to a diminished response.

In summary, in newborn infants hypoxia leads to an increase in ventilation, followed by a fall. The magnitude of the increase is similar in all three states (REM, QS and awake), though the late fall is less marked or absent in QS. The initial increase is greater and more sustained in term than pre-term infants, but is much smaller in the first 24 hours of age at all gestations (Rigatto, 1979).

Central Chemoreceptor

The respiratory response to CO_2 is mediated partly by direct effects on the peripheral chemoreceptors, and partly by the effect of CO_2 on medullary extracellular fluid pH. Little information is available on the peripheral chemoreceptors' role in CO_2 responses in the newborn period, so the two will be considered together.

Newborn infants respond to inhaled CO_2 and when this response is expressed in terms of body weight their sensitivity is similar to that of adults (Avery *et al.*, 1963). The effects of gestational age and sleep on CO_2 responses are controversial. Frantz *et al.* (1976) showed an increasing steady-state response to inhaled 4 per cent CO_2 with increasing gestational and postnatal age in pre-term infants. Their results were similar whether expressed in terms of minute ventilation or in terms of the pressure generated in the first breath on airway occlusion, suggesting that a changing vagal effect with increasing age was not responsible for the increasing response. Davi *et al.* (1979) found no evidence of increasing CO_2 response with increasing gestational age, and similar responses in REM and QS at all gestations. The response curve in REM was shifted to the left (i.e. higher ventilation at the same level of $PaCO_2$) reflecting the higher metabolic rate in REM than in QS. Using a similar steady state CO_2 stimulus, Bryan *et al.* (1976) showed greater responses to CO_2 in QS than in REM. Haddad *et al.* (1980) reported a longitudinal study of CO_2 responses in term infants from one week to four months of age, and showed no difference in response between QS and REM. All these studies showed that the increase in ventilation with inspired CO_2 is primarily by an increase in tidal volume, with little change in inspiratory time or respiratory frequency.

The discrepancies between these studies may be partly explained by the fact that REM was considered as a homogeneous state. Sullivan *et al.* (1979) have described two distinct forms of REM sleep in dogs – phasic and tonic. They found that in tonic REM CO_2 responsiveness was similar to that in QS, whereas in phasic REM the response was depressed. From 34 weeks gestational age when QS first appears in pre-term newborn infants, to three months past term when there is a steady decrease in the proportion of REM and indeterminate sleep, and a rise in the proportion of QS (Parmalee *et al.*, 1967). No information is available on the incidence of phasic versus tonic REM in human infants, though clearly changes in their relative proportions might alter CO_2 responses.

Newborn guinea pigs and newborn rabbits, which are relatively mature and immature respectively at birth both give appropriate respiratory responses to change in CSF pH (Wennergren & Wennergren, 1980). Bureau *et al.* (1979) showed that changes in arterial pH produced larger and more rapid fluctuations in the CSF in newborn lambs than in mature sheep. This suggests that the central chemoreceptor may play a greater role in short-term respiratory modulation in the newborn than in the adult. The same authors suggest that the response of the newborn lamb to such pH changes is less than that of an adult. They did not, however, maintain a constant $PaCO_2$, so the changes in pH observed were accompanied by changing peripheral chemoreceptor stimulation by the changing $PaCO_2$, and no conclusions can be drawn on central chemoreceptor sensitivity from their results.

In summary, the newborn infant responds to CO_2 in a similar way to the adult. There is possibly an increase in response with increasing gestational age, whilst the effect of sleep state remains controversial.

Vagal, Trigeminal and Proprioceptive Reflexes

Trigeminal Reflexes
Stimulation of the area innervated by the trigeminal nerve, i.e. the face, nasal and nasopharyngeal mucosa, has important cardiorespiratory effects (Haddad & Mellins, 1977). Effects include a fall in respiration and heart rates, and redistribution of cardiac output to the heart and brain, with reduced flow to kidneys, gut and skeletal muscle. There is considerable inter-species variation in these effects; their relationship to the diving reflex of aquatic mammals is not well understood. Little is known of the role of trigeminal reflexes in newborn infants, though clinical use is made of the cardiovascular effects of facial cooling as a means of terminating paroxysms of supraventricular tachycardia in newborn infants. Gentle stimulation of the malar region in pre-term and term infants gives rise to significant respiratory slowing in many infants, in both REM and QS (Fleming *et al.*, 1982). Thus, use of a face mask to record ventilation in newborn infants may itself alter respiratory pattern.

Laryngeal Reflexes
Chemoreceptors in the upper larynx are involved in important reflexes in newborn lambs (Downing & Lee, 1975). Instillation of water or cow's milk into this area leads to apnoea and cardiovascular changes. Instillation of ewe's milk produces no such effects. The reflex disappears with maturation and its role in human infants is unknown.

In dogs, laryngeal stimulation leads to respiratory slowing in QS, but often prolonged apnoea in REM sleep (Sullivan *et al.*, 1978).

Vagal Stretch Receptors
Gradual cooling of the vagus nerve in dogs produces sequential blockade

of afferent fibres — first those from slowly adapting stretch receptors, followed by those from irritant receptors. Use of this technique has led to a better understanding of the role of stretch receptors in normal breathing. Afferent impulses generated in the stretch receptors during inspiration render the inspiratory neurones refractory and hence delay the start of the next inspiration (see Chapter 3). The stretch receptors are mainly responsible for determining expiratory time. Inspiratory time is affected by other vagal afferents, probably including irritant receptors (Haddad & Mellins, 1977).

Two techniques have been used to quantify stretch-receptor activity in newborn infants — lung inflation (Cross *et al.*, 1960) and end-expiratory airway occlusion (Thach *et al.*, 1978) — the two methods described by Hering & Breuer. Whilst allowing static quantification of stretch-receptor activity, these two methods do not give information on the dynamic role of stretch receptors. Both methods show greater stretch receptor activity in the newborn infant than the adult. With end-expiratory airway occlusion the prolongation of inspiratory time in the first breath after occlusion is a measure of stretch receptor activity. Confusion over the most appropriate way to quantify this effect has led to statements that stretch-receptor activity increases (Thach *et al.*, 1978) or decreases (Kirkpatrick *et al.*, 1976) with increasing gestational and post-natal age. The reflex is more marked in QS than REM in human infants (Finer, Abroms & Taeusch, 1976) and in dogs (Phillipson, 1977).

Irritant Receptors

Deflation of the lung or inhalation of irritant gases in many animals produce marked hyperventilation. This response is mediated via small myelinated vagal fibres from epithelial irritant receptors or rapidly adapting stretch receptors. These receptors are distributed throughout the airways, and also respond to direct tactile stimulation. Stimulation of the tracheal mucosa in intubated infants elicits a response which changes with age (Fleming *et al.*, 1978). Prior to 35 weeks gestation such stimulation is commonly followed by respiratory slowing or apnoea, whereas in term infants there is marked hyperventilation. The inhibitory response in many pre-term infants may be due to lack of vagal myelination. Unmyelinated fibres are unable to transmit the high-frequency spike trains of irritant receptor discharge. In effect they act as a low-pass filter, changing the character of the centrally received signals, perhaps converting a facilitatory high frequency signal into an inhibitory low frequency one.

Whatever the mechanism involved, the importance of this response is that endotracheal intubation and suctioning, two procedures which usually produce vigorous respiratory efforts, may lead to apnoea in pre-term infants. In addition, the reduction in lung volume which commonly accompanies brief apnoea may lead to deflation-mediated irritant stimulation and prolongation of the apnoea in pre-term infants.

Head's Paradoxical Reflex
This consists of a sudden deep inspiration which is produced by artificially inflating the lungs. It is present in term newborn infants and is thought to be mediated by irritant receptors (Cross *et al.*, 1960). This reflex may be involved in sighing, which occurs frequently in newborn infants, and is thought to be important in maintaining lung compliance through recruitment of atelectatic alveoli (Thach & Taeusch, 1976).

Intercostalphrenic Reflex
Inward movement of the rib cage on inspiration (i.e. distortion) stimulates intercostal muscle spindles leading to phrenic inhibition via a supraspinal reflex, the intercostal-phrenic inhibitory reflex (Knill & Bryan, 1976). The lack of intercostal muscle activity in REM sleep leads to marked inspiratory distortion of the rib cage in newborn infants, particularly pre-term infants in whom the rib cage is relatively compliant. The infant is thus in effect 'sucking in ribs instead of fresh air', and the intercostal phrenic reflex may variably shorten inspiratory time, further reducing effective ventilation (Bryan & Bryan, 1978).

Sleep State
Respiration in REM in infants is more rapid and irregular than in QS, and is accompanied by inhibition of postural muscle tone and monosynaptic reflexes (Finer *et al.*, 1976; Bolton & Herman, 1974). Vagal reflexes are also less active in REM, though other reflexes, notably the intercostal-phrenic, may be important in determining respiratory rate (Bryan & Bryan, 1978).

The immediate ventilatory response to hypoxia in REM is comparable to that in QS, though is less well maintained, perhaps because of hypoxic central depression (Rigatto *et al.*, 1979). CO_2 responsiveness in REM sleep is controversial though probably comparable to that in QS.

Perhaps the most important feature of respiration in REM sleep is the influence of phasic activity in the behavioural control system, leading to the irregular breathing pattern. This pattern whilst influenced by chemoreceptor and reflex activity is not dependent on such input for its continuation.

In contrast in QS respiration is more regular, is under tight metabolic and reflex control, and is dependent on these outputs. This important distinction is illustrated by infants with absent or grossly depressed chemoreceptor function, who maintain adequate ventilation in REM or when awake, but hypoventilate severely in QS (Fleming *et al.*, 1980).

Stability of Control

The control of respiration has many features similar to those of control problems encountered in engineering. Application of engineering control theory might therefore give useful information on the dynamic control of respiration. Such an approach has been used by Hathorn (1978). Using the techniques of Fourier transforms and autocorrelation he showed evidence of oscillations in tidal volume and respiratory frequency in newborn infants. The oscillations had a period of 8–12 seconds and were more marked in REM than in QS. Cross-variance analysis showed that the oscillations in tidal volume were usually out of phase with those in frequency, particularly in QS, i.e. an increase in tidal volume was usually accompanied by a fall in frequency and vice versa. In pre-term infants with periodic breathing this relationship was reversed, the oscillations being in phase with each other. The separate and out-of-phase control of tidal volume and frequency may be a feature of mature respiratory control. Certainly, in engineering terms such a system would lead to precise and stable control, with little tendency to 'hunt' widely around the desired setting. Lack of this precise out-of-phase control may be a factor in the irregularity of respiration in REM sleep and pre-term infants.

Hathorn noted that the pattern of breathing often changes after a sigh and therefore only studied stable inter-sigh periods. As discussed earlier, sighing is common in newborn infants, is vagally mediated, and is important in minimising atelectasis. Whatever the cause of a sigh, one predictable effect of a sudden deep breath will be to produce a transient increase in CO_2 washout, with a transient fall in $PaCO_2$, and perhaps a rise in PaO_2. If the oscillations of tidal volume and frequency are related to feed-back from chemoreceptors then such oscillations should be accentuated by the transient changes in blood gases in the period after a sigh (see Chapter 7). Examination of the respiratory pattern after sighs in term infants in QS shows a pattern which changes with increasing post-natal age (Fleming & Levine, 1982). In the first 24–48 hours there is a fall in both tidal volume and frequency with a gradual return to baseline values over 25–40 seconds. Beyond two to three days of age sighs are commonly followed by a clear pattern of damped oscillation, particularly involving tidal volume. The tidal volume oscillates with a period of 12-20 seconds whilst the oscillations in frequency have a period of 30-40 seconds and are of smaller amplitude. Ventilation (the product of tidal volume and frequency) oscillates with a pattern and period similar to that of tidal volume. These observations support the concept of separate control of tidal volume and frequency. The period of the oscillation in tidal volume is consistent with the effect of a feed-back loop with a lag of 6-10 seconds which is consistent with the response time of the peripheral chemoreceptors (Hathorn, 1978). The lack of such oscillations in the first 24-48 hours of life may be due to decreased chemoreceptor activity, or to increased right to left shunting (see sections on peripheral chemoreceptors). Such oscillations are consistently found after sighs in infants up to about three months of age. Beyond that age the oscillations, though present, are more highly damped, being extinguished after only one or two cycles (Figure 8.2). This changing pattern of dynamic response to spontaneous disturbances with increasing age suggests that the two aspects of the control system, sensitivity and stability, may develop at different times. The infant under three months of age responds promptly to disturbances, but the response is underdamped, leading to an overshoot and 'hunting' about the desired value of ventilation. Beyond this age the response is more nearly critically damped, with a rapid response and minimal overshoot. Thus stability of control is not established until around 3-4 months of age.

Detailed study of the relationship between the development of sensitivity and of stability may lead to a better understanding of the dynamic control of respiration in the undisturbed individual.

Clinical Aspects

A detailed account of clinical aspects of respiratory control in the newborn infant is beyond the scope of this review, but two problems of particular importance will be discussed: apnoea of prematurity, and sudden infant death syndrome.

Apnoea of Prematurity

As described in the previous sections many of the important protective feed-back systems are not fully developed prior to term. Hypoxia, particularly if accompanied by hypothermia, leads to marked respiratory depression in pre-term infants, and may be accompanied by a failure to respond to CO_2 accumulation (Rigatto *et al.*, 1975b). The inhibitory

Figure 8.2: Five Sections of Respiratory Recordings, from One Infant, at Ages from 23 h to 151 days, to Show the Change, with Increasing Age, in the Pattern of Respiration after a Sigh. At 23 h the sigh is followed by a brief apnoea and then a slow rise in ventilation back to baseline values. By 47 h there is a clear single oscillation in tidal volume before returning to baseline values. The following two recordings show a clear pattern of damped oscillation in tidal volume and frequency, which, at 94 days, takes the form of overt periodic breathing. In addition to the oscillations becoming more obvious, the period of the oscillations in the tidal volume decreases, from 25 s at 47 h, to 14 s at 38 days, and 12.5 s at 94 days. By 151 days the oscillations have again become highly damped, and are extinguished after only two cycles. The period remains short, at 12.5 s. Thus by 151 days the infant's respiratory control system shows a prompt and highly damped (but *not* critically damped) response to spontaneous disturbance.

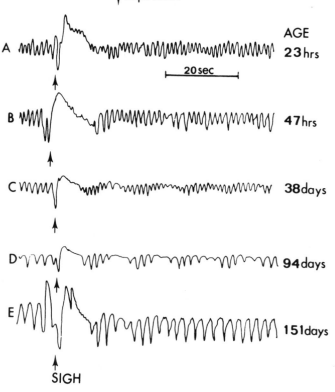

BABY 26 : Impedance Pneumograms

Hering–Breuer inflation reflex is strongly present, and stimulation of irritant receptors may lead to paradoxical respiratory inhibition (Fleming *et al.*, 1978). The pre-term infant spends most sleeping time in REM, when inspiratory rib cage distortion leads to increased diaphragmatic work (Fleming *et al.*, 1979) and shortened inspiratory time (via intercostal phrenic reflex). An increase in work is poorly tolerated by the immature diaphragm which may fatigue and further compromise effective gas exchange (Bryan & Bryan, 1978).

There is a marked inhibitory bias in the control of respiration in the pre-term infant. Indeed there are so many potentially inhibitory factors acting that perhaps the surprising feature is absence of apnoea in many infants! Effective management of the apnoea of prematurity depends on an understanding of the physiology and appropriate preventive measures, e.g. maintenance of normoxia and a neutral thermal environment; maintenance of FRC and reduction of inspiratory distortion by continuous positive airway pressure; increasing sensitivity to CO_2 by the use of caffeine or theophylline. Pathophysiology and management of apnoea of prematurity have been well reviewed by Kattwinkel (1977).

The Sudden Infant Death Syndrome (SIDS)

The possible role of abnormal control of respiration in causation of SIDS has been extensively investigated. There is evidence of pre-existing hypoxia in many such infants who die suddenly and unexpectedly, and family studies have shown decreased responsiveness to both hypoxia and CO_2 in parents and siblings of some such children, though other studies have failed to demonstrate such an association. Sleep apnoea and recurrent airway obstruction during sleep have also been variably reported in such families. Reviewing over 140 papers on SIDS published in six years (1974–79) Valdes-Dapena (1980) states 'it would now seem that idiopathic protracted apnoea, especially during sleep, may be part of the pathogenetic mechanism in *some* instances of S.I.D.S.'. Clearly this is an area in which speculation has far outstripped scientific observation. Hopefully the results of a number of large prospective studies on respiratory pattern in newborn infants being carried out now on both sides of the Atlantic will help to answer some of the questions.

References

Avery, M.E., Chernick, V., Dutton, R.E. & Permutt, S. (1963), 'Ventilatory Response to Inspired CO_2 in Infants and Adults', *J. Appl. Physiol.*, *18*, 895-903

Barcroft, J. & Barron, D.H. (1936) 'The Genesis of Respiratory Movements in the Foetus of the Sheep', *J. Physiol.*, *88*, 56-61

Barcroft, J. & Barron, D.H. (1937) 'Movements in the Mid-foetal Life in the Sheep Embryo', *J. Physiol.*, *91*, 329-51

Biscoe, T.J., Purves, M.J. & Sampson, S.R. (1969) 'Types of Nervous Activity which may be Recorded from the Carotid Sinus Nerve in the Sheep Foetus', *J. Physiol.*, *202*, 1-23

Boddy, K. & Dawes, G.S. (1975) 'Foetal Breathing', *Br. Med. Bull.*, *31*, 3-7

Boddy, K., Dawes, G.S., Fisher, R., Pinter, S. & Robinson, J.S. (1974) 'Foetal Respiratory Movements, Electrocortical and Cardiovascular Responses to Hypoxaemia and Hypercarbia in Sheep', *J. Physiol.*, *243*, 599-618

Boddy, K. & Mantell, C.D. (1972) 'Observations of Foetal Breathing Movements Transmitted through Maternal Abdominal Wall', *Lancet*, *ii*, 1219-20

Bolton, D.P.G. & Herman, S. (1974) 'Ventilation and Sleep State in the Newborn', *J. Physiol.*, *240*, 67-77

Bowes, G., Adamson, T.M., Ritchie, B.C., Dowling, Margaret, Wilkinson, M.H. & Maloney, J.E. (1981) 'Development of Patterns of Respiratory Activity in Unanaesthetised Fetal Sheep *in utero*', *J. Appl. Physiol.: Resp. Environ. Exercise Physiol.*, *50*, 693-700

Brady, J.P. & Ceruti, E. (1966) 'Chemoreceptor Reflexes in the Newborn Infant', *J. Physiol.*, *184*, 631

Bryan, A.C. & Bryan, M.H. (1978) 'Control of Respiration in the Newborn', *Clin. Perinatol.*, *5*, 269-81

Bryan, M.H., Bryan, A.C., Hagan, R. & Gulston, G. (1976) 'CO_2 Response in Sleep State in Infants', *Clin. Res.*, *24*, 698A

Bureau, M.A., Begin, R. & Berthiaume, Y. (1979) 'Central Chemical Regulation of Respiration in Term Newborn', *J. Appl. Physiol.: Resp. Environ. Exercise Physiol.*, *47*, 1212-7

Bystrzycka, E., Nail, B.S. & Purves, M.J. (1975) 'Central and Peripheral Neural Respiratory Activity in the Mature Sheep Foetus and Newborn Lamb', *Resp. Physiol.*, *25*, 199-215

Ceruti, E. (1966) 'Chemoreceptor Reflexes in the Newborn Infant. Effect of Cooling on the Response to Hypoxia', *Paediatrics*, *37*, 37-55

Chapman, R.L.K., Dawes, G.S., Rurak, D.W. & Wilds, R.L. (1977) 'Foetal Breathing and Nerve Stimulation *in utero*', *J. Physiol.*, *272*, 13-14P

Chapman, R.L.K., Dawes, G.S., Rurak, D.W. & Wilds, P.L. (1980) 'Breathing Movements in Fetal Lambs and the Effect of Hypercapnia', *J. Physiol.*, *302*, 19-29

Cross, K.W., Klaus, M., Tooley, W.H. & Weisser, K. (1960) 'The Response of the New-born Baby to Inflation of the Lungs', *J. Physiol.*, *151*, 551-65

Davi, M., Sankaran, K., MacCallum, M., Cates, D. & Rigatto, H. (1979) 'Effect of Sleep State on Chest Distortion and on the Ventilatory Response to CO_2 in Neonates', *Pediatr. Res.*, *13*, 982-6

Davis, M.E. & Potter, E.L. (1946) 'Intra-uterine Respiration in the Human Foetus', *J. Am. Med. Assoc.*, *131*, 1194-201

Dawes, G.S., Duncan, S., Lewis, B.V., Merlet, C.L., Owen-Thomas, J.B. & Reeves, R.T. (1969) 'Hypoxaemia and Aortic Chemoreceptor Function in Foetal Lambs', *J. Physiol.*, *201*, 105-16

Dawes, G.S., Fox, H.E., Leduc, B.M., Liggins, G.C. & Richards, R.T. (1972)

'Respiratory Movements and Rapid Eye Movement Sleep in the Foetal Lamb', *J. Physiol.*, *220*, 119-43

Downing, S.W. & Lee, J.C. (1975) 'Laryngeal Chemosensitivity', *Pediatrics*, *55*, 640

Fagenholz, S.A., O'Connell, K. & Shannon, D.C. (1976) 'Chemoreceptor Function and Sleep State in Apnea', *Pediatrics*, *58*, 31-6

Finer, N.N., Abroms, I.F. & Taeusch, H.W. (1976) 'Ventilation and Sleep States in Newborn Infants', *J. Pediatr.*, *89*, 100-8

Fleming, P.J., Bryan, A.C. & Bryan, M.H. (1978) 'Functional Immaturity of Pulmonary Irritant Receptors and Apnea in Newborn Preterm Infants', *Pediatrics*, *61*, 515-8

Fleming, P.J., Cade, D., Bryan, M.H. & Bryan, A.C. (1980) 'Congenital Central Hypoventilation and Sleep State', *Pediatrics*, *66*, 425-8

Fleming, P.J. & Levine, M.R. (1982) 'Observations on Respiratory Stability in the Newborn Infant', *J. Physiol.*, *326*, 58-9P

Fleming, P.J., Levine, M.R., Goncalves, A. & Purves, M.J. (1982) 'The Effects of Trigeminal Stimulation when Using a Facemask to Record Respiration in Newborn Infants', *Paediatric Research*, *16*, 1031-4

Fleming, P.J., Muller, M., Bryan, M.H. & Bryan, A.C. (1979) 'The Effect of Abdominal Loading on Ribcage Distortion in Premature Infants', *Pediatrics*, *64*, 425-8

Folgering, H., Ponte, J., Sadig, T. (1982) 'Adrenergic Mechanisms and Chemoreception in the Carotid Body of the Cat and Rabbit', *J. Physiol.*, *325*, 1-21

Franz, Ivan D. III, Adler, S.M., Thach, B.T. & Taeusch, H.W. (1976) 'Maturational Effects on Respiratory Responses to CO_2 in Premature Infants', *J. Appl. Physiol.*, *41*, 41-5

Frantz, I.D. III, Adler, S.M., Abroms, J.F. & Thach, B.T. (1976) 'Respiratory Response to Airway Occlusion in Infants: Sleep State and Maturation', *J. Appl. Physiol.*, *41*, 434-638

Haddad, G.C., Leistner, H.L., Epstein, R.A., Epstein, M.A.F., Grodin, W.K. & Mellins, R.B. (1980) 'CO_2 Induced Changes in Ventilation and Ventilatory Pattern in Normal Sleeping Infants', *J. Appl. Physiol.: Resp. Environ. Exercise Physiol.*, *48*, 684-8

Haddad, G.C. & Mellins, R.B. (1977) 'The Role of Airway Receptors in the Control of Respiration in Infants, A Review', *J. Pediat.*, *91*, 281-6

Harned, H.S., Jr., Griffin, C.A. III, Berryhill, J.S. Jr., Mackinney, L.G. & Sygioka, K. (1967) 'Role of Carotid Chemoreceptors in the Initiation of Effective Breathing of the Lamb at Term', *Pediatrics*, *39*, 329-36

Hathorn, M.K.S. (1978) 'Analysis of Periodic Changes in Ventilation in Newborn Infants', *J. Physiol.*, *285*, 85-99

Hodson, W.A., Fenner, A., Brumley, G., Cerick, V. & Avery, M.E. (1968) 'Cerebrospinal Fluid and Blood Acid-Base Relationships in Fetal and Neonatal Lambs and Pregnant Ewes', *Resp. Physiol.*, *4*, 322-32

Jansen, A.H., Purves, M.J. & Tan, E.D. (1980) 'The Role of the Sympathetic Nerves in the Activation of the Carotid Body Chemoreceptors', *J. Develop. Physiol.*, *2*, 305-21

Jones, Colin T. (1980) 'Circulating Catecholamines in the Fetus, their Origin, Actions and Significance', in H. Parvez & S. Parvez (eds), *Biogenic Amines in Development*, Elsevier/North-Holland Biomed. Press, Amsterdam

Jost, R.G., Quilligan, E.J., Yeh, S.Y & Anderson, G.G. (1972) 'Intra-Uterine Electroencephlogram of the Sheep Fetus', *Am. J. Obstet. Gynecol.*, *114*, 535-9

Kattwinkel, J. (1977) 'Neonatal Apnea: Pathogenesis and Therapy', *J. Pediatr.*, *90*, 342-7

296   *Control of Respiration in the Fetus and Newborn*

Klaus, M., Fanaroff, A. & Martin, R.J. (1979) in M. Klaus & A. Fanaroff (eds), *Respiratory Problems In Care of the High-risk Neonate*, Saunders, London, pp. 173-204

Kirkpatrick, S.M.L., Olinsky, A., Bryan, M.H. & Bryan, A.C. (1976) 'Effect of Premature Delivery on the Maturation of the Hering-Breuer Inspiratory Inhibitory Reflex in Human Infants', *J. Pediatr.*, *88*, 1010-4

Knill, R. & Bryan, A.C. (1976) 'An Intercostal-phrenic Inhibitory Reflex in Human Newborn Infants', *J. Appl. Physiol.*, *40*, 352-6

Maloney, J.E., Adamson, T.M., Brodecky, V., Cranage, S., Lambert, T.F. & Ritchie, B.C. (1975) 'Diaphragmatic Activity and Lung Liquid Flow in the Unanaesthetised Fetal Sheep', *J. Appl. Physiol.*, *39*, 423-8

Mantell, C.D. (1976) 'Breathing Movements in the Human Foetus', *Am. J. Obstet. Gynecol.*, *125*, 550-3

Martin, C.B., Murata, Y., Petrie, R.H. & Parer, J.T. (1974) 'Respiratory Movements in Fetal Rhesus Monkeys', *Am. J. Gynecol.*, *119*, 939-48

Meschia, G., Cotter, J.R., Breathnach, C.S. & Barron, D.H. (1965) 'The Haemoglobin, Oxygen, Carbon Dioxide and Hydrogen Ion Concentrations in the Umbilical Bloods of Sheep and Goats as Sampled by Indwelling Catheters', *Q. J. Exp. Physiol.*, *50*, 185-95

Mitchell, R.A. & Berger, A.J. (1975) 'Neural Regulation of Respiration', *Am. Rev. Resp. Dis.*, *111*, 206-24

Moss, Immanuala, R. & Scarpelli, E.M. (1979) 'Generation and Regulation of Breathing *in utero*: Fetal CO_2 Response Test', *J. Appl. Physiol.: Resp. Environ. Exercise Physiol.*, *47*, 527-31

Parmelee, A.H. Jr., Wenner, W.H., Akiyama, Y., Schultz, M. & Stern, E. (1967) 'Sleep States in Premature Infants', *Dev. Med. Child Neurol.*, *9*, 70-7

Persson, Hans E. (1973) 'Development of Somatosensory Cortical Function: an Electrophysiological Study in Pre-natal Sheep', *Acta Physiol. Scand.*, *394*, (Suppl.), 1-64

Phillipson, E.A. (1977) 'Regulation of Breathing During Sleep', *Am. Rev. Resp. Dis.*, *115*, (Suppl.) 217-24

Phillipson, E.A. (1978) 'Respiratory Adaptations in Sleep', *Am. Rev. Physiol.*, *40*, 133-56

Phillipson, E.A., McClean, P.A., Sullivan, C.E. & Zamel, N. (1978) 'Interaction of Metabolic and Behavioural Respiratory Control during Hypercapnia and Speech', *Am. Rev. Resp. Dis.*, *117*, 903-9

Plum, F. (1970) 'Neurological Integration of Behavioural and Metabolic Control of Breathing', in R. Porter (ed.), *The Control of Breathing. Hering-Breuer Centenary Symposium*, Churchill, London, pp. 159-75

Ponte, J. & Purves, M.J. (1973) 'Types of Afferent Nervous Activity which may be Measured in the Vagus Nerve of the Sheep Foetus', *J. Physiol.*, *229*, 51-76

Purves, M.J. (1966) 'The Effect of a Single Breath of Oxygen in Respiration in the Newborn Lamb', *Resp. Physiol.*, *1*, 297-307

Reuss, M.L. & Rudolph, A.M. (1980) 'Distribution and Recirculation of Umbilical and Systemic Venous Blood Flow in Fetal Lambs during Hypoxia', *J. Develop. Physiol.*, *2*, 71-84

Rigatto, A. (1979) 'A Critical Analysis of the Development of Peripheral and Central Respiratory Chemosensitivity during the Neonatal Period', in C. von Euler (ed.), *Central Nervous Control Mechanisms in Breathing*, Pergamon, Oxford, pp. 137-48

Rigatto, H., Brady, J.P. & Torre, Verduzco R. del (1975a) 'Chemoreceptor Reflexes in Preterm Infants. I. The Effect of Gestational Age on the Ventilatory Response to Inhalation of 100% and 15% Oxygen', *Pediatrics*, *55*, 604-13

Rigatto, H., Torre, Verduzco R. de la & Cates, D.B. (1975b) 'Effects of O_2 on the Ventilatory Response to CO_2 in Preterm Infants', *J. Appl. Physiol., 39*, 896-9

Rigatto, A., Kalapesi, Z., Leahy, F.N., MacCallum, M. & Cates, D. (1979) 'Ventilatory Response to 100% and 15% Oxygen During Wakefulness and Sleep in Preterm Infants', *Pediatr. Res., 13*, 504

Scarpelli, E.M., Condorelli, S. & Cosmi, E.V. (1977) 'Cutaneous Stimulation and Generation of Breathing in the Fetus', *Pediatr. Res., 11*, 24-8

Snyder, F.F. & Rosenfeld, M. (1937) 'Direct Observation of Intrauterine Respiratory Movements of the Foetus and the Role of CO_2 and O_2 in their Regulation', *Am. J. Physiol., 119*, 153-65

Steele, A.G. & Windle, W.F. (1939) 'Some Correlations Between Respiratory Movements and Blood Gases in Cat Foetuses', *J. Physiol., 94*, 531-8

Sullivan, C.E., Murphy, E., Kozar, L.F. & Phillipson, E.A. (1978) 'Waking and Ventilatory Responses to Laryngeal Stimulation in Sleeping Dogs', *J. Appl. Physiol.: Resp. Environ. Exercise Physiol., 45*, 681-9

Sullivan, C.E., Murphy, E., Kozar, K.F. & Phillipson, E.A. (1979) 'Ventilatory Responses to CO_2 and Lung Inflation in Tonic Versus Phasic REM Sleep', *J. Appl. Physiol.: Resp. Environ. Exercise Physiol., 47*, 1304-10

Tenney, S.M. & Ou, L.C. (1979) 'Ventilatory Response of Decorticate and Decerebrate Cats to Hypoxia and CO_2', *Resp. Physiol., 29*, 81

Thach, B.T. & Taeusch, H.W. (1976) 'Sighing in Newborn Human Infants: Role of Inflation-Augmenting Reflex', *J. Appl. Physiol., 41*, 502-7

Thach, B.T., Frantz, I.D., Adler, S.M. & Taeusch, H.W. (1978) 'Maturation of Reflexes Influencing Inspiratory Duration in Human Infants', *J. Appl. Physiol.: Resp. Environ. Exercise Physiol., 45*, 203-11

Towell, M.E. & Salvador, H.S. (1974) 'Intrauterine Asphyxia and Respiratory Movements on the Foetal Goat', *Am. J. Obstet. Gynecol., 118*, 1124-31

Valdes-Dapena, M.A. (1980) 'Sudden Infant Death Syndrome: A Review of the Medical Literature 1974-1979', *Pediatrics, 66*, 597-614

Wennergren, G. & Wennergren, M. (1980) 'Respiratory Effects Elicited in Newborn Animals via the Central Chemoreceptors', *Acta Physiol. Scand., 108*, 309-11

9 INITIATION AND CONTROL OF VENTILATORY ADAPTATION TO CHRONIC HYPOXIA OF HIGH ALTITUDE

Sukhamay Lahiri, P. Barnard and R. Zhang

Introduction

Life begins with hypoxia *in utero* (Barcroft, 1925). The newborn after birth is exposed to an environment of higher PO_2 — the higher the altitude of birth the less drastic is the change of environment. Thus, the first acclimatisation in life is from low to high oxygen pressure. Anatomical and physiological development in this neonatal period may make the responses and adaptation different from those observed in the adult man and animals. The ventilatory response to changes in the inspired PO_2 in the newborn is small which is in part due to small changes in the stimulus level in the arterial blood due to arterio-venous shunts. The fact that ventilation is depressed even by mild hypoxia in the newborn infant after a brief initial stimulation (e.g. Brady & Ceruti, 1966) shows that the hypoxic stimulus does reach the cells and tissues. High altitude newborn infants exhibit the same lack of oxygen effect on ventilation. A sustained response to hypoxia develops within a few weeks after birth both at sea level and high altitudes (see Lahiri, 1977).

Most literature on ventilatory adaptation to chronic hypoxia deals with adult humans and some with animals like the goat, pony, sheep, cat, dog, rabbit and rat which are exposed to hypoxia by design. A good part also deals with native human residents and animals of high altitude. Thus the time frame of hypoxic exposure leading to initiation and maintenance of response or loss of response is a critical consideration.

The intensity of hypoxic stimulus in another critical element. An acute PIO_2 decrease from 150 Torr to 100 Torr ($P_AO_2 \simeq 60$ Torr) does not cause a measurable increase in ventilation in the steady state, but a prolonged exposure to the same exposure increases it gradually by a process known as ventilatory acclimatisation. What initiates this ventilatory response still remains a question, but the phenomenon also suggests that a ventilatory response with an initial alkalosis may not be what initiates the acclimatisation. It is well known though that carotid body chemosensory input at this level of PaO_2 is considerably increased (see Lahiri & Gelfand, 1981). It is as if some factors prevented the input being translated into neural respiratory output during mild acute hypoxia. In the course of time this barrier is removed and the central mechanism becomes responsive to the chemosensory input. The sensory input itself may be responsible for the time-dependent change.

The next range of hypoxic stimulus to which a ventilatory response occurs promptly by chemoreflex mechanism followed by a sustained adaptation lies between PIO_2 of 100 Torr and 70 Torr, the latter corresponding approximately to permanent residence at the highest altitude. Beyond that altitude physical and mental deterioration dominate, although ventilatory acclimatisation may continue. However, ventilatory acclimatisation decreases with further increase of altitude (see later).

In principle, the time-dependent increase in ventilation during hypoxic exposure could result from a change in the stimulus level or from an increase in the response of the sensory receptor and the integrative respiratory system or from a change in the mechanical property of the system, or from a combination of all these changes. In developing this chapter we will consider these questions as well as the physical environmental factors which provide the stimulus in the first place. Finally, the inevitable question of respiratory adaptation to chronic hypoxia in the context of human diseases will be discussed.

Physical Environmental Factors

Barometric Pressure

With the increase of altitude the proportion of constituent gases in the air on the surface of the earth does not change significantly but the air pressure decreases as the length of air column and its weight decrease. As a consequence, partial pressure of the constituent gases and the number of molecules in a given volume of expanded ambient air are reduced. The actual barometric pressure at a given altitude varies according to the geographical location, particularly latitude, due to the gravitational effect and to air temperature (see Pugh, 1964; West *et al.*, 1983). The barometric pressure in the mountains is higher than that predicted by the International Civil Aviation Organisation. For example, the predicted pressure on the Everest summit is 236 Torr, and the actual measured value is 253 Torr (West *et al.*, 1983). Thus, the inspired PO_2 is significantly higher than expected. Nonetheless, because of the physical reason a given volume of alveolar ventilation delivers a smaller number of oxygen molecules to the lung at higher altitudes, and the partial pressure of oxygen pressure in the alveoli (PAO_2) falls. The physiological system and process responsible for delivering oxygen to the lungs and blood react to remedy and restore the situation.

Temperature, Humidity and Alveolar Gas Pressures

With the increase of altitude there is a graded decrease in temperature and absolute humidity. This raises the question of equilibration of temperature and humidity between the inhaled and alveolar air. The potential that temperature may not equilibrate particularly at a very high altitude during high rate of ventilation seems real. This cold dry air at high altitude causes severe cough reflex during exercise indicating failure of temperature equilibration in the airway. If the alveolar air temperature is lower than $38\,^{\circ}C$, the water vapour pressure will be lower allowing a rise in the inspired and alveolar PO_2. This effect would be greater at higher altitudes. Dejours (1982) has discussed this aspect recently.

Elements of Control System in Hypoxic Adaptation

The receptors, sensory input, and the output of the central nervous system to the respiratory apparatus constitute the controller, and the respiratory apparatus is the plant which carries out the ventilatory function of the lungs. The mechanical events elicit several reflex actions initiated by the mechanoreceptors; the chemical events in the arterial blood following ventilation determine peripheral, central and other chemoreceptor activity which contribute to the control of ventilation. In the chronic state, blood and tissue stimulus levels also change. These include acid-base, blood O_2 content and O_2 equilibrium, cellular levels and release of excitatory or inhibitory substances such as dopamine, and circulating hormones of relevance. Our plan is to review the litera-ture in terms of controller and plant of the breathing system. Most literature is focused on the chemical aspects of the controller.

The subject has been reviewed quite frequently in recent years in various forms. The reader is referred to these articles: Berger, Mitchell & Severinghaus (1977); Cerretelli (1980); Dempsey & Forster (1982); Frisancho (1981); Lahiri (1977); Lenfant & Sullivan (1971); Mitchell (1966); which also provide the viewpoint of the reviewers.

Figure 9.1: Relationships Between Barometric Pressure and Alveolar O_2 and CO_2 Concentrations and Pressures. The important point of this illustration is that while the alveolar gas pressures decrease with the barometric pressure, alveolar gas concentrations do not. (See text for an explanation.)

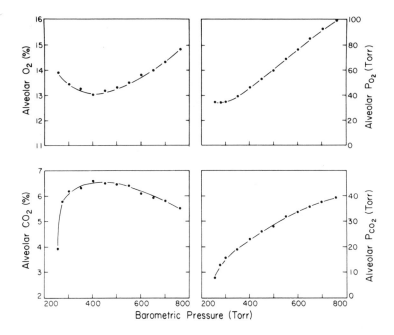

Effect of Barometric Pressure on Alveolar Gas Concentrations and Pressures

Because of the physical effect of barometric pressure on gas volume and pressure and the response of the control system to the reduced oxygen pressure and gas density, alveolar gas pressures and concentrations may not show a clear correspondence. This can be appreciated from Figure 9.1 constructed from the compiled data on the acclimatised human subjects (see Lahiri, 1968; West *et al.*, 1982). With the fall of barometric pressure over the range of 760 Torr to 400 Torr alveolar PO_2 decreases practically linearly from 100 Torr to 45 Torr, but the alveolar O_2 concentration decreases from 14.8 per cent to 13.3 per cent only. The alveolar PCO_2 decreases from 39.5 Torr to 25 Torr but the alveolar CO_2 concentration increases from 5.5 per cent to 6.8 per cent. The data below 400 Torr barometric pressure are sparse, but they seem to show a turning point with a severe decrease in $PACO_2$ and a lesser decrease in PAO_2. Alveolar O_2 concentration increases whereas CO_2 concentration decreases. These data illustrate the point that except at extreme altitude variation of O_2 and CO_2 concentrations lie within 1 per cent despite large decreases in partial pressure.

Because alveolar O_2 and CO_2 concentrations at altitude changed little compared with their partial pressures, it may appear that the controlled variables are the alveolar gas concentrations. However, the idea can be dismissed by considering the effect of low inspired O_2 concentrations on alveolar CO_2 concentration at sea level (760 Torr) – alveolar CO_2 concentration decreases almost linearly with the decrease of PIO_2 because of hyperventilation. On the other hand, raised barometric pressure from 1 atmosphere to 30 atmosphere decreases alveolar CO_2 concentrations exponentially from 5.6 per cent to 0.17 per cent because of the mechanical compression of air.

Characteristics of Ventilatory Responses

The resting ventilation response to hypoxia is well described by the Rahn-Fenn-Otis diagram relating alveolar PO_2 to alveolar PCO_2 (Rahn & Fenn, 1962; Rahn & Otis, 1949). This description takes the metabolic effect of hypoxia into account as well so that $PACO_2$ and PAO_2 values are independent of any change in the resting metabolic rates. The responses of alveolar gases to acute and short-term chronic hypoxia are shown in Figure 9.2. The altitude diagonals at 5400m and 8840m for R.Q. values of 0.8 and 1.0 are also shown. During steady-state acute hypoxia to the level of 60 Torr there is no decrease in $PACO_2$, indicating no change in the alveolar ventilation; below a PAO_2 of 60 Torr $PACO_2$ decreases because of the increase in ventilation. Prolongation of hypoxia at a given altitude, say 5400m, increases ventilation over several days, decreasing $PACO_2$ and increasing PAO_2 along the R.Q. diagonal until the acclimatised point is reached in a few days. The rate of acclimatisation varies between species (see Dempsey & Forster, 1982). The lower curve in Figure 9.2 describes the alveolar values found in the newly acclimatised subjects. Further acclimatisation may occur by raising R.Q., for example from 0.8 to 1.0, which increases both PAO_2 and $PACO_2$. This type of acclimatisation is smaller at higher altitudes as can be seen from the diagonals at 8840m (near Everest summit). Return to sea level normoxia does not return the steady-state ventilation to the initial sea level value immediately — the effect of chronic hypoxia endures for a few days (Lahiri, 1972).

Figure 9.2: Effects of Acute and Chronic Exposure to Low Barometric Pressures on Alveolar PO_2 and PCO_2 Relationship. R.Q. diagonals of 0.8 and 1.0 at 5400 m and 8840 m are also shown to help understand the effect of R.Q. increase on the alveolar gases at any altitude. The solid line for acute and the broken line for chronic hypoxia, representing the compiled data, are drawn.

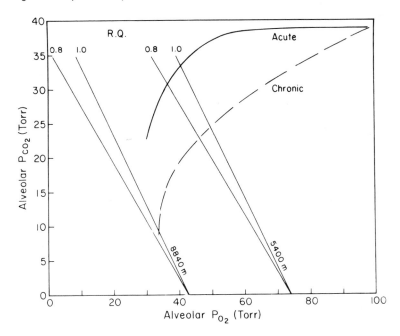

The mechanism of the ventilatory response to acute hypoxia is more defined than that to chronic hypoxia. The acute response is initiated by the chemosensory discharge from carotid and aortic bodies. The discharge rates of these chemoreceptors increase hyperbolically with the decrease of PaO_2, and the ventilatory response approximately corresponds to the chemosensory input (Lahiri & DeLaney, 1975). Denervation of these receptors abolishes the normal ventilatory response (see Dempsey & Forster, 1982) often leading to depression particularly observed in anaesthetised animals (Lahiri, 1976). Inhalation of 100 per cent O_2 increases ventilation above the normoxic level (Miller & Tenney, 1975; Nishino & Lahiri, 1981). Severe hypoxia in the chemodenervated animals may lead to tachypnoea mediated by the changes in brain monoamine metabolism (Gautier & Bonora, 1980). Hypoxia also increases cerebral blood flow (see Dempsey & Forster, 1982) which may decrease the central stimulus level. Also, because of the normally high aerobic metabolic rate, brain anerobic metabolism may increase with hypoxia, causing central metabolic acidosis (Siesjo, 1978). Thus, hypoxia causes a variety of effects which may influence ventilation during acute and hence chronic hypoxia.

An increase in ventilation during acute hypoxia also opposes its full effect by decreasing PCO_2-H^+ in arterial blood and cerebral fluid. Artificially maintained PCO_2 during acute hypoxia thus increases ventilation further (see Cunningham, 1973). This is partly because of the fact that the arterial chemosensory response to hypoxia is augmented by PCO_2-H^+ (see Lahiri, Smatresk & Mulligan, 1983) and partly because a prevention of respiratory alkalosis also prevents its inhibitory effect on chemosensory mechanism (see Leusen, 1972). At a given inspired PO_2, however, increases in ventilation would increase PaO_2 at an artificially maintained $PaCO_2$, contributing to a decrease in the stimulus level of the arterial chemoreceptors.

The foregoing observations on the responses to acute hypoxia no doubt set the stage for responses to long-term hypoxia but its mechanism is still much debated. The thesis in its simplest form consists of restoration of normal acidity in the receptor environment by metabolic means following the initial respiratory alkalosis. The most striking support was provided by Severinghaus, Mitchell, Richardson & Singer (1963) who reported that the ventilatory acclimatisation to chronic hypoxia coincided with restoration of CSF pH from alkaline to nearly normal values which in turn restored the chemoreceptor stimulation so that the full effect of the peripheral chemosensory input was accomplished. Acute normoxia decreased peripheral chemoreceptor input, reducing ventilation and consequently raising $PaCO_2$. This increased PCO_2 led to an increased acidity of CSF in equilibrium with the central chemoreceptor environment which kept ventilation higher than the original normoxic value. Also, they argued that the pH compensation through cerebral [HCO_3^-] reduction resulted in a reduction of PCO_2 threshold and an increased slope of ventilatory response to CO_2. Crawford & Severinghaus (1978) partly substantiated these claims but conceded that CSF pH was not restored to sea level normal value during acclimatisation to high altitude. This latter reinvestigation was prompted by refutation of their claim of CSF pH homeostasis by several groups of investigators (Dempsey, Forster & doPico, 1974;

Lahiri & Milledge, 1967; Weiskopf, Gabel & Fencl, 1976), the staunchest of which were Dempsey, Forster and their colleagues. They reported that arterial and CSF pH increased along with ventilation during acclimatisation, and concluded that these pH increases resulted from the increased ventilation and damped ventilatory acclimatisation rather than caused it. This conclusion also refutes any claim which is tied to restoration of CSF pH.

The assumption that CSF pH is in equilibrium with the ISF pH of the chemosensory elements in the CNS during ventilatory acclimatisation has been questioned. Fencl, Gabel & Wolfe (1979) investigated the question in goats by ventriculocisternal perfusion technique measuring transependymal flux before and five days after exposure to 446 Torr. They concluded that unlike chronic normoxia there was a large $[HCO_3^-]$ gradient from CSF to cerebral ISF due to an increased lactic acid concentration in the ISF, and that this acidity stimulated the intracranial chemoreceptors and was responsible for ventilatory acclimatisation. An increased lactate and other metabolic acid level of the brain during chronic hypoxia (Weyne, Van Leuven & Leusen, 1977) is consistent with this claim. However, ventilatory acclimatisation requires that the cerebral acidity should increase with time, establishing a positive correlation between cerebral ISF acidity and ventilation. However, it has been reported that cerebral acid metabolite level is maximal during the first two hours of acute hypoxia and then declined with time or changed little (Lahiri, Edelman, Cherniack & Fishman, 1981; Musch, Bateman, Smith & Dempsey, 1980; Weyne *et al.*, 1977). Thus, cerebral acidity may provide a greater ventilatory stimulus at a given time during hypoxia but does not explain the time-dependent increase in ventilation. Therefore other factors which are elaborated during chronic hypoxia and stimulate breathing directly or sensitise the system to the known stimuli seem to be involved.

Role of Peripheral Chemoreceptors

Although peripheral chemoreceptors are known to be responsible for the ventilatory response to acute hypoxia, they are not *a priori* essential for ventilatory acclimatisation. However, the balance of evidence suggests that they are critical. Practically all the evidence comes from animal experiments which involved exposure to chronic hypoxia before and after denervation of carotid and aortic bodies (Bouverot, Candas & Libert, 1973; Forster *et al.*, 1976; Lahiri *et al.*, 1981). Denervation is followed in a few hours or days by an increase in $PaCO_2$ and a decrease in PaO_2. When these animals were exposed to chronic hypoxia either they did not increase their ventilation or showed a slow, attenuated response. Most of the denervated animals did not show a normal increase in ventilation, although cerebral metabolic acidosis occurred. This central metabolic acidosis is, however, expected to decline with time because of normal acid-base homeostasis. In any case, a lack of normal ventilatory response to central metabolic acidosis during hypoxia may suggest that either the peripheral chemosensory input is needed for the full central acidosis effect on ventilation, or the concomitant hypoxia suppresses the acidosis effect.

The mechanism of the effect of peripheral chemoreceptors inducing ventilatory acclimatisation despite a decrease in the arterial stimulus level (PaO_2 and $PaCO_2$-H^+) in the normal subjects is unclear. There are two ways to go about it. One, a time-dependent increase in the arterial chemoreceptor response to arterial chemical stimuli; or, a central time-dependent increase in the effect of the chemosensory input. The hypoxic chemoreflex effect in man has been found to be unchanged in short-term hypoxia (see Lahiri, 1977) but ventilation data alone do not speak for the arterial chemoreceptor activity. Measurement of carotid chemoreceptor activity from a few chemosensory fibres of otherwise intact carotid sinus nerve in the chronically hypoxic cats was found to be comparable to those of normoxic cats (Smatresk, Lahiri, Pokorski & Barnard, 1981). Thus, ventilatory acclimatisation is not accounted for by an increased chemosensory input from the peripheral chemoreceptors.

The observation that acute hypoxia of less severe intensity than 60 Torr does not elicit an immediate steady-state response but leads to the time-dependent acclimatisation strongly suggests a slow 'trophic' effect of chemosensory input. The mechanism of this effect which perhaps is the basis of ventilatory acclimatisation is not clear. By the same token it is also not clear why acute denervation of the arterial chemoreceptor in the normoxic man and animals results in a decrease in ventilation (see Dempsey & Forster, 1982; Lahiri, 1977; Unger & Bouverot, 1980) if there is no stimulatory effect of acute mild hypoxia. As suggested before, it is possible that other extra-chemoreceptor effects of hypoxia prevent the response. With time these inhibitory effects are overcome giving rise to ventilatory acclimatisation.

Central Neural Process

Suprapontine facilitatory and inhibitory responses have been implicated as the determinants of the ventilatory acclimatisation to chronic hypoxia (see Tenney & Ou, 1977), but how hypoxia exerts the differential effects has not been suggested. There are also observations indicating that the state of CNS excitability is increased, but its mechanism and how it relates to acclimatisation are unclear (see Dempsey & Forster, 1982). Whether these changes are dependent on the intact chemosensory input from the arterial chemoreceptor is not established.

Chemical changes in the CNS induced by hypoxia have been a focus of investigation. The two amino acids, tyrosine and tryptophan, which have played critical roles along with oxygen in biological evolution (Dickerson, 1978), provide the source materials for catecholamines and indolamines, respectively, which are known transmitters in the CNS and elsewhere (see Eldridge & Millhorn, 1981). The rate-limiting enzymes tyrosine hydroxylase and tryptophan hydroxylase have high K_m values for oxygen and are affected by PO_2. However, the responses are different in different tissues like carotid bodies and central nervous system (Davis, 1975; Hanbauer, Karoum, Hellstrom & Lahiri, 1981). Metabolism and levels of these amines in the CNS have been linked with respiratory control because of their presence in the respiratory neurones and because of their effects on respiratory neurones when applied externally. However, the findings are contradictory, and a real understanding of the function of central neurotransmitters in the respiratory control has not evolved (see Dempsey & Foster, 1982; Eldridge & Millhorn, 1981). Because of the lack of understanding of the basic mechanisms, a role of these neurotransmitters in ventilatory acclimatisation cannot be clearly formulated at this point in time. Also, if the observation that ventilatory acclimatisation to chronic hypoxia does not occur in the chemodenervated animals were to be taken into account one is confronted at once with the question whether these neurotransmitters are the mediators of the slow, time-dependent chemosensory effect of hypoxia. These questions have not been answered.

Role of Carbon Dioxide

Since hyocapnia is an inevitable consequence of ventilatory acclimatisation to chronic hypoxia, attempts have been made to ascertain its role in the process. If hypocapnia was a key to the ventilatory acclimatisation, then prevention of hypocapnia by CO_2 inhalation should prevent ventilatory acclimatisation by raising ventilation to the acclimatised level promptly in the acute state, and should not decrease ventilatory PCO_2 threshold and increase CO_2 sensitivity. Eger *et al.* (1968) compared in man the effect of hypoxia with and without controlled CO_2 on the intercept of the CO_2 response curves, and found that a significant descrease in the CO_2 intercept (threshold) occurred during hypoxia even without hypocapnia. Hypocapnia alone also decreased the intercept. Cruz *et al.* (1980) reported a similar study in man but their subjects did not show the initial hypocapnia and therefore the question of its prevention was not addressed in the study. However, hypercapnia superimposed on hypoxia raised the acute response, minimising the time-dependent effect of hypoxia, but a decrease in the intercept of the CO_2 response curve was not prevented, although this fact escaped the authors' attention. Also, it should be appreciated that a sufficient hypercapnia could cause a time-dependent decrease rather than an increase in ventilation during hypoxia (Carson & Lahiri, 1979). These results confirm that CO_2 has a role but do not exclude the role of chronic hypoxia in acclimatisation. Similarly, metabolic manipulation of plasma $[HCO_3^-]$ during chronic hypoxia also did not prevent acclimatisation (see Dempsey & Forster, 1982).

Resemblance Between the Ventilatory Effects of Hypoxia and Metabolic Acidosis

Regulation of extracellular (blood and interstitial fluid) and intracellular $[H^+]$ which influence the chemosensitive elements in the carotid and aortic bodies and in the central nervous system during acute and chronic hypoxia, resembles that in metabolic acidosis in some respects. The literature on acid-base regulation and ventilation is relevant and is worthwhile to review briefly.

There are two diametrically opposite views on the mechanism of how metabolic acidosis stimulates the arterial chemoreceptors, and the increased chemosensory input to the central respiratory neurones increases ventilation which in turn causes central alkalosis. This alkalosis opposes the full effect of the peripheral chemosensory input (Mitchell, 1966). During prolonged chronic acidosis the central alkalosis is compensated by decreasing $[HCO_3^-]$, and ventilation increases further in a time-dependent fashion. This effect of metabolic acidosis is analogous to the effect of hypoxia. Accordingly, peripheral chemoreceptors are responsible for the initial response, and the central chemosensors make the secondary adjustment.

The proponents of the other view (Pappenheimer, Fencl, and their colleagues) discount any significant contribution of peripheral chemoreceptors and claim that the effect is totally central due to an increase in $[H^+]$ at the chemosensitive site away from arterial blood. A decrease in plasma $[HCO_3^-]$ is followed by a decrease in the central ISF $[HCO_3^-]$ causing ISF acidosis. This acidity is expected to stimulate ventilation and lower PCO_2, making the bulk CSF alkaline which would oppose the full effect of the ISF acidosis. With time CSF $[HCO_3^-]$ would decrease with a consequent further increase in ventilation. However, the required data showing the time-dependent changes are not clearly established.

The peripheral chemoreceptor theory relies heavily on the observations that mild metabolic acidosis does not stimulate ventilation and cause CSF acid-base changes in chemodenervated animals (see Dempsey & Forster, 1982; Mitchell, 1966). Also, it is well known that metabolic acidosis stimulates carotid chemoreceptor activity (Eyzaguirre & Lewin, 1961; Hornbein & Roos, 1963) which is likely to stimulate ventilation. On the other hand, there is evidence that a more severe metabolic acidosis stimulates breathing in the chemodenervated animals (Javaheri, Herrera & Kazemi, 1979). This is consistent with the proposal of Pappenheimer, Fencl and colleagues (Pappenheimer *et al.*, 1965; Fencl, Miller & Pappenheimer, 1966). They showed that in the steady-state CSF $[H^+]$-ventilation relationship in chronic metabolic acidosis and alkalosis in the unanaesthetised goats is the same as that during steady-state CO_2 inhalation. Their observations, however, cannot exclude a contribution of peripheral arterial chemoreceptors.

The question really hangs on the relative contribution of the two chemosensitive systems under the conditions of the experimental study. These have been attempted by separately controlling the environment of the two systems (Dutton, Hodson, Davies & Chernick, 1967; see Dempsey & Forster, 1982). Despite a quantitative disagreement, it is clear that both systems contribute to the regulation of ventilation during metabolic and respiratory acid-base perturbation. Another way to examine the question is quantitatively to relate ventilation to arterial chemoreceptor activity at two different levels of $[H^+]$ at a constant steady-state $PaCO_2$. The rationale is that if the effect of H^+ was due entirely to the stimulation of arterial chemoreceptors then ventilation would be a single function of the chemoreceptor activity. It has been found that at two different levels of $[H^+]$ ventilation is not defined by a single function of carotid chemoreceptor activity indicating that H^+ has an additional stimulatory effect on ventilation (see Lahiri & Gelfand, 1981; Pokorski & Lahiri, 1982).

Thus, acute metabolic acidosis and acute hypoxia appear to have similar effects on ventilation and CSF pH. With acidosis, however, arterial $[H^+]$ is expected to decrease and CSF $[H^+]$ to increase during chronic state. The general characteristics of the ventilatory response are a time-dependent increase with a consequent decrease in arterial and cerebral PCO_2 (see Dempsey & Forster, 1982, for a detailed account and references). The time-dependent increase in ventilation indicates that a sustained increased sensory input from the arterial chemoreceptors may produce a time-dependent central effect simultaneously with, but separate from the central $[H^+]$ effect described earlier.

Long-term Chronic Hypoxia

Adult high-altitude residents in the Andes and Himalayas have been found to show a blunted ventilatory response to hypoxia and stimulation during hypoxia (see Cerretelli, 1980; Dempsey & Forster, 1982; Lahiri, 1977). Rocky Mountain highlanders also show a similar but a less blunted response presumably because of their lower altitude residence than the other two populations. Ventilation of these high-altitude residents, however, is greater than that in the sea level natives during acute hypoxia of the same intensity. As a result their alveolar PO_2–PCO_2 points lie in the intermediate range between curves for acute and short-term chronic hypoxia of the sojourners (see Figure 9.2). The question whether the blunted response is due to a lowered PaO_2 threshold for ventilation (Tenney & Ou, 1977) has been considered, but both threshold and sensitivity may contribute to the phenomenon.

The characteristic in the adult high-altitude residents is not reversed by a short-term exposure to sea-level atmosphere (Lahiri *et al.*, 1969; Sorensen & Severinghaus, 1968). However, the characteristic is acquired. Lahiri, Brody, Motoyama & Velasquez (1978) reported that infants and young children did not show the characteristic blunted response but developed it in later life. Young individuals who were relocated at sea level did not show it either. On the other hand, those adults who changed residence from low to high altitude developed the characteristic only partially in a few years. The acquired attenuation of ventilatory responses to hypoxia was also found in the residents of European ancestry in the Rockies of Colorado (Dempsey *et al.*, 1972; Weil *et al.*, 1971).

The observation that sea-level residents who are hypoxic before and after birth because of congenital cyanotic heart disease show a blunted ventilatory response to transient hypoxia before the corrective surgery but not after suggest the labile nature of the characteristic (Blesa, Lahiri, Rashkind & Fishman, 1977; Edelman *et al.*, 1970; see Lahiri, 1977).

A loss of ventilatory response to hypoxia during long-term acclimatisation has not been found systematically in all the species studied. It is possible that exposure might not have been long enough to induce the characteristics. Bisgard & Vogel (1971) described a decreased ventilatory response to hypoxia in the calf at high altitude but they were not devoid of carotid chemoreceptor drive because chemodenervation increased their $PaCO_2$ dramatically. Tenney & Ou (1977) reported a loss of ventilatory drive due to hypoxia in the cats maintained at 5,500 m for several weeks. More recently Barnard *et al.* (1981) confirmed these observations for some but not all cats chronically exposed to PIO_2 of 70 Torr.

The mechanism of the loss of hypoxic ventilatory response in man and animal is not well understood. The stimulus levels in the arterial blood and brain in terms of PO_2 and H^+ are no less than in short-term acclimatisation (Blayo, Marc-Vergnes & Pocidalo, 1973; Lahiri & Milledge, 1967). Indeed these stimulus levels are slightly higher in the highlanders presumably because of relative hypoventilation. Red blood cell concentrations are high in the highlanders and also in the acclimatised lowlanders (see Lahiri, 1977). Further, a large range of haematocrit has little effect on carotid chemoreceptor responses to hypoxia (see Lahiri *et al.*, 1983). Some highlanders, particularly South Americans, who show a totally depressed ventilatory response to hypoxia and hyperventilate on breathing 100 per cent O_2 also show the highest haematocrits. It seems, however, that these high haematocrits are the result rather than the cause of hypoventilation.

Because the hypoxic chemoreflex originates in the arterial chemoreceptors, carotid and aortic bodies are the natural target for investigation. This intuitive suspicion is strengthened by the observations that carotid bodies are visibly enlarged in the chronically hypoxic man at high altitude (see Eyzaguirre, Fitzgerald, Lahiri & Zapata, 1983; Fitzgerald & Lahiri, 1983). In order to investigate a role of the arterial chemoreceptor, attempts have been made to develop an animal model. Chronically hypoxic rats have been found to show carotid body hypertrophy and hyperplasia of Type I and Type II cells in several weeks (Laidler & Kay, 1978; see McDonald, 1981) and increases in their carotid body catecholamine concentrations (Hanbauer *et al.*, 1981). Barer, Edwards & Jolly (1976) reported that the ventilatory response to hypoxia in such rats was blunted but Oslon & Dempsey (1978) did not confirm this finding. Dopamine may decrease chemosensory responses to hypoxia in rats (Zapata & Zuazo, 1980) as it does in other species (Nishino & Lahiri, 1981; Lahiri, Nishino, Mokashi & Mulligan, 1980; Welsh, Heistad & Abboud, 1978) but the actual chemosensory responses in chronically hypoxic rats are not known. On the other hand, carotid chemosensory responses in chronically hypoxic cats, which showed a blunted ventilatory response to hypoxia were comparable to those in the normal cats (Barnard *et al.*, 1981; Smatresk *et al.*, 1981). The carotid bodies of these cats were not particularly enlarged (unpublished observations), but their catecholamine levels are not known. Thus, we do not have all the necessary data in one species to come to a conclusion regarding catecholamines and chemosensory function of carotid and aortic bodies and ventilatory response to chronic hypoxia.

Response and Adaptation of the Respiratory Plant

A change in the respiratory mechanics during acute and chronic hypoxia could change ventilation despite a constant neural respiratory output. This mechanical change is expected to alter mechanoreceptor input from the lungs, respiratory airways and the thoracic cage. As a consequence the V_T/T_I relationship would change. Few studies so far have not produced unequivocal results (Brody *et al.*, 1977; Cruz, 1973; Gautier, Milic-Emili, Miserochi & Siafakas, 1980; Tenney, Rahn, Stroud & Mithoffer, 1953). A low air density no doubt contributes to increased ventilatory volumes but it is not a determinant of ventilatory acclimatisation.

Ventilation During Exercise

Metabolic rate is a critical determinant of pulmonary ventilation. Under normoxic condition, an increased metabolic rate during exercise is followed by a proportional increase in alveolar ventilation so that the initial alveolar and arterial PCO_2 remain constant. This relationship deviates as the level of exercise approaches maximum level leading to hyperventilation and lowering alveolar PCO_2 because of an increased blood lactate level, increased body temperature, and some other factors. This principle also applies during hypoxia – ventilation increases in proportion to the metabolic rate so as to maintain the initial effect of hypoxia on arterial PCO_2 constant. This means that the initial effect of hypoxia on ventilation during rest increases by a multiplied factor of metabolic rate during exercise. All these results follow from the metabolic hyperbola describing the relationship between metabolic rate, ventilation and alveolar PCO_2 : $\dot{V}_E = K\ \dot{V}CO_2 / PACO_2$. At higher metabolic rates the gain of the hyperbola (V_E vs. $PACO_2$) is increased, and the effect of hypoxia, acute or chronic is augmented. Because the relationship between hypoxia and ventilation is also hyperbolic the interaction between increasing hypoxia and exercise is expected to result in a large increase in ventilation at extreme altitudes.

The situation is further complicated by the fact that arterial PO_2 and O_2 saturation decline with exercise at a given low inspired PO_2 providing further stimulus to breathing. This happens during both acute and chronic hypoxia (see Lahiri, 1977). An additional unidentified stimulus to breathing during hypoxic exercise is present during exercise which lowers arterial PCO_2 giving rise to alkalosis even when arterial PO_2 is maintained constant (Asmussen & Nielsen, 1958; see Cunningham, 1973; Masson & Lahiri, 1974). The mechanism of all these effects of hypoxia on ventilation during exercise is not fully understood but a great number of these effects are dependent on the arterial chemoreceptor responses and chemosensory drive. Cunningham with his colleagues contributed to the understanding of these mechanisms (see Chapter 6).

Because exercise augments the effect of hypoxia on ventilation it has been utilised as a tool to investigate ventilatory responses to short- and long-term chronic hypoxia. This application has confirmed unequivocally that the adult natives of high altitude who show blunted resting ventilatory responses to hypoxia also ventilate less at a given hypoxic exercise (see Lahiri, 1977). This effect, however, follows from the relationship of metabolic hyperbola but the physiological mechanisms are not clear.

Ventilation at Extreme Altitudes

Measurement of ventilation and alveolar gases at extreme altitudes have been made on recent Himalayan expeditions. Sea-level man has climbed Mount Everest without the use of oxygen supplements (see Dejours, 1982; Messner, 1979; West & Wagner, 1980). Acclimatisation at lower altitudes allowed them to tolerate PIO_2 of 43 Torr at a barometric pressure of 253 Torr on the Everest summit. Recently a subject, who reached the summit breathing supplemental oxygen but collected alveolar samples breathing ambient air showed an average $PACO_2$ of 8 Torr and PAO_2 of 35 Torr (West *et al.*, 1983). This means that his near-resting alveolar ventilation was about 40 litres min^{-1} (BTPS) and total ventilation of about 52 litres min^{-1} (BTPS). Doubling his metabolic rate would increase his ventilation at least to 104 litres min^{-1} approaching the sea level maximum value. Because of lower air density and possible changes in the pulmonary mechanics the subject may be able to sustain this or higher levels of ventilation, but the oxygen cost of breathing may be prohibitive at these extreme altitudes.

In any case it is not realistic for a subject to maintain ventilation at the rate of 100 litres min^{-1} for a long time and perform physical work effectively. Thus ventilation itself could become a limiting factor for exercise at great altitudes. Pugh *et al.* (1964) reported such a limitation at 7450 m, and West (1983) and Schoene (1983) (also personal communication) confirmed this limitation by lowering PIO_2 acutely in acclimatised subjects at 6300 m. This predicts that the maximum oxygen consumption near the Everest summit would not be very much greater than double the resting value.

It seems that the stimulus to resting ventilation at these great altitudes is derived entirely from the peripheral chemoreceptors because the estimated arterial pH calculated from the measured buffer base and alveolar PCO_2 is about 7.60. This alkalosis is likely to persist during exercise because blood lactic acid level is diminished during chronic hypoxia (see Cerretelli, 1980; Maher, Jones & Hartley, 1974; Lahiri & Samaja, unpublished observation). It is interesting to note that the climbers with supplemental oxygen breathing also experience dyspnoea and exhaustion during their slow climb suggesting that an extra-peripheral chemoreceptor drive exists.

Oxygen Delivery, Tissue PO_2 and Respiratory Drive

Pulmonary ventilation is only one of the several physiological responses to oxygen demand and CO_2 production maintaining tissue homeostasis of PO_2 and $[H^+]$. Other major responses to hypoxia involve O_2 transport and metabolism. These include cardiac output and blood flow to the organs, red blood cells carrying O_2 and position of the O_2 dissociation curves. All these responses vary directly with the intensity of hypoxia. Acute hypoxia increases cardiac output at a given oxygen uptake but the maximum oxygen uptake is diminished. During chronic hypoxia cardiac output returns to the normal normoxic relationship but the maximal value remains low. Cerebral blood flow follows the same pattern (see Lahiri, 1977). These responses do not parallel ventilatory responses, and therefore are not causally connected.

The organic phosphates in red blood cells increase during a few hours to days of respiratory alkalosis and deoxygenation. These phosphates, the major one of which is 2,3-disphosphoglycerate in man, bind with haemoglobin, lowering affinity for oxygen. However, alkalosis *per se* increases the affinity, and the net result is a practically unchanged *in vivo* oxygen–haemoglobin dissociation curve (Lahiri, 1975; see Lenfant & Sullivan, 1971; Winslow *et al.*, 1981).

In the course of a few weeks red cell mass increases raising blood haematocrit and O_2 content at a given PaO_2. Thus, in a steady acclimatised state both PaO_2 and O_2 carrying capacity increase so that the arterial O_2 content is practically restored to the initial normoxic sea level value (see Lahiri, 1977). However, the maintained high haemacrit must mean that the sensor PO_2 which initiated the response is still lower than the sea-level value.

There has been much discussion regarding the physiological meaning of these changes in terms of arterial O_2 delivery to the tissues, work of the heart, performance capacity, etc., but a ventilatory effect is not apparent. One can visualise the effects on two counts: an increased blood buffering of CO_2-H^+ because of an increased haemoglobin and hence a decrease of arterial CO_2-H^+ stimulus change and a possible reduction of arterial chemoreceptor response to hypoxia. An increased haematocrit and O_2 content of arterial blood is expected to decrease aortic chemoreceptor responses to hypoxia whereas a less intense effect may occur on carotid body chemoreceptors (see Lahiri *et al.*, 1983), but experimental data showing the effect of an altered O_2 transport on these chemoreceptors in chronic hypoxia are not available.

Sleep and Ventilatory Acclimatisation

Quiet sleep results in hypoventilation raising $PaCO_2$ and lowering PaO_2 which in turn increase stimulus to peripheral as well as central chemoreceptors (see Phillipson, 1978). Ventilatory response to mild hypercapnia is depressed, but there is some uncertainty regarding hypoxic sensitivity (Phillipson, 1978; see Cherniack, 1981). However, it seems that the response to CO_2 is more depressed than that to hypoxia. Sino-aortic denervation in the cat further increases the depressant effect of sleep on ventilation (Guazzi & Freis, 1969).

Figure 9.3: Stethograph Tracing of Breathing at 14,100 Feet (Pike's Peak). Tracing 2, evening of arrival. Natural periodic breathing was enhanced after six forced breaths. Tracing 3, four days later. Oxygen breathing decreased the intensity of periodic breathing but it continued to cycle with longer period. Withdrawal of O_2 was followed by the reappearance of the enhanced periodic breathing.

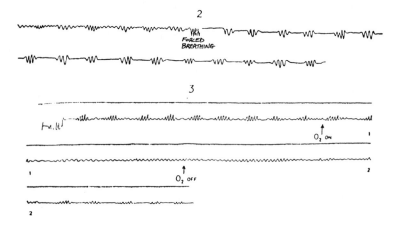

Source: Douglas, Haldane, Henderson & Schneider, 1913.

Slight period oscillations in breathing in normal subjects during sleep at sea level is discernible. It is, however, striking at high altitudes. Periodic breathing with apnoea during drowsiness and quiet sleep is a common feature in the sojourners at high altitude (Mosso, 1898; Douglas *et al.*, 1913). A similar phenomenon can be elicited during sleep in some subjects with acute hypoxia (Berssenbrugge, Dempsey, Skatrud & Iber, 1980; see Cherniack, 1981). This periodic breathing pattern is eliminated by removing acute as well as chronic hypoxia. An example of the effect of breathing 100 per cent O_2 during periodic breathing on Pike's Peak (14,100 ft) is shown in Figure 9.3. Clearly, periodic breathing with apnoea was replaced by more regular breathing but periodicity with a longer cycle was still visible. We studied the effect of chronic hypoxia on ventilation during sleep at 5400 m in the newcomers and the Sherpa high-altitude residents in the Himalayas (Lahiri, 1983). The newcomers were acclimatised for about eight weeks before the study. All of them showed periodic breathing with apnoea during quiet sleep, an example of which is shown in Figure 9.4. O_2 breathing prolonged the apnoea for the first few breaths which were followed by the appearance of more regular breathing but the cycling with a longer periodic duration continued as illustrated in Figure 9.3. It is interesting to note that the trough and crest of the SaO_2 per cent waves were lower and higher respectively than the mean SaO_2 per cent during waking state. Thus the subjects are exposed to varying PaO_2 levels during sleep rather than a steady low PaO_2 only.

Figure 9.4: Effect of Raised PIO_2 on Periodic Breathing with Apnoea During Quiet Sleep in a Sojourner after 62 days at High Altitude. The experimental record was made at 5400 m ($PB \cong 400$ Torr, Everest Base Camp). Tracings from the top are Respitrace records of tracheal air flow (V_T, inspiration up), arterial O_2, saturation (SaO_2 %) by earoimetry, time marker in seconds, electrooculogram, electrocardiogram, time marker in seconds. Oxygen breathing and the associated increase in SaO_2 % were followed simultaneously by an immediate lengthening of apnoeic period and decreased tidal volumes. Subsequently, after a delay, apnoea disappeared but the cycling continued at a suppressed level with a longer period. (This experimental record shows only part of the O_2 effect.)

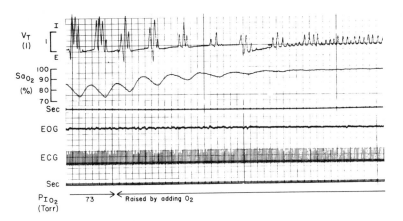

Inhalation of CO_2 during periodic breathing eliminates the apnoeic period, making the breathing pattern more uniform, but the periodicity does not disappear altogether unless the sleep stage also changed or the subject is awakened. The effect of CO_2 inhalation is in part due to the fact that it dampens blood gas oscillations (Band *et al.*, 1978) and hence the chemosensory discharges and the corresponding ventilation.

A dominant role of arterial chemoreceptors with high nonlinear gain in periodic breathing is indicated by the foregoing results. Cherniack (1981) emphasised this possible role of arterial chemoreceptors and small O_2 stores. Accordingly subjects with low or blunted hypoxic ventilatory drive would show least periodic breathing during sleep. The Sherpas in our study (Lahiri, 1983) who showed a blunted ventilatory response to hypoxia also failed to show periodic breathing, an observation consistent with the model. According to this model, a further increase of hypoxic stimulus should augment periodic breathing further, but in fact breathing is stimulated and apnoeic period disappears, although breathing continues to cycle. The role of sleep in this model is to increase the central threshold of response to the CO_2 stimulus, predisposing the subject to apnoea when the existing stimulus goes below this threshold. The effects of sleep are complex. A decrease in the controller gain, that is CO_2 sensitivity, is not expected to contribute to the instability but certainly lessens the stimulus to breathe. On the other hand, hypoventilation during quiet sleep lowers PaO_2 as well, augmenting the peripheral chemosensory input which would lower the threshold for the central CO_2 response. The overall gain of the controller may increase due to increasing hypoxia, and ventilation may become more dependent on the peripheral chemosensory input. The augmented rate sensitivity of the arterial chemoreceptors to the combination of PO_2-PCO_2 stimuli would make the chemosensory input oscillate more sharply with the oscillations of blood gases, and hence the breathing pattern. The balance of all these effects is to induce periodic breathing. This model requires that the system remains sensitive to peripheral chemoreceptor input while the central CO_2-H^+ related response is diminished during sleep or any other condition inducing periodic breathing. It does not, however, explain the residual periodicity when oxygen is breathed. A study of breathing pattern in sleep in subjects without intact arterial chemoreceptors is of great interest.

The time-course of development of periodic breathing at high altitude is not well documented but it has been suggested that its intensity and the frequency of apnoea are diminished with time.

However, it was found to be sustained at 5400 m exceeding eight weeks (Lahiri, 1983). It is possible that at higher altitude the periodic breathing lasts longer and the higher the altitude the more intense is the periodic breathing with apnoea during sleep.

The subjective experience of the sojourners at high altitude is that of a disturbed sleep. How the disturbed sleep pattern contributes to respiratory changes at high altitude requires further study.

Ventilatory Acclimatisation in Fetuses and Newborns

Breathing movements in the fetuses *in utero* are inhibited rather than stimulated by hypoxaemia. Although breathing after birth can be stimulated by carotid body chemosensory input (see Lahiri, 1977), it is not essential for the onset and sustained breathing (Jansen, Ioffe, Russell & Cherniack, 1981). Newborn infants shortly after birth show little chemoreflex response to hypoxia both at sea level and at high altitude (Lahiri *et al.*, 1978). The mechanism of this insensitivity is not clear but one common factor is that the fetuses at any altitude are hypoxaemic. A few days after birth the full-term infants begin to show sustained response to hypoxaemia at both low and high altitudes. This pattern of ventilatory response to hypoxaemia appears to correspond to carotid body chemosensory response to hypoxaemia in fetal and newborn lambs (Biscoe, Purves & Sampson, 1969).

Maladaptation to Chronic Hypoxia

There are numerous diseases which lead to hypoxaemia but there are a few which stem from hypoxaemia. The most striking ones are known as acute and chronic mountain sickness. Acute mountain sickness develops after rapidly moving to high altitude. The usual symptoms are excruciating headache, nausea, vomiting, loss of appetite, lack of sound sleep, dyspnoea, pulmonary congestion and sometimes pulmonary oedema. Within hours and days acclimatisation occurs and the symptoms are very much reduced and may disappear, although their recurrence is not unusual. Acute mountain sickness lasts longer at higher altitudes. At altitudes above 5500 m the symptoms persist at a reduced level.

Chronic mountain sickness, as described by Monge & Monge (1966), is found among the high-altitude residents of the Andes. The characteristic features are extremely high haematocrit (e.g. 80 per cent at 4540 m), cyanosis, engorgement of blood vessels, pulmonary hypertension, hypertrophy of the right heart and many other circulatory symptoms. All these subjects show very poor ventilatory response to hypoxia, and hypoventilate at high altitude and are unable to perform physical exercise. They conspicuously hyperventilate on breathing 100 per cent O_2 (see Lahiri, 1977). These diseases are not common at sea level but pathophysiology of the disease provides the foundation of medical practice of the diseases due to hypoxia.

Conclusion

The basic mechanisms of the time-dependent increase in ventilation during short-term chronic hypoxia and blunting of the peripheral chemo-reflex response during long-term chronic hypoxia remain unresolved. A sustained peripheral chemoreceptor input seems to be critical for ventilatory acclimatisation. It is curious that cerebral acidosis during hypoxia without intact carotid sinus nerves does not stimulate ventilation. Blunting of the chemoreflex response is not due to blunting of the arterial chemoreceptor response. On the other hand, efferent inhibitory effect on the carotid body chemosensory discharge is increased during chronic hypoxia presumably associated with an increased dopamine concentration. The structure and function of carotid body are altered but their role in the ventilatory adaptation is only partially understood. A role of central nervous mechanisms for both ventilatory acclimatisation and blunting of chemoreflex response also emerges but it seems to be dependent on the interaction with the peripheral chemosensory input. Understanding of the role of central neurotransmitters in the process is critical. Since the central chemo-receptors are not clearly identified, their stimulus-response cannot be ascertained quantitatively making the study indirect and uncertain. Use of ventilation as a measure of central chemoreceptor activity offers the conflict of logic because ventilation itself is the subject of investigation.

The central nervous system with numerous synaptic connections has the inherent adaptive property. A sustained hypoxic stimulus can change this adaptive property of the central nervous system, altering the input-output relationship. Altered characteristics of respiratory control in chronic hypoxia are in part due to this central plasticity. A study of the adaptive system controlling respiration, as revealed by the state of chronic hypoxia, is multifaceted and requires a multidisciplinary approach.

Acknowledgement

We are grateful to Ms J. Callaghan, Ms K. Hart, Ms B. Pauly and Mr A. Mokashi for their help. Supported in part by the NIH Grants HL-19737 and HL-26533.

References

Asmussen, E. & Nielsen, M. (1958) 'Pulmonary Ventilation and Effect of Oxygen Breathing in Heavy Exercise', *Acta Physiol. Scand.*, *43*, 365-78

Band, D.M., McClelland, M., Phillips, D.L., Saunders, K.B. & Wolff, C.B. (1978) 'Sensitivity of the Carotid Body to Within-breath Changes in Arterial PCO_2', *J. Appl. Physiol.: Resp. Environ. Exercise Physiol.*, *45*, 768-77

Barcroft, J. (1925) *The Respiratory Function of the Blood. Part I. Lessons from High Altitude*, The University Press, Cambridge

Barer, G.R., Edwards, C.W. & Jolly, A.J. (1976) 'Changes in the Carotid Body and Ventilatory Response to Hypoxia in Chronically Hypoxic Rats', *Clin. Sci. Molec. Med.*, *50*, 311-3

Barnard, P.A., Smatresk, N., Pokorski, M., Zhang, R. & Lahiri, S. (1981) 'Carotid Chemoreceptor and Ventilatory Responses in Chronically Hypoxic Cats', *Physiologist*, *24*, 114 (abstract)

Berger, A.J., Mitchell, R.A. & Severinghaus, J.W. (1977) 'Regulation of Respiration', *N. Engl. J. Med.*, *297*, 92-7

Berssenbrugge, A., Dempsey, J., Skatrud, J. & Iber, C. (1980) 'The Effect of Chronic Hypoxia on Breathing Pattern in Wakefulness and Sleep', *Physiologist*, *23*, 727 (abstract)

Biscoe, T.J., Purves, M.J. & Sampson, S.R. (1969) 'The Types of Nervous Activity which may be Recorded from the Carotid Sinus Nerve in the Sheep Fetus', *J. Physiol.*, *202*, 1-23

Bisgard, G.E. & Vogel, J.H.K. (1971) 'Hypoventilation and Pulmonary Hypertension in Calves after Carotid Body Excision', *J. Appl. Physiol.*, *31*, 431-7

Blayo, M.C., Marc-Vergnes, J.P. & Pocidalo, J.J. (1973) 'pH, PCO_2, and PO_2 of Cisternal Cerebrospinal Fluid in High Altitude Natives', *Resp. Physiol.*, *19*, 298-331

Blesa, M.K., Lahiri, S., Rashkind, W.J. & Fishman, A.P. (1977) 'Normalisation of the Blunted Ventilatory Response to Acute Hypoxia in Congenital Cyanotic Heart Disease', *N. Engl. J. Med.*, *296*, 237-41

Bouverot, P., Candas, V. & Libert, P. (1973) 'Role of Arterial Chemoreceptors in Ventilatory Adaptation to Hypoxia in Awake Dogs and Rabbits', *Resp. Physiol.*, *17*, 209-19

Brady, J.P. & Ceruti, E. (1966) 'Chemoreceptor Reflexes in the Newborn Infant: Effects of Varying Degrees of Hypoxia on Heart Rate and Ventilation in a Warm Environment', *J. Physiol.*, *184*, 631-45

Brody, J.S., Lahiri, S., Simpser, M., Motoyama, E.K. & Velasquez, T. (1977) 'Lung Elasticity and Airdynamics in Peruvian Natives to High Altitude', *J. Appl. Physiol.: Resp. Environ. Exercise Physiol.*, *42*, 245-51

Carson, G. & Lahiri, S. (1979) 'Effects of Hypercapnia on Ventilatory Acclimatization to Hypoxia', *Fed. Proc.*, *38*, 1390 (abstract)

Cerretelli, P. (1980) in J.B. West (ed.), *Pulmonary Gas Exchange*, Academic Press, New York, pp. 98-148

Cherniack, N.S. (1981) 'Respiratory Dyschythmias during Sleep', *N. Engl. J. Med.*, *35*, 325-30

Crawford, R.D. & Severinghaus, J.W. (1978) 'CSF pH and Ventilatory Acclimatization to Altitude', *J. Appl. Physiol.: Resp. Environ. Exercise Physiol.*, *45*, 275-83

Cruz, J.C. (1973) 'Mechanics of Breathing in High Altitude and Sea Level Subjects', *Resp. Physiol.*, *17*, 146-61

Cruz, J.C., Reeves, J.T., Grover, R.F., Maher, J.T., McCullough, R.C., Cymerman, A. & Denniston, J.C. (1980) 'Ventilatory Acclimatization to High Altitude is Prevented by CO₂ Breathing', *Respiration*, *39*, 121-30

Cunningham, D.J.C. (1973) 'The Control System Regulating Breathing in Man', *Q. Rev. Biophys.*, *6*, 433-83

Davis, J.N. (1975) 'Adaptation of Brain Monoamine Synthesis to Hypoxia in the Rat', *J. Appl. Physiol.*, *39*, 215-20

Dejours, P. (1982) in C.R. Taylor, K. Johansen & L. Bolis (eds), *A Comparison to Animal Physiology*, Cambridge University Press, New York (in press)

Dempsey, J.A. & Forster, H.V. (1982) 'Mediation of Ventilatory Adaptations', *Physiol. Rev.*, *62*, 262-346

Dempsey, J.A., Forster, H.V., Birnbaum, M.L., Reddan, W.G., Thoden, J. & Grover, R.F. (1972) 'Control of Exercise Hypercapnia under Varying Durations of Exposure to Moderate Hypoxia', *Resp. Physiol.*, *16*, 213-31

Dempsey, J.A., Forster, H.V. & doPico, G.A. (1974) 'Ventilatory Acclimatization to Moderate Hypoxemia in Man', *J. Clin. Invest.*, *53*, 1091-100

Dickerson, R.E. (1978) 'Chemical Evolution and the Origin', *Sci. Am.*, *239*, 70-86

Douglas, C.G., Haldane, J.S., Henderson, Y. & Schneider, E.C. (1913) 'Physiological Observations Made on Pike's Peak, Colorado, with Special Reference to Adaptations to Low Barometric Pressures', *Phil. Trans. R. Soc.*, Series B, *203*, 185-381

Dutton, R.E., Hodson, W.A., Davies, D.G. & Chernick, V. (1967) 'Ventilatory Adaptation to a Step Change in PCO₂ at the Carotid Bodies', *J. Appl. Physiol.*, *23*, 195-202

Edelman, N.H., Lahiri, S., Braudo, L., Cherniack, N.S. & Fishman, A.P. (1970) 'The Blunted Ventilatory Response to Hypoxia in Cyanotic Heart Disease', *N. Eng. J. Med.*, *282*, 405-11

Eger, E.I., II, Kellog, R.H., Mines, A.H., Lima-Ostos, M., Morrill, C.G. & Kent, D.W. (1968) 'Influence of CO₂ on Ventilatory Acclimatization to Altitude', *J. Appl. Physiol.*, *24*, 607-15

Eldridge, F.L. & Millhorn, D.E. (1981) 'Central Regulation of Respiration by Endogenous Neurotransmitters and Neuromodulators', *Ann. Rev. Physiol.*, *43*, 121-35

Eyzaguirre, C., Fitzgerald, R.S., Lahiri, S. & Zapata, P. (1983) 'Arterial Chemoreceptors' in J.T. Shepherd and F.M. Abboud (eds), *Circulation, Handbook of Physiology*, The American Physiological Society, Washington, D.C. (in press)

Eyzaguirre, C. & Lewin, J. (1961) 'Chemoreceptor Activity of the Carotid Body of the Cat', *J. Physiol.*, *159*, 222-37

Fencl, V., Gabel, R.A. & Wolfe, D. (1979) 'Composition of Cerebral Fluids in Goats Adapted to High Altitude', *J. Appl. Physiol.: Resp. Environ. Exercise Physiol.*, *47*, 508-13

Fencl, V., Miller, T.B. & Pappenheimer, J.R. (1966) 'Studies on the Respiratory Responses to Disturbances of Acid-Base Balance, with Deductions Concerning the Ionic Composition of Cerebral Interstitial Fluid', *Am. J. Physiol.*, *210*, 459-72

Fitzgerald, R.S. & Lahiri, S. (1983) in N.S. Cherniack & J.G. Widdicombe (eds), *Respiration, Handbook of Physiology*, The American Physiological Society,

Washington, D.C.

Forster, H.V., Bisgard, G.E., Rasmussen, B., Orr, J.A., Buss, D.D. & Manohar, M. (1976) 'Ventilatory Control in Peripheral Chemoreceptor Denervated Ponies during Chronic Hypoxemia', *J. Appl. Physiol.*, *41*, 878-85

Frisancho, A.R. (1981) *Human Adaptation, A Functional Interpretation*, University of Michigan Press, Ann Arbor, pp. 101-68

Gautier, H. & Bonora, M. (1980) 'Possible Alterations in Brain Monoamine Metabolism during Hypoxia-induced Tachypnea in Cats', *J. Appl. Physiol.: Resp. Environ. Exercise Physiol.*, *49*, 767-77

Gautier, H., Milic-Emili, J., Miserochi, G. & Siafakas, N.M. (1980) 'Patterns of Breathing and Mouth Occlusion Pressure during Acclimatization to High Altitude', *Resp. Physiol.*, *41*, 365-77

Guazzi, M. & Freis, E.D. (1969) 'Sino-aortic Reflexes and Arterial pH, PCO_2, and PO_2 in Wakefulness and Sleep', *Am. J. Physiol.*, *217*, 1623-7

Hanbauer, I., Karoum, F., Hellstrom, S. & Lahiri, S. (1981) 'Effects of Hypoxia Lasting up to One Month on the Catecholamine Content in Rat Carotid Body', *Neuroscience*, *6*, 81-6

Hornbein, T.H. & Roos, A. (1963) 'Specificity of H Ion Concentration as a Carotid Chemoreceptor Stimulus', *J. Appl. Physiol.*, *18*, 580-4

Jansen, A.H., Ioffe, S., Russell, B.J. & Chernick, V. (1981) 'Effect of Carotid Chemoreceptor Denervation on Breathing *in Utero* and after Birth', *J. Appl. Physiol.*, *51*, 630-3

Javaheri, S., Herrera, L. & Kazemi, H. (1979) 'Ventilatory Drive in Acute Metabolic Acidosis', *J. Appl. Physiol.: Resp. Environ. Exercise Physiol.*, *46*, 913-8

Lahiri, S. (1968) 'Alveolar Gas Pressures in Man with Life-Time Hypoxia', *Resp. Physiol.*, *4*, 373-86

Lahiri, S. (1972) 'Dynamic Aspects of Regulation of Ventilation in Man during Acclimatization to High Altitude', *Resp. Physiol.*, *16*, 245-58

Lahiri, S. (1975) 'Blood Oxygen Affinity and Alveolar Ventilation in Relation to Body Weight in Mammals', *Am. J. Physiol.*, *229*, 529-36

Lahiri, S. (1976) in A.S. Paintal (ed.), *Morphology and Mechanisms of Chemoreceptors*, University of Delhi, Delhi, pp. 138-46

Lahiri, S. (1977) 'Physiological Responses and Adaptation to High Altitudes' in D. Robertshaw (ed.) *Environmental Physiology II*, University Park Press, Baltimore, pp. 217-51

Lahiri, S. (1983) in J.B. West & S. Lahiri (eds), *Man at High Altitude*, The American Physiological Society, Washington, D.C. (in press)

Lahiri, S., Brody, J.S., Motoyama, E.K. & Velasquez, T.M. (1978) 'Regulation of Breathing in Newborns at High Altitude', *J. Appl. Physiol.: Resp. Environ. Exercise Physiol.*, *44*, 673-8

Lahiri, S. & DeLaney, R.G. (1975) 'Relationship Between Carotid Chemoreceptor Activity and Ventilation in the Cat', *Resp. Physiol.*, *24*, 267-86

Lahiri, S., Edelman, N.H., Cherniack, N.S. & Fishman, A.P. (1981) 'Role of Carotid Chemoreflex in Respiratory Acclimatization to Hypoxemia in Goat and Sheep', *Resp. Physiol.*, *46*, 367-82

Lahiri, S. & Gelfand, R. (1981) in T.F. Hornbein (ed.), *Regulation of Breathing*, vol. II, Dekkar, New York, pp. 773-844

Lahiri, S., Kao, F.F., Velasquez, T., Martinez, C. & Pezzia, W. (1969) 'Irreversible Blunted Respiratory Sensitivity to Hypoxia in Man Born at High Altitude', *Resp. Physiol.*, *6*, 360-74

Lahiri, S., Manet, K. & Sherpa, M.G. (1983) 'Dependence of High Altitude Sleep Apnea on Ventilatory Sensitivity to Hypoxia', *Resp. Physiol.* (in press)

Lahiri, S. & Milledge, J.S. (1967) 'Acid-base in Sherpa Altitude Residents and Lowlanders at 4,880m', *Resp. Physiol.*, *2*, 323-34

Lahiri, S., Nishino, T., Mokashi, A. & Mulligan, E. (1980) 'Interaction of Dopamine and Haloperidol with O_2 and CO_2 Chemoreception in Carotid Body', *J. Appl. Physiol.: Resp. Environ. Exercise Physiol.*, *49*, 45-51

Lahiri, S., Smatresk, N. & Mulligan, E. (1983) in H. Acker & R.G. O'Regan (eds), *Physiology of Peripheral Arterial Chemoreceptors*, Elsevier/North-Holland, Amsterdam (in press)

Laidler, P. & Kay, J.M. (1978) 'A Quantitative Study of Some Ultrastructural Features of Type I Cells in the Carotid Bodies of Rats Living at a Simulated Altitude of 4,300 meters', *Neurocytol.*, *7*, 183-92

Lenfant, C. & Sullivan, K. (1971) 'Adaptation to High Altitude', *N. Engl. J. Med.*, *284*, 1298-309

Leusen, I. (1972) 'Regulation of Cerebrospinal Fluid Composition with Reference to Breathing', *Physiol. Rev.*, *52*, 1-56

Maher, J.T., Jones, L.G. & Hartley, L.H. (1974) 'Effects of High Altitude Exposure on Submaximal Exercise Capacity in Man', *J. App. Physiol.*, *37*, 895-8

Masson, R.G. & Lahiri, S. (1974) 'Chemical Control of Ventilation during Hypoxic Exercise', *Resp. Physiol.*, *22*, 241-62

Messner, R. (1979) *Everest: Expedition to the Ultimate*, Kaye and Ward, London

Milledge, J.S. & Lahiri, S. (1967) 'Respiratory Control in Lowlanders and Sherpa Highlanders at Altitude', *Resp. Physiol.*, *2*, 310-22

Miller, J.D. & Tenney, S.M. (1975) 'Hypoxia-induced Tachypnea in Carotid-deafferented Cats', *Resp. Physiol.*, *23*, 31-9

Mitchell, R.A. (1966) 'Cerebrospinal Fluid and the Regulation of Respiration', in C.G. Caro (ed.), *Advances in Respiratory Physiology*, Baltimore, Williams and Wilkins, pp. 1-47

Monge, C.M. & Monge, C.C. (1966) *High Altitude Diseases*, Thomas, Springfield, Ill.

Mosso, A. (1898) in T. Fisher (ed.), *Life of Man in the High Alps*, Unwin, London, pp. 42-7

Musch, T.I., Bateman, N.T., Smith, C.A. & Dempsey, J.A. (1980) 'Lactate Production in Rat Brain during Acclimatization to Hypoxia', *Fed. Proc.*, *39*, 830 (abstract)

Nishino, T. & Lahiri, S. (1981) 'Effects of Dopamine on Chemoreflex in Breathing', *J. Appl. Physiol.: Resp. Environ. Exercise Physiol.*, *50*, 892-7

Olson, E.B., Jr, & Dempsey, J.A. (1978) 'Rat as a Model for Human-like Ventilatory Adaptation to Chronic Hypoxia', *J. Appl. Physiol.: Resp. Environ. Exercise Physiol.*, *44*, 763-9

Pappenheimer, J.R., Fencl, V., Heisey, S.R. & Held, R. (1965) 'Role of Cerebral Fluids in Control of Respiration as Studied in Unanesthetized Goats', *Am. J. Physiol.*, *208*, 436-50

Phillipson, E.A. (1978) 'Control of Breathing during Sleep', *Am. Rev. Resp. Dis.*, *118*, 909-39

Pokorski, M. & Lahiri, S. (1982) 'Inhibition of Aortic Chemoreceptor Responses by Metabolic Alkalosis in the Cat', *J. Appl. Physiol.: Resp. Environ. Exercise Physiol.*, (in press)

Pugh, L.G.C.E. (1964) in D.B. Dill (ed.) *Adaptation to the Environment, Handbook of Physiology*, The American Physiological Society, Washington, D.C., pp. 861-8

Pugh, L.G.C.E., Gill, M.B., Lahiri, S., Milledge, J.S., Ward, M.P. & West, J.B. (1964) 'Muscular Exercise at Great Altitudes', *J. Appl. Physiol.*, *19*, 431-40

Rahn, H. & Fenn, W.O. (1962) *A Graphical Analysis of the Respiratory Gas Exchange: The O_2-CO_2 Diagram*, The American Physiological Society, Washington, D.C.

Rahn, H. & Otis, A.B. (1949) 'Man's Respiratory Response during and after

Acclimatization to High Altitude', *Am. J. Physiol.*, *157*, 445-62

Schoene, R.S. (1983) in J.B. West and S. Lahiri (eds), *Man at High Altitude*, The American Physiological Society, Washington, D.C. (in press)

Severinghaus, J.W., Mitchell, R.A., Richardson, B.W. & Singer, M.M. (1963) 'Respiratory Control at High Altitude Suggesting Active Transport Regulation of CSF pH', *J. Appl. Physiol.*, *18*, 1155-66

Siesjo, B.K. (1978) *Brain Energy Metabolism*, John Wiley and Sons, New York

Smatresk, N., Lahiri, S., Pokorski, M. & Barnard, P. (1981) 'Augmented Efferent Inhibition of Carotid Chemoreceptor Activity in Chronically Hypoxic Cats', *Physiologist*, *24*, 114 (abstract)

Sorensen, S.C. & Severinghaus, J.W. (1968) 'Irreversible Respiratory Insensitivity to Acute Hypoxia in Man Born at High Altitude', *J. Appl. Physiol.*, *25*, 217-20

Tenney, S.M. & Ou, L.C. (1977) 'Hypoxic Ventilatory Response of Cats at High Altitude: An Interpretation of "Blunting"', *Resp. Physiol.*, *30*, 310-22

Tenney, S.M., Rahn, H., Stroud, R.C. & Mithoffer, J.C. (1953) 'Adaptation to High Altitude Changes in Lung Volume in the First Seven Days at Mt. Evans, Colorado', *J. Appl. Physiol.*, *5*, 607-13

Ungar, A. & Bouverot, P. (1980) 'The Ventilatory Responses of Conscious Dogs to Isocapnic Oxygen Tests. A Method of Exploring the Central Component of Respiratory Drive and its Dependence on O_2 and CO_2', *Resp. Physiol.*, *39*, 183-97

Weil, J.V., Byrne-Quinn, E., Sodal, I.E., Filley, G.F. & Grover, R.F. (1971) 'Acquired Alternation of Chemoreceptor Function in Chronically Hypoxic Man at High Altitude', *J. Clin. Invest.*, *50*, 186-95

Weiskopf, R.B., Gabel, R.A. & Fencl, V. (1976) 'Alkaline Shift in Lumbar and Intracranial CSF in Man after 5 Days at High Altitude', *J. Appl. Physiol.*, *41*, 93-7

Welsh, M.J., Heistad, D.D. & Abboud, F.M. (1978) 'Depression of Ventilation by Dopamine in Man', *J. Clin. Invest.*, *61*, 708-13

West, J.B. (1983) in J.B. West & S. Lahiri (eds), *Man at High Altitude*, The American Physiological Society, Washington, D.C. (in press)

West, J.B., Pizzo, C.J., Milledge, J.S., Maret, K.H. & Peters, R.M. (1982) 'Barometric Pressure and Alveolar Gas Composition on the Summit of Mt. Everest', *Fed. Proc.*, *41*, 1109 (abstract)

West, J.B. & Wagner, P.D. (1980) 'Predicted Gas Exchange on the Summit of Mt. Everest', *Resp. Physiol.*, *42*, 1-16

Weyne, J., Van Leuven, F. & Leusen, I. (1977) 'Brain Amino Acids in Conscious Rats in Chronic Normocapnic and Hypocapnic Hypoxemia', *Resp. Physiol.*, *31*, 231-9

Winslow, R.M., Monge, C.C., Statham, N.J., Gibson, C.G., Charache, S., Whittenbury, J., Moran, O. & Berger, R.L. (1981) 'Variability of Oxygen Affinity of Blood: Human Subjects Native to High Altitude', *J. Appl. Physiol.: Resp. Environ. Exercise Physiol.*, *51*, 1411-6

Zapata, P. & Zuazo, A. (1980) 'Respiratory Effects of Dopamine-induced Inhibition of Chemosensory Inflow', *Resp. Physiol.*, *40*, 79-92

INDEX

accessibility, pulmonary 79, 85
acclimatisation 258-60, 298-9, 302-5
acetazolamide 8-9
acetylcholine 8, 29-33, 52, 62, 66-7
acidity effects 41-51, 58
acidosis 42, 48-51, 308-10
adaptation indices 83-4
adenosine diphosphate (ADP) 181-2, 185
adenosine triphosphate (ATP) 8, 34, 181-5
age-dependence 117-18, 137, 277, 283, 288, 310, 319
airway resistance 118-20, 213, 312
alkalosis 298, 308, 314-15
alloxan 91, 95-6
almitrine 18, 19
alveolar ventilation 233-5, 255-8
anaemia 188
anaerobic thresholds 223, 265
anaesthesia 7, 44, 116-17
angina pectoris 191-2
aortic bodies 1-2
aortic chemoreceptors 9, 14-15
apnoea 16, 113, 137, 216, 291-3, 317
arrhythmia 137
arterial blood pressure, fall 136-7
arterial oscillations, CO_2 253-5
asthma 123, 125, 131-3, 222, 262-7
atropine 52, 62, 123, 130
autocoids, lung 130-2

barometric pressures 299-302
baroreceptors 5, 137-8
beta-adrenergic mechanisms 218-19, 282
bicarbonate exchanges 68-72
bipuvacaine 80, 85
blood
    components affecting ventilation 41
    flow 1, 12-13, 27, 92-5, 100
    transmitting chemoreceptors 68-72, 235-40
body fluids, boiling 187
bradykinin 85, 86, 130

brain
    electrical activity 278
    oxygen partial pressures 162-5
breath-holding 206-8
breathlessness 209-10
bronchial receptors 86-7, 130-1
bronchitis 16-17, 37, 209-11, 260-9
bronchoconstriction 10, 123, 129-30
α-bungarotoxin 31

C fibres 89
calcium, affecting pH sensitivity 62-3
capillary, structure 166-9
capsaicin 35-6, 86, 98, 129, 135
carageenin 103
carbon dioxide
    acclimatisation effects 307-8
    analysed 233
    chemoreceptor response 8-9, 81, 85, 96-7, 128
    molecules 41-2
    partial pressures 7, 165-6
    pressures related to oxygen pressures 222
    sensitivity 7, 44, 117, 214, 230-2, 247
    solubility in blood 177-8
carbon monoxide poisoning 11, 188-9
carbonic anhydrase inhibitor 8-9, 230
carotid bodies 1-4, 6, 14-15, 18, 20-2
    carbon dioxide changes 228-32, 240-5
    high-altitude hypertrophy 15, 311-12
catecholamines 33, 282, 312
cell types, chemosensitive 18, 20-2, 60
central chemoreceptors
    defined 41-2
    newborn 285-7
central inspiratory activity (CIA) 110
central nervous system (CNS), hypoxia effects 6, 9, 194, 307
centrifugal activity 24
cerebral circulation 193-4